Constructed Wetlands *for the* Treatment *of* Landfill Leachates

T0203860

edited by

George Mulamoottil
Edward A. McBean
Frank Rovers

CRC Press
Taylor & Francis Group
Boca Raton London New York

CRC Press is an imprint of the
Taylor & Francis Group, an **informa** business

Front cover: *Typhalatigolia* in foreground at the Grønmo MSW landfill in Norway; back cover: aerial view of the Bølstad landfill treatment system in Norway. Photographs courtesy of W. S. Warner.

CRC Press
Taylor & Francis Group
6000 Broken Sound Parkway NW, Suite 300
Boca Raton, FL 33487-2742

First issued in paperback 2019

ISBN-13: 978-1-56670-342-0 (hbk)
ISBN-13: 978-0-367-40030-9 (pbk)

Library of Congress Cataloging-in-Publication Data

Constructed wetlands for the treatment of landfill leachates / edited
   by George Mulamoottil, Edward A. McBean, and Frank Rovers.
       p.   cm.
    Papers presented at an international symposium held in June of 1997.
    Includes bibliographical references and index.
    ISBN 1-56670-342-5 (alk. paper)
    1. Constructed wetlands--Congresses. 2. Sanitary landfills--
Leaching--Congresses. 3. Constructed wetlands--Case studies--
Congresses. I. Mulamoottil, George. II. McBean, Edward A. III. Rovers, Frank.
TD756.5.C657   1998
628.3'9--dc21

                                                             98-23910
                                                               CIP

Library of Congress Card Number 98-23910

Visit the Taylor & Francis Web site at
http://www.taylorandfrancis.com

and the CRC Press Web site at
http://www.crcpress.com

Preface

The improvements in the design of landfills result in extending the contaminating life span of these facilities. The disposal of landfill leachates is of concern because they have the potential to degrade the environment. The treatment and disposal of landfill leachates is becoming a major environmental issue, especially with regulatory agencies and environmentalists. Constructed wetlands are increasingly being employed to treat landfill leachate, and the use of natural systems in waste management seems to be gaining in popularity as a result of their sustainability and cost savings. At present, there are several constructed wetland treatment facilities in operation around the world. This symposium was organized to consolidate the results and experiences in the use of this ecotechnology for landfill leachate treatment. The symposium also provided an open forum to discuss the issues and concerns and to generate recommendations for future work.

There were 56 participants who attended the symposium, and they brought together a wide spectrum of interests and experiences. The theme of the symposium, wetlands for the treatment of landfill leachates, was of interest to professionals of diverse backgrounds and attracted scientists, engineers, policy makers, and practitioners. The attendees included members from the private and public sectors. The papers included in this book were presented at the symposium, except for one. We believe that this book has brought, under one cover, all that we know about constructed wetlands and leachate treatment. This book will have served its purpose if researchers continue to communicate effectively on new findings on the subject, and if government agencies allocate more funds to understand and improve this exciting and emerging natural treatment technology.

A number of individuals helped in the preparation of this book. Janet Ozaruk assisted in organizing and hosting the symposium. We take this opportunity to recognize her diligence, enthusiasm, and attention to detail. Conestoga-Rovers and Associates, Waterloo, Ontario provided some of the logistical support. Sharon Rew assisted in typesetting the introduction and panel discussion sections. Manjit Kerr-Upal assisted with the manuscript preparation and some editorial work, and we thank her for a job well done. We are also thankful to all symposium participants, session chairs, and authors of papers for their contributions, for without their efforts the symposium would not have been a success.

George Mulamoottil
Edward A. McBean
Frank Rovers

The Editors

George Mulamoottil (Ph.D.) is a Professor in the School of Urban and Regional Planning and the Department of Geography, a member of the Wetlands Research Centre, and, at present, Associate Dean of Graduate Studies at the University of Waterloo. He has been involved in several areas of environmental research and has worked in Indonesia, Thailand, India, and Kenya. Dr. Mulamoottil's research interests have focused on the planning and management of water resources. Most of his recent contributions are related to wetlands and storm water management. Dr. Mulamoottil has worked as a civil servant and has been active in environmental consulting in Ontario for over 29 years.

Edward McBean (Ph.D.) is an Associate of Conestoga-Rovers and Associates and President of CRA Engineering, Inc. Dr. McBean's previous experience includes more than 20 years as a faculty member at the University of Waterloo and the University of California, Davis. His research and professional work have focused on the fate and transport of contaminants as they move through the environment. He is the author of more than 300 papers in the technical literature and has authored/edited eight books.

Frank Rovers (M.A.Sc.) is President of Conestoga-Rovers and Associates, an environmental engineering company with more than 850 employees located in 29 offices. Mr. Rovers has been involved for more than 25 years in a large number of environmental engineering problems dealing with a wide spectrum of environmental quality issues. He is the author of numerous technical journal articles dealing with landfills and fate and transport of contaminants in the environment.

Contributors

Mike Barr
Aspinwall and Company
Walford Manor
Baschurch, Shrewsbury
Shropshire, SY4 2HH, United Kingdom

John M. Bernard
Department of Biology
Ithaca College
Ithaca, New York, USA 14850

Tom Brown
Soil Enrichment Systems, Inc.
10800 Weston Road
Vaughan, Ontario, Canada L4L 1A6

Martin Carville
Aspinwall and Company
Walford Manor
Baschurch, Shrewsbury
Shropshire, SY4 2HH, United Kingdom

Normand Castonguay
Castonguay Technologies
21 Bradley St.
Stittsville, Ontario, Canada K2S 1M9

William F. DeBusk
Soil and Water Science Department
University of Florida
P.O. Box 110510
Gainesville, Florida, USA 32611-0512

David A. V. Eckhardt
U.S. Geological Survey
Water Resources Division
903 Hanshaw Road
Ithaca, New York, USA 14850-1573

Leta Fernandes
Department of Civil Engineering
University of Ottawa
Ottawa, Ontario, Canada K1N 6N5

Gwyn Harris
Aspinwall and Company
Walford Manor
Baschurch, Shrewsbury
Shropshire, SY4 2HH United Kingdom

James Higgins
Soil Enrichment Systems, Inc.
10800 Weston Road
Vaughan, Ontario, Canada L4L 1A6

Petter D. Jenssen
Department of Engineering
Agricultural University of Norway
1432 Ås, Norway

Keith D. Johnson
Wetland Sciences, Inc.
3025 Gulf Breeze Parkway
Gulf Breeze, Florida, USA 32561

Joseph Julik
Minnesota Pollution Control Agency
520 Lafayette Road North
St. Paul, Minnesota, USA 55155

Robert H. Kadlec
Wetland Management Services
6996 Westbourne Drive
Chelsea, Michigan, USA 48118-9527

Francine Kelly-Hooper
Environmental Consultant
680 Irwin Crescent
Newmarket, Ontario, Canada L3Y 5A2

François La Forge
Conestoga Rovers & Associates
179 Colonnade Road
Suite 400
Nepean, Ontario, Canada K2E 7J4

Steve Last
Aspinwall and Company
Walford Manor
Baschurch, Shrewsbury
Shropshire, SY4 2HH United Kingdom

Joseph Loer
8378 88th Street, South
Cottage Grove, Minnesota, USA 55016

Trond Mæhlum
Jordforsk
Centre for Soil and Environmental Research
N-1432 Ås, Norway

Craig D. Martin
Wetland Sciences, Inc.
3025 Gulf Breeze Parkway
Gulf Breeze, Florida, USA 32561

Edward A. McBean
Conestoga Rovers & Associates
651 Colby Dr.
Waterloo, Ontario, Canada N2V 1C2

William C. McCrory
McCrory & Williams
Consulting Engineers
Daphne, Alabama, USA

Erik L. Melear
Solid Waste Department
Orange County Utilities Department
5901 Church Street
Orlando, Florida, USA 32829

Gerald A. Moshiri
Wetland Sciences, Inc.
3025 Gulf Breeze Parkway
Gulf Breeze, Florida, USA 32561

George Mulamoottil
School of Urban and Regional Planning
Faculty of Environmental Studies
University of Waterloo
200 University Ave. W.
Waterloo, Ontario, Canada N2L 3G1

John H. Peverly
Department of Agronomy
Purdue University
West Lafayette, Indiana, USA 47907

Sharon Rew
School of Urban and Regional Planning
Faculty of Environmental Studies
University of Waterloo
200 University Ave. W.
Waterloo, Ontario, Canada N2L 3G1

Howard Robinson
Aspinwall and Company
Walford Manor
Baschurch, Shrewsbury
Shropshire, SY4 2HH United Kingdom

Frank Rovers
Conestoga-Rovers & Associates
651 Colby Dr.
Waterloo, Ontario, Canada N2V 1C2

William E. Sanford
Department of Earth Resources
Colorado State University
Fort Collins, Colorado, USA 80523-1482

Majid Sartaj
Civil Engineering Department
University of Ottawa
Ottawa, Ontario, Canada K1N 6N5

Katrin Scholz-Barth
North American Wetland Engineering, P.A.
20920 Keewahtin Ave.
Forest Lake, Minnesota, USA 55025

Larry N. Schwartz
Camp Dresser & McKee, Inc.
1950 Summit Park Drive
Suite 300
Orlando, Florida, USA 32810

Per Stålnacke
Jordforsk
Centre for Soil and Environmental Research
N-1432 Ås, Norway

Jan M. Surface
U.S. Geological Survey
903 Hanshaw Road
Ithaca, New York, USA 14850

Mostafa A. Warith
Gore & Storrie Limited
Ottawa, Ontario, Canada K2C 3W7

William S. Warner
Jordforsk
Centre for Soil and Environmental Research
N-1432 Ås, Norway

Douglas Wetzstein
Minnesota Pollution Control Agency
520 Lafayette Road North
St. Paul, Minnesota, USA 55155

Lee P. Wiseman
Camp Dresser & McKee, Inc.
1950 Summit Park Drive
Suite 300
Orlando, Florida, USA 32810

Contents

Introduction

The mass of solid wastes produced globally is increasing at a rapid pace. Although improvements are being made in reducing, reusing, and recycling of wastes, protecting the environment and human health continues to be a challenge. According to Kopka (1990), 80% of the approximately 160 million tons of garbage produced in the U.S. during 1986 was disposed in landfills. In a related issue, during a 10-year period (1978 to 1988), 14,000 landfills were closed in the U.S., many of which were unlined and only some included collection and treatment of leachate. The leachate may migrate from the refuse and contaminate the surface waters and groundwaters, potentially impacting aquatic ecosystems and human health. As the potential for environmental contamination from landfills continues, it is imperative that better design and operating procedures are implemented.

A report on the status of waste disposal sites in Ontario by the Ontario Ministry of Environment and Energy (1991) is indicative of the ongoing concerns with landfills. The Ministry classified the Ontario sites into those that have the potential to impact human health and those that have the potential to impact the environment. As of 1990, there were 1358 active sites and 2334 closed sites. The closed sites include uncertified sites, which may be grouped as older illegal operations. As there is very little information available on the waste type and physical setting of the closed sites, any attempt to remediate the facility will face difficult problems.

A major concern also exists in Europe on problems related to landfills. In some countries, the situation is serious and requires immediate remedial action. For example, Bulc et al. (1997) reported the existence of 60,000 illegal landfill sites and only 43 registered landfill facilities in Slovenia. In developing countries, numerous examples exist where the solid waste management practices are uncontrolled, chaotic, and very expensive to remediate.

The number of approved and illegal landfill sites, from a global perspective, is enormous. At many of these sites, leachates find their way into the environment because there is no provision for their treatment. In some cases, leachate is conveyed to sewage treatment plants where leachate treatment occurs along with domestic sewage. The degree of treatment effectiveness of certain chemical refractories in this type of treatment environment is variable. Furthermore, these technologies are capital-intensive and require ongoing costly operation and maintenance.

Constructed wetlands, because of their ecosystem characteristics that are similar to natural wetlands, are generally accepted as an ecotechnology for treatment of various types of wastewaters. The ecosystem characteristics are attributed to a combination of factors, such as high plant productivity, large adsorptive surfaces on sediments and plants, an aerobic–anaerobic interface, and an active microbial population (Urbanic-Bercic, 1994). Since these properties result in high rates of biological activity, they provide an opportunity to transform many of the common pollutants in, for example, conventional municipal wastewater, to less harmful products or essential nutrients that can be used by the biota (Kadlec and Knight, 1996). As a consequence, in recent years constructed wetlands are increasingly being utilized to emulate natural wetlands in treating wastewater. At present, several treatment facilities are in operation using both natural and constructed wetlands. The wastewaters that are being treated by wetlands include drainage from acid mines, urban storm water runoff, livestock waters, secondary-treated sewage effluents, and landfill leachates (Bobberteen and Nickerson, 1991). However, as the list of piloted and full-scale constructed wetland projects grows, there is a need to collect more data on their performance and utility.

The potential to expand the use of constructed wetlands to the treatment of leachates beyond the more general treatment of wastewaters, is relevant in today's context. Leachates from landfills vary in their hazardous characteristics and include numerous priority pollutants and phenolics. The treatment of leachates by natural systems seems to be environmentally sustainable for the treatment of many constituents. According to Kadlec and Knight (1996), both subsurface-flow and surface-flow wetlands are emerging ecotechnologies with the potential to treat landfill leachates.

The use of constructed wetlands has considerable promise for the control of a large number of organic compounds, which are the subject of landfill leachate regulation. However, the development of this technology using constructed wetlands to treat landfill leachates has to be supplemented by investigations on the breakdown and pathways of the contaminants. To improve the design of constructed wetlands for better performance, it is essential that practitioners understand the movement, breakdown, and accumulation of contaminants in different compartments of a wetland including those of the substrate, microbial population, plants, and animals. Extensive and long-term studies are required to collect such information from differing climatic and environmental conditions. As well, investigations on the growth and nutrient uptake, including trace and heavy metal adsorption by plants and their tolerance/adaptability to toxicity of leachates are important areas of needed research.

Elevated levels of iron, copper, zinc, lead, and cadmium were found in the roots of the common reed in constructed wetlands treating landfill leachate, but these metals were not accumulated in the shoots or rhizomes (Peverly et al., 1995). Accumulation of metals in different species of wetland plants may also show variations that may influence the selection of plants to be established in constructed wetlands. It is expected that wetland plants will be exposed to slightly higher temperatures in the winter in locations where leachates are discharged. The increase in temperature may have some beneficial effect on the plants resulting in improved treatment performance. Further, by providing oxygen, the plants have a positive effect on the aerobic biodegradation of the organics and have the potential to change the redox reactions causing the solubilization/precipitation of metals. However, for situations where there is a concentration of leachates by evapotranspiration, there could be negative effects on the biota.

As the use of constructed wetlands to treat landfill leachates is a relatively new ecotechnology, the data on their performance are being accumulated in different parts of the world. More data are needed on the effectiveness of different wetland systems, such as the subsurface-flow and surface-flow wetlands, and some combinations of the system, such as peat infiltration or extended aeration. Related to the performance of the system is the development of improved design guidelines, which are still rudimentary at present. Kadlec and Knight (1996) have also commented that design data are meager.

The performance of constructed wetlands in treating leachates in cold climatic conditions is not adequately documented because of the expected reduction in the functioning of the wetlands as a treatment system due to the dormancy of plants and microbial populations. The lack of information is not surprising since the performance of treatment wetlands in handling other types of wastewater on a year-round basis using greenhouse technology is only now being compiled (Smith and Mulamoottil, 1998). A recent paper by Jenssen et al. (1994) has pointed out that, based on a theoretical model developed to predict the influence of insulation, the threat of hydraulic failure in subsurface-flow wetlands in colder regions can be mitigated, if not eliminated.

Germane to the use of constructed wetlands for the treatment of landfill leachate is to develop cost comparisons of the technology with other treatment technologies. Since many landfill sites are located in isolated and rural areas, connection of a leachate collection system to sanitary sewers may pose a substantial problem. When reliance is made on the pumping and hauling of leachates to sewage treatment plants, there will be an appreciable recurring expenditure. For these sites, once the landfill site is closed, even with no revenue generated, the conveyance of the leachates to sewage treatment plants will have to be continued at potentially significant monetary costs (Rew, 1996).

In response to the issues briefly outlined, there is a need to consolidate the existing information and outline future research needs. The International Symposium "Wetlands for Treatment of Landfill Leachates" was organized for June 24 and 25, 1997 to facilitate such an activity. This meeting was held in Romulus (Detroit, MI), and provided an opportunity to present the results and experiences obtained to date, and to provide a forum for researchers and practitioners to address the issues and concerns related to this ecotechnological approach.

The papers included in the book were presented at the symposium except for the contribution (Chapter 10) by Sartaj et al., who were unable to attend. It is our hope that, by bringing together the experts in the field to the symposium and by sharing their experience, we are able to assemble in this book what is known on the use of constructed wetland systems for leachate treatment. Brief outlines of the papers are presented beginning with those describing the broad general contributions of constructed wetlands to leachate treatment. This will be followed by papers related to case studies of constructed wetlands and integrated constructed wetland systems, plants, and landfill leachate, problems associated with iron in wetlands, and cost comparisons of alternative leachate treatment systems.

McBean and Rovers (Chapter 1), based on landfill monitoring results, discuss the magnitudes of leachate concentrations and their variations from site to site and over time. The authors further describe the extent of uncertainties that may be expected as related to leachate characteristics. These factors are taken into consideration to predict typical ranges of leachate concentrations and the contaminating life span of landfills. The information so generated reflects the design input information for constructed wetlands.

The reduction of contaminants in leachate by wetland treatment and the processes involved in the different contaminant transformations are described by Kadlec (Chapter 2). Use of appropriate design methods is recommended to deal with issues related to the potential exposure of contaminants to wildlife. Compared with other technologies, Kadlec points out that the stochastic component of wetland performance is larger. A balanced view of the positive and negative aspects in the use of constructed wetlands for leachate treatment is presented.

The paper by Mæhlum (Chapter 3) discusses the various issues, concerns, and successes of the use of constructed wetlands to treat landfill leachates in cold climates. Integrated wetland systems seem to function well in colder regions and meet effluent criteria for most of the water quality parameters. Mæhlum provides supporting evidence for better performance of subsurface-flow wetlands, which afford greater thermal protection.

Sanford (Chapter 4) clearly demonstrates the significant role of substrate type in the functioning of constructed wetlands as leachate treatment systems. The clogging of substrates, decrease of hydraulic conductivity, overland flow of leachate, and stratification due to density differences in rainwater and leachate within the substrate are described in his paper. According to Sanford, density stratification has the potential to reduce residence time which can interfere with treatment efficiency.

The performance of surface-flow constructed wetland treating leachate contaminated groundwater in Albania, FL is presented in the paper by Johnson et al. (Chapter 5). The treated effluent met federal and state water quality standards, including standards for several heavy metals, for discharge into receiving waters. The paper indicates that the quality of the treated leachate effluent remained unchanged even though the original design loading volumes have been exceeded frequently during the past 20 years. The importance of using site-specific information in the design of constructed wetlands for leachate treatment is emphasised by the authors.

Landfill leachate composition and leachate management strategies in the U.K. are briefly presented by Robinson et al. (Chapter 6). This is followed by a description of a constructed wetland system used to treat leachate from Monument Hill Landfill Site in southern England. In the U.K. there are several dozen reed beds in operation, and they are mostly established for the polishing of effluents after aerobic biological pretreatment. The removal of contaminants was most effective during warm summer months. The timing of the contaminant removal was very beneficial because of the high sensitivity of the receiving stream in the summer. The case study is placed within the context of the leachate treatment practice in the U.K.

Schwartz et al. (Chapter 7) describe the effective treatment of landfill leachate mixed with storm water in the North Wide Cypress Swamp, FL. The heavy metals and nutrients are retained in the wetland. The benthic organisms and fish in the wetland were indicative of good water quality conditions. The authors report that the biological communities adapted to new hydrologic regimes.

A study by La Forge et al. (Chapter 8) documented attenuation of landfill leachate in a natural marsh in eastern Ontario, Canada. The prediction of the migration potential of zinc and cadmium was made possible by combining two computer models. The initial results point out the capacity of the marsh soil to reduce substantially the movement of contaminants from leachate.

The paper by Mæhlum et al. (Chapter 9) discusses the adoption of pretreatment of leachates before the treatment in constructed wetlands. The Norwegian experience on the use of integrated wetland systems at Esval and Bølstad is documented. Detention times of more than 20 days is recommended during cold seasons.

The performance of a constructed wetland integrated system in treating leachate from a landfill receiving industrial, commercial, and institutional wastes is documented in the paper by Sartaj et al. (Chapter 10). The system has provision for spray irrigation of leachate over a peat filter followed by surface-flow wetlands. The characteristics of the leachate are different from most landfills in Ontario. The authors report that the integrated wetland treatment system is very effective.

DeBusk (Chapter 11) reports on landfill leachate treatment by a constructed wetland system that incorporates the use of a primary treatment pond, wetland cells in series, and a detention pond with a final discharge to a percolation pond for groundwater recharge. The wetland system provides a high level of treatment of leachate based on 6 years of operation.

Loer et al. (Chapter 12) describe increasing dissolved oxygen in the leachate and precipitating solids and heavy metals as the first component of an integrated natural system for leachate treatment. The other components of the treatment system consisted of a sedimentation basin and a free-water-surface constructed wetland. The effluent from the constructed wetland is then infiltrated to the surficial aquifer.

In the paper by Eckhardt et al. (Chapter 13), results of a dual system involving a surface-flow and a subsurface-flow wetland to treat landfill leachate in Munroe County, NY are presented. The surface-flow wetland provided an effective pretreatment site for certain constituents, while the subsurface-flow wetland functioned as an environment for plant uptake, microbial activity, and geochemical processes, all of which assisted in contaminant removal. The authors report highest removal rates for organic constituents and lowest for inorganic ions.

Besides describing the growth pattern of wetland plants growing in landfill leachate, Bernard (Chapter 14) points out the capabilities of plants to grow well in polluted water, and their adaptability to modify their structural characteristics. Harvesting plants seems to be an appropriate strategy, and Bernard makes specific recommendations in this regard. A comprehensive coverage of the literature on wetland plants useful for researchers and practitioners can be found in this paper with 57 citations.

Higgins and Brown (Chapter 15) present an evaluation of surface- and subsurface-flow constructed wetlands to leachate. Based primarily on cost calculations and ease of operation, the authors recommend surface-flow wetlands to treat landfarm leachate at the refinery in Sarnia, Ontario.

Kelly-Hooper (Chapter 16) provides a brief review of the role of invertebrates in constructed wetlands. The invertebrates, because of their feeding activities involving shredding and filtering, assist in water quality improvement and act as an important food source for other animals. As a major constituent of municipal landfill leachate, iron, when exposed to oxygenated water, settles out and forms a blanketing layer on the sediment. Kelly-Hooper has discussed the effects of ferric hydroxide on benthic invertebrates and points out the changes in their species composition as indicative of iron saturation in the sediment. The use of sedimentation ponds as pretreatment is recommended before the landfill leachate is conveyed to constructed wetlands.

Rew and Mulamoottil (Chapter 17) describe a cost comparison of three leachate treatment alternatives for the Glanbrook Landfill Site in the Regional Municipality of Hamilton-Wentworth in Ontario.

The symposium assessed the status of this ecotechnology and outlined future research needs. The papers have demonstrated and documented the success of constructed wetlands to treat landfill leachate. However, as discussed in the panel discussion summary, much remains to be done. Rapid strides are being made in designing with the natural environment by using ecosystems, communities, and organisms. The success of this ecotechnology will contribute toward a sustainable environment.

REFERENCES

Bobberteen, S. and Nickerson, J., 1991. Use of created cattail (*Typha*) wetlands in mitigation strategies; *Environmental Management* 15:785–795.

Bulc, T., Vrhousek, D., and Kukanja, V., 1997. The use of constructed wetlands for landfill leachate treatment, *Water Science and Technology* 35:301–306.

Jenssen, P. D., Mæhlum, T., and Warner, W. S., 1994. The influence of cold climate upon constructed wetlands: performance of treating domestic wastewater and landfill leachate in Norway, in Collins, E., Ed., *Onsite Wastewater Treatment: Proceedings of the Seventh International Symposium on Individual and Small Community Sewage Systems*, December 11–13, 1994, American Society of Agricultural Engineers, MI.

Kadlec, R. and Knight, R., 1996. *Treatment Wetlands*, Lewis Publishers, Boca Raton, FL.

Kopka, J. K., 1990. Evaluation of constructed wetlands for the treatment of Municipal Solid Waste Landfill Leachate; M.Sc. thesis, Cornell University, Ithaca, NY.

Ministry of Environment and Energy, 1991. Waste Disposal Site Inventory, Toronto, Ontario.

Peverly, J. H., Surface, J. M., and Wang, T., 1995. Growth and trace metal absorption by *Phragmites australis* in wetlands constructed for landfill leachate treatment, *Ecological Engineering* 5:21–35.

Rew, S., 1996. The Feasibility of Using Constructed Wetlands to Treat Landfill Leachates, B.A. thesis, University of Waterloo, Waterloo, Ontario.

Smith, S. and Mulamoottil, G., 1998. Year-round Performance of Constructed Wetlands to Treat Stormwater in Aurora, Ontario. Report under preparation for Ontario Ministry of Environment and Energy, Toronto.

Urbanic-Bercic, O., 1994. Investigation into the use of constructed reedbeds for municipal waste dump leachate treatment, *Water Science and Technology* 29:289–294.

CHAPTER 1

Landfill Leachate Characteristics as Inputs for the Design of Wetlands Used as Treatment Systems

Edward A. McBean and Frank Rovers

CONTENTS

ABSTRACT: The principles of decomposition in landfills are reviewed to establish bases for quantifying the concentrations of individual constituents that are expected in leachate. Indications of the expected values and ranges of concentrations are provided. The trends in landfilling are examined, and implications for the contaminating life span of landfills are presented. The need to view leachate treatment as a long-term commitment and the resulting interest in cost avoidance are major determining factors in reaching a decision on leachate management strategy. The paper describes the typical concentrations within leachate and provides indications of circumstances when deviations from these concentrations may occur. Wetlands provide an important opportunity to contribute in this activity, but the magnitudes of the concentrations and the implications of temporal variabilities in leachate quantities and qualities must be understood, before incorporating a flexible design in the establishment of constructed wetlands.

1.1 INTRODUCTION

There is general societal acceptance of the principles of recycling, reuse, and reducing in solid waste management. However, the utilization of solid waste landfills as a partial means to manage solid wastes is widespread and will continue for the foreseeable future. Hence, the consequences of the past and ongoing use of landfills and, more specifically, the disposal of landfill-generated leachates are widely recognized as matters of concern because of the potential for negative impacts to the environment. As protection of the environment is vital to the well-being of present and future generations, any improvements that could be made in solid waste management practices will be welcomed by society at large and by all levels of government.

The dimensions of the problems that may arise from landfills became apparent in the early 1970s with the identification of environmental degradation caused by migration of landfill-generated gas and leachate (see, for example, McBean et al., 1995, for more details regarding the historical documentation). During the past 25 years, major improvements have been made in our understanding of the physical/chemical/biological processes that occur within refuse in landfills. The causative factors that result in the migration of contaminants of concern from landfills into adjacent environmental media are also better understood. The discernment of landfills as a complex system provided the impetus to develop comprehensive control strategies, which are now used in current landfill designs. These control strategies have included the implementation of leachate collection systems. However, leachate collected by these systems must be dealt with, as part of the long-term commitment for solid waste management.

Changes to landfill designs have included the trend toward reducing infiltration into the landfilled wastes, thus acting toward the entombment of the wastes. Decreasing the moisture content of the wastes results in significant increases in the contaminating life span of landfilled refuse. Consequently, the reduction in infiltration will result in a leachate collection system that must be maintained for a lengthier period and causes increases in the long-term operation and maintenance costs. Cost avoidance is a major determining factor in reaching a decision as to which of the leachate management strategies is adopted at specific sites. Selection of the control strategies most appropriate for a particular situation requires a good understanding of existing conditions and how these conditions may change in the future.

Temporal variability in terms of the quality and quantity of leachate is an acknowledged phenomenon. Our knowledge base to predict these variations is improving. It is interesting to note that as a landfill site becomes larger, the temporal changes in the leachate concentrations tend to be smoothed out as a result of the different stages of decomposition at different areas within the landfill.

This paper describes landfill leachate characteristics as evidenced by the monitoring results obtained at different landfills. Case study examples are described that demonstrate site-specific characteristics related to the quantity and quality of landfill leachates. The purpose of these characterizations is to indicate some of the fundamentals of the decomposition processes that occur in landfills. An improved understanding of the decomposition processes and how these affect leachate characteristics will assist system designers in predicting changes in leachate quality that may occur over time at a particular site. These findings are intended as inputs to improve our design basis of constructed wetlands leachate management systems. Further, from site-specific information representative of various landfill situations, uncertainty ranges in concentrations are presented. These uncertainty ranges are relevant in the planning for use of constructed wetlands as part of a landfill leachate management strategy.

1.2 BACKGROUND

Landfill design engineers now incorporate substantial liner and leachate recovery systems to divert leachate percolating through the solid wastes toward collection networks to prevent off-site leachate migration. The diverted leachate is then conveyed to one or more locations for treatment and disposal. These leachate collection and liner systems are universally used in current design of solid waste landfills.

Once the collection of the leachate has become part of a solid waste management system, a series of leachate treatment options must be examined. Examples of the available leachate treatment options are depicted in Figure 1.1. The identification of the preferred option in specific circumstances is a function of the costs, both operating and capital, and the limitations imposed on the quality and quantity of discharge. Costs, including those of conveyance, surcharges imposed on discharges to the publicly owned treatment works (POTW), and land values, influence the selection of a leachate management option. A frequently utilized option, when available, involves discharge to a POTW. However, because of stipulations on maximum concentrations of constituents allowed to be discharged to the POTW and/or economic surcharges imposed for discharge to the POTW, at many locations partial on-site treatment is first utilized, before discharge to a POTW. Table 1.1 provides examples of average sewer surcharges imposed on six facilities in Wisconsin to indicate the magnitude of costs. The average for the six locations, as applied to a 20-ha site with 30 cm of percolation, results in sewer surcharges of $240,000/year.

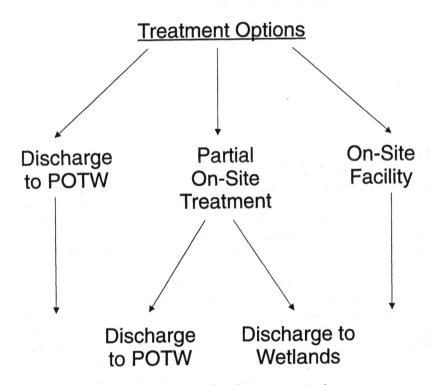

Figure 1.1 Schematic of leachate treatment options.

The array of potential methods for the management of landfill leachate is further elaborated upon in Table 1.2, where various combinations and permutations for leachate treatment are listed. The selection of the most-effective leachate strategy is influenced by the leachate management costs, the computation of which must include the implications of future temporal variations in leachate quantities and qualities. Wetlands provide a useful natural treatment system, either for leachate from old wastes and/or for partial treatment of leachates from newer wastes. However, significant variations in the quality and quantity of leachate occur over time, and, consequently, wetland treatment systems must be designed accordingly.

Table 1.1 **Examples of Sewer Surcharges Utilized in Wisconsin**

	Flow ($/m³)	BOD ($/kg)
Average	0.27	0.42
Range	0.05–0.51	0.18–0.88

Example: 20-ha site with 30 cm of percolation
BOD = 10,000 mg/L
average, $240,000/year

Table 1.2 **Alternative Methods for Management of Leachate**

1. Spray irrigation on adjacent grassland
2. Recirculation of leachate through the landfill
3. Disposal off site to sewer for treatment as an admixture with domestic sewage
4. Physical-chemical treatment
5. Anaerobic biological treatment
6. Aerobic biological treatment and
7. Wetlands

1.3 VARIABILITIES IN LEACHATE QUANTITY AND QUALITY

1.3.1 Leachate Quantity

Chemical and biological reactions occur as infiltrating water percolates through refuse. The products of the complex combination of ongoing reactions within the refuse are transported by the infiltrating water. Besides the chemical and biological reactions, physical processes such as sorption and dissolution also take place during the passage of water through the waste. The sum total of all of these processes and reactions is a leachate with dimensional attributes varying over time.

The transport of leachate through refuse material is analogous to the transport of contaminants in groundwater through heterogeneous, variably saturated soils. Primary transport will occur through preferential flow pathways, and secondary transport will occur as a result of storage diffusion from dead-end pores and advection/dispersion in low-permeability regions.

In addition to the leachate generation induced by precipitation, it is also produced as a result of biochemical processes that convert solid materials to liquid form. The leachate generated from the biochemical reactions is characterized by very high concentrations of organic and inorganic contaminants. Water percolating through the landfill surface from infiltration will actually dilute contaminants in the leachate, as well as aid in the formation of new leachate (Bagchi, 1990).

Percolate or leachate will collect a variety of dissolved organic and inorganic contaminants resulting from the dissolution or decomposition of refuse. The biodegradability of organic matter in leachate is generally inversely proportional to the molecular weight of the various components of the organic matter; lower molecular weights indicate a higher degree of biodegradability. Chian and deWalle (1977) state that organic matter in landfills ranges from small, volatile acids with low molecular weight to large, refractory (i.e., recalcitrant) fulvic- and humic-like compounds of intermediate and high molecular weights, respectively. Relatively unstabilized (or acid-phase) leachate is dominated by up to 90% short-chain, readily biodegradable volatile fatty acids (VFAs), which primarily consist of acetic, propionic, and butyric acids (Lu et al., 1985). These changes and uncertainties in the extent of the variability, as they occur over time, are important considerations in the design of wetlands for the treatment of leachate. The temporal variability of leachate quantities and qualities is influenced by features including the meteorology, the cover overlying the refuse, the hydrogeologic conditions in the vicinity of the site, and numerous characteristics of the refuse such as age, composition, and degree of compaction.

The features of the water balance, such as precipitation, interception and surface runoff, evapotranspiration by vegetation, and infiltration at a landfill site, are schematically depicted in Figure 1.2. In the majority of well-designed landfills, the leachate is designed to percolate into the leachate collection tile system. However, the processes that generate the leachate are not uniform.

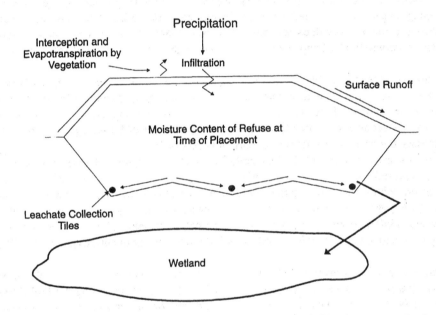

Figure 1.2 Water balance components and direction of leachate to a wetland.

Field capacity is an important feature in landfills and is defined as the maximum moisture content that can be maintained without producing continuous downward percolation due to gravitational forces. During precipitation events, water infiltrating into the refuse will increase the moisture content until field capacity is reached. However, solid wastes because of their heterogeneity and variable moisture content, will show differences in permeability and porosity within a landfill (Murray et al., 1995). As a result, preferential leachate flow pathways occur since leachate quantities will change over time as a result of, for example, changes in the attainment of field capacity within the refuse and/or changes in the percolation rate through the landfill surface as a result of changes in the surface cover of the landfill. Estimates of leachate quantity are usually obtained using the Hydrologic Evaluation Landfill Program (HELP) model (see Schroeder et al., 1994).

Although the concepts are relatively straightforward, there are definite limits to the accuracy of prediction of leachate quantities using the HELP model. An important element of the problem of prediction is related to field capacity. Examples of field capacities reported in the technical literature are listed in Table 1.3. The considerable variability in magnitudes is apparent, and

Table 1.3 Examples of Field Capacities for Municipal Solid Wastes

Rovers and Farquhar (1973)	30
Wigh (1979)	37
Walsh and Kinman (1979)	42
Walsh and Kinman (1981)	40
Fungaroli and Steiner (1979)	34
Oweis and Kehra (1986)	20–35
Zeiss and Major (1993)	13.6
Schroeder et al., (1994)	29.2
Schroeder et al., (1994) (with channeling)	7.3

Note: All values in m/m.

particularly noteworthy are the last few entries, which list lower magnitudes of field capacity. These decreasing magnitudes are a reflection of channeling in the refuse. The existence of channeling indicates that the leachate does not move as a "front" through the refuse. Instead, channels develop in the refuse and the first appearance of leachate is the result of channelized flow through the refuse.

The formation of these preferential leachate flow pathways is likely to occur, particularly during the initial stages of waste placement, before significant settlement and compaction of the refuse occurs. Ongoing settlement of the refuse material compresses the skeleton of the refuse and significantly reduces variations in porosity and permeability between different regions of the refuse, thereby decreasing the preferential nature of some flow pathways.

The growth toward attainment of field capacity is depicted in Figure 1.3 based on the experimental results reported by Fungaroli and Steiner (1979). Successive additions of water to the lysimeter in accord with a seasonal pattern of rainfall are depicted by the heavy solid line. Initially, most of the added water was retained by the refuse during the first year of the experiment. For subsequent years, additional water added produced lysimeter volumes as indicated in Figure 1.3. By the third year, the quantities of water added essentially mirrored the quantities of water collected as leachate.

The characteristics of field capacity indicate that not all parts of a landfill site reach field capacity simultaneously. Further, the uncertainty of prediction of leachate quantities using the HELP model is provided in Table 1.4. The values listed in the table indicate the percentage differences between the HELP model predictions of leachate quantities and field measurements. Significant differences are readily apparent. Seasonality of leachate quantities and changes in leachate production with time complicate the problem of prediction of leachate quantities.

1.3.2 Leachate Quality Variations

A series of phases is discernible in the decomposition of solid waste. Although the phases are variously defined (e.g., Rovers and Farquhar, 1973), there is general agreement on changes, as shown in Figure 1.4. Phase I, or the hydrolysis and acidification phase, involving aerobic decomposition, is typically brief, and lasts for less than a month. Once the available oxygen within the waste is utilized, except in the vicinity of the landfill surface, aerobic decomposition terminates.

Phase II begins with the initiation of activities of anaerobic and facultative organisms (involving acetogenic bacteria). They hydrolyze and ferment cellulose and other putrescible materials, producing simpler, soluble compounds such as VFAs (which produce a high biochemical oxygen demand, BOD, value) and ammonia. Phase II can last for years, or even decades. Leachates produced during this stage are characterized by high BOD values (commonly greater than 10,000 mg/L) and

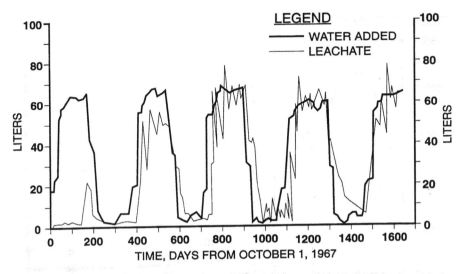

Figure 1.3 Volume of water added and lysimeter leachate quantities over time. (After Fungaroli and Steiner, 1979.)

Table 1.4 Percentage Differences between the HELP Model Predictions and Field Measurements

Landfills	%
Brown County	
Daily cover	65
Final cover	−29
Eau Claire County	
Daily sand cover	138
Interim sludge cover (uncompacted)	−96
Interim sludge cover (clayey loam)	−31
Final cover	127
Marathon County	
Daily cover	29
Final cover	−33

high ratios of BOD to chemical oxygen demand, COD (commonly greater than 0.7), indicating that high proportions of soluble organic materials are readily biodegradable. Other typical characteristics of phase II leachates are acidic pH levels (typically, 5 to 6), strong, unpleasant odors, and high concentrations of ammonia, in the range of 500 to 1000 mg/L. The aggressive chemical nature of this leachate assists in dissolution of other components of the waste, and produces high levels of iron, manganese, zinc, calcium, and magnesium in the leachate.

The anaerobic phase II is characterized by acid fermentation, reducing redox potential, high concentrations of readily degradable organic acids and inorganic ions, such as chloride, sulfate, calcium magnesium, and sodium, as well as ammonia and carbon dioxide (McBean et al., 1995). The partial reduction of sulfate to sulfides during this phase will result in decreasing concentrations of sulfate. Further, the generated sulfides may form precipitates with iron, manganese, and heavy metals that were dissolved during the acid fermentation.

Phase II also involves slower-growing methanogenic bacteria gradually becoming established and consuming simple organic compounds, with the production of a mixture of carbon dioxide, methane, and other trace gaseous constituents that constitute landfill gas. The transition from phase II to phase III decomposition can take many years, may not be completed for decades, and is sometimes never completed. In phase III, bacteria gradually become established that are able to

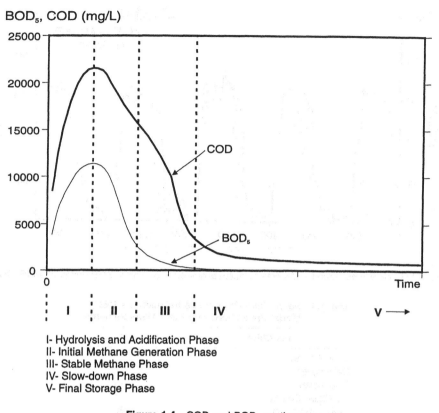

BOD$_5$, COD (mg/L)

I- Hydrolysis and Acidification Phase
II- Initial Methane Generation Phase
III- Stable Methane Phase
IV- Slow-down Phase
V- Final Storage Phase

Figure 1.4 COD and BOD$_5$ vs. time.

remove the soluble organic compounds, mainly fatty acids, which are largely responsible for the characteristics of phase II leachates. There is a depletion of both COD and BOD over time in phase III.

Leachates generated during phase III are often referred to as "stabilized," but in the life cycle of a landfill at this stage the landfill is biologically at its most active level. A dynamic equilibrium is eventually established between acetogenic and methanogenic bacteria, and wastes continue to actively decompose. Leachates produced during phase III are characterized by relatively low BOD values and low ratios of BOD to COD. However, ammonia nitrogen continues to be released by the first-stage acetogenic process and will be present at high levels in the leachate. Inorganic substances such as iron, sodium, potassium, sulfate and chloride may continue to dissolve and leach from the landfill refuse for many years.

The composition of phase III leachate is characterized by neutral pH levels, strongly reducing redox potential, low concentrations of VFAs and higher concentrations of refractory organic matter. The methanogenic phase is the longest and most important phase of waste stabilization, with phases IV and V occurring as the refuse becomes depleted of degradable organics.

Indications of the predominant organic acids in leachate over time are depicted in Figure 1.5. The result is that leachate is typically composed of hundreds of organic and inorganic contaminants. Christensen et al. (1994) categorize landfill contaminants into four groups:

1. Anthropogenic organic compounds originating from household or industrial chemicals. These compounds include, among others, aromatic hydrocarbons, phenols, and pesticides;
2. Organic matter that primarily consists of degradation products of solid organic matter in the waste material, in addition to anthropogenic compounds that typically have concentrations lower than 1 mg/L. (Christensen et al., 1994);

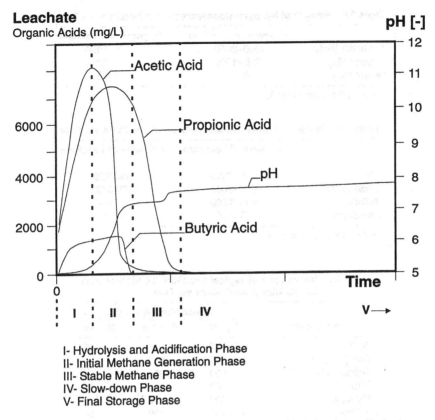

Figure 1.5   Change of concentrations of organic acids in leachate with time.

3. Inorganic species including calcium, magnesium, sodium potassium, ammonia, iron, manganese, chloride, sulfate, and bicarbonate; and,
4. Heavy metals including cadmium, chromium, copper, lead, nickel, and zinc.

Information presented so far indicates that the leachate coming from a single location within a landfill is highly variable over time. In addition, the variability in the refuse from location to location influences observed variabilities of leachate quality. It is important to note that some locations within a landfill will be at one phase of decomposition, while others will be at very different stages of decomposition. As a result, while much of the research focus has been on the use of isolated lysimeters, dealing with landfills as a whole is much more difficult because of the spatial heterogeneity of refuse in terms of the stages of decomposition.

Factors that influence leachate composition include refuse type and composition, refuse density, pretreatment, placement sequence and depth, moisture infiltration, ambient air temperature, landfill management practices, and time. The recognition and understanding of these influencing factors has greatly assisted the leachate characterization process. However, uncertainty is still inherent in the prediction of temporal trends in leachate strength in the highly heterogeneous landfill environment. Examples of leachate strength variability over time for nitrogen are provided in Table 1.5, for anion concentrations in Table 1.6 and some typical leachate concentrations for an array of constituents in Table 1.7. The ranges presented in these tables demonstrate that there can be significant degrees of variability of leachate strength from site to site.

The decline in chloride concentrations in landfill leachate can be primarily attributed to the long-term flushing from the refuse material. Chloride is not subject to such reactive transport processes as sorption or decay over time and space. The concentrations of organic compounds

Table 1.5   Ranges of Nitrogen Concentrations in Leachate vs. Time

	1–2 years of Leachate	10 years of Leachate
Ammonia (NH_3)	1000–2000	500–1000
Organic N_{org}	500–1000	10–50
Nitrate NO_3	0	0–10

Note: All values in mg/L.

Table 1.6   Ranges of Anion Concentrations in Leachate vs. Time

	1–2 years of Leachate	4–5 years of Leachate
Chloride	1000–3000	500–2000
Bicarbonate	1000–3000	1000–2000
Sulfate	500–1000	50–500
Phosphate	50–150	10–50

Note: All values in mg/L.

Table 1.7   Indications of Typical Leachate Concentrations for Various Constituents vs. Time

Constituent	Concentration (mg/L)		
	1 year	5 years	15 years
BOD	20,000	2,000	50
TKN	2,000	400	70
Ammonia-N	1,500	350	60
TDS	20,000	5,000	2,000
Chloride	2,000	1,500	500
Sulfate	1,000	400	50
Phosphate	150	50	—
Calcium	2,500	900	300
Sodium and potassium	2,000	700	100
Iron and magnesium	700	600	100
Aluminum and zinc	150	50	—

Note: BOD, biochemical oxygen demand; TKN, total Kjeldahl-N; TDS, total dissolved solids.

decrease more rapidly than chloride concentrations with age of the refuse, as indicated by the higher decay rate for total organic carbon (TOC) estimated by Lu et al. (1985). The more rapid decline in concentrations of organic compounds is primarily due to the additional process of biodegradation.

Table 1.8 lists examples of COD and BOD concentrations for a number of field sites. As the age of the site increases, the concentrations of BOD and COD decrease. With the passage of decades, the BOD, and to a lesser extent the COD, decrease to low concentrations. Table 1.9 indicates some examples of ranges of metal concentrations for municipal solid waste leachate. Decreases of metal concentrations over time are apparent as magnitudes typically decrease by one or more orders of magnitude.

Lu et al. (1985) have generalized the trends in the behavior of leachate contaminant concentrations as showing an initial rise to peak levels within the first few years of leachate generation, followed by a decline in concentrations at a constant exponential rate. There are two approaches indicated in the literature for predicting leachate contaminant concentrations. Wigh (1979) and other researchers have proposed semiempirical equations to describe the concentration history of various contaminants in leachate. The second approach for predicting leachate contaminant con-

Table 1.8 Examples of COD and BOD₅ Concentrations for Field Sites[a]

Field Site	Age (yr)	BOD (mg/L)	COD (mg/L)
Germany (several)	12	—	2,500
Guelph (Cureton et al., 1991)	15	—	14,300
Illinois (Dupage Co.)	15	—	1,340
Muskoka (Cureton et al., 1991)	20	—	2,727
England (Rainham)	24	30	1,300
Canada (Waterloo)	35	2	—

Note: COD, chemical oxygen demand; BOD₅, 5-day biochemical oxygen demand.

Table 1.9 Examples of Ranges or Median Values of Metal Concentrations for Municipal Solid Waste Leachate

Metal	Concentration (mg/L)	
	2 years	10 years
Iron (Fe)	500–1000	100–500
Calcium (Ca)	500–1000	100–500
Magnesium (Mg)	135	74–927
Manganese (Mn)	3.7	0.03–79
Arsenic (As)	0.0135	0.0002–0.98
Cadmium (Cd)	0.0135	0.0007–0.15
Chromium (Cr)	0.06	0.005–1.9
Copper (Cu)	0.054	0.003–2.8
Mercury (Hg)	0.006	0.0001–0.009
Nickel (Ni)	0.17	0.02–2.23
Lead (Pb)	0.063	0.005–1.6
Zinc (Zn)	0.68	0.03–350

centrations involves the fitting of empirical equations to leachate concentration curves based on data obtained from landfills or test lysimeters. The disadvantage of these empirical models is that they are applicable only to landfills of similar refuse composition, depth, and infiltration rate. The appropriateness of these models for application to other sites must be carefully assessed.

A better understanding of the changes of leachate concentration with time can be made by analyzing data from ongoing landfill monitoring. Information available from three landfills are briefly described as follows.

1.3.2.1 The Breitenau Landfill

This landfill represents a research reactor landfill consisting of 25,600 t of refuse (Lechner et al., 1993). Because 75 cm of raw compost was added to the refuse, significant moisture was available to promote microbial degradation. The ratio of BOD and COD concentrations changed during 4 years and Figure 1.6 shows the changes in the COD and BOD concentrations over time. Initially, the BOD/COD ratio was approximately 0.5 but, as the BOD concentration decreased through methanogenesis, both the COD and BOD decreased accordingly and the ratio became approximately 10 to 1. A continuing long tailing off of concentrations is clearly seen in the figure.

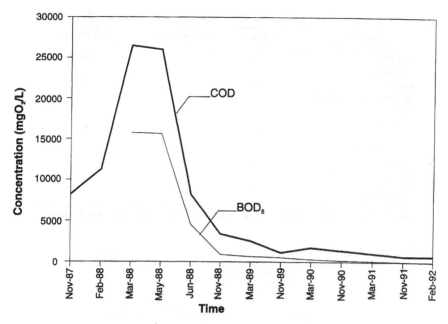

Figure 1.6   COD and BOD$_5$ vs. time for the Breitenau landfill leachate.

The corresponding changes in the pH as a function of time at the Breitenau landfill are depicted in Figure 1.7, where an initial dip in the pH value is evident. This is followed by a recovery to a pH of approximately 8. As the fatty acids are consumed by microorganisms, the pH increases and, since the acids are consumed, the BOD concentration of the leachate decreases. These temporal changes in BOD, COD, and pH follow exactly the temporal patterns referred to earlier.

Similarly, the concentrations of nitrogen subspecies vs. time are indicated in Figure 1.8. Organic and ammonia nitrogen attain concentrations of the order of 1500 mg/L, while nitrate nitrogen remains at very low levels. In Figure 1.9 concentrations of calcium, potassium, and magnesium vs. time are plotted for the Breitenau landfill. The results demonstrate dissolution, with long-term tailing off of concentrations; the results are in agreement with the attenuation of the various constituents as shown in Table 1.7. Due to the uniformity of the refuse resulting from the addition of raw compost, even a higher die-off rate is noted in their concentrations. The contents of the Breitenau landfill simulate behavior of a large lysimeter because field capacity due to the addition of the raw compost occurred from the outset.

1.3.2.2   The Greenlane Landfill

Greenlane landfill is a municipal solid waste landfill located in southwestern Ontario. The COD and BOD concentrations in the leachate vs. time for the Greenlane landfill are depicted in Figure 1.10 over a 1-year period. Perturbations of concentrations to an order of magnitude are apparent, indicating that single-grab samples of constituent concentrations may not be representative of average conditions. In this landfill, there is a lesser demonstration of the characteristic hydrolysis/acidification/methanogenesis shape as shown in Figure 1.4, because of the averaging effect from the size of the landfill and relatively brief period of monitoring. However, the temporal fluctuations are demonstrative of the types of variabilities witnessed at other sites, and the magnitudes of the COD and BOD values are indicative of expectations, given the pH values illustrated in Figure 1.11. The BOD concentrations in the 1000 to 6000 mg/L range indicate methanogenesis, with the expectation being

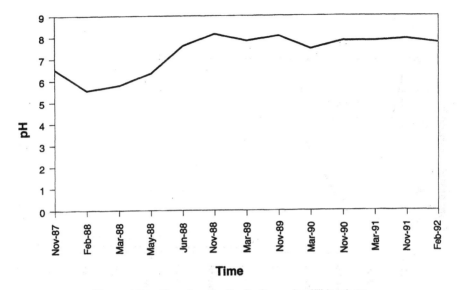

Figure 1.7 pH vs. time for the Breitenau landfill leachate.

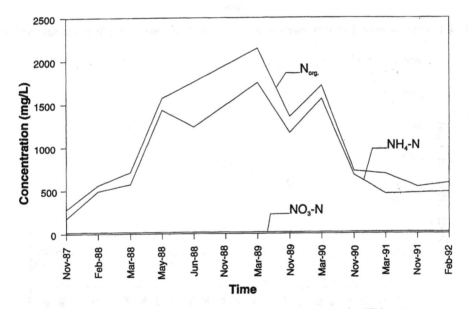

Figure 1.8 Nitrogen concentrations vs. time for the Breitenau landfill leachate.

that the pH would be buffered, to be in the range of 6.5 to 7.5. Similarly, during the 1-year period of monitoring, the COD and BOD concentrations decreased while the pH levels increased.

1.3.2.3 The Azusa Landfill

The Azusa landfill is located in Azusa, CA. Landfilling at the site commenced in 1952 and continued until 1995. It is relevant to note that this site received only approximately 45 cm of precipitation per year, and hence the percolation rate is low relative to landfills in many other regions. Therefore, similar to the trends noted at many landfills toward entombment of wastes, the contaminating life span of the refuse is very long when moisture levels and microbial degradation

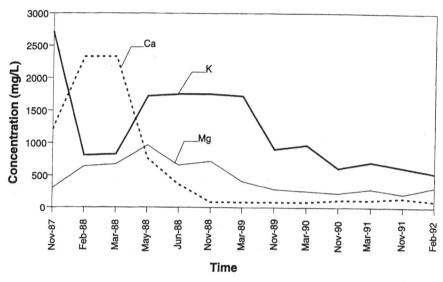

Figure 1.9   Calcium, potassium, and magnesium concentrations vs. time for the Breitenau landfill leachate.

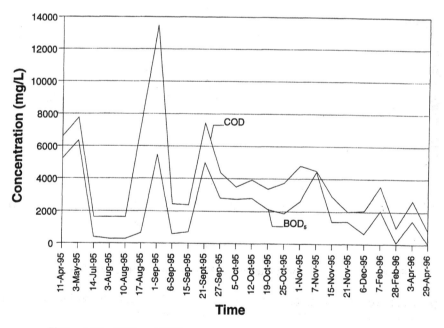

Figure 1.10   COD and BOD$_5$ vs. time for the Greenlane landfill leachate.

rates are low and COD and BOD concentrations are high. The site has a series of gas collection probes from which leachate was collected (Geoscience Support Services, Inc., and Conestoga–Rovers and Associates, 1995) and provided information on the quality of leachate from different portions of the landfill. The results of the monitoring at different locations for COD are illustrated

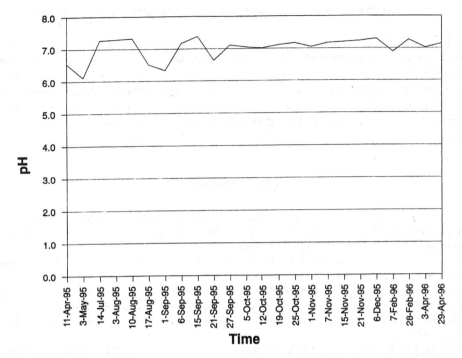

Figure 1.11   pH vs. time for the Greenlane landfill leachate.

in Figure 1.12; they illustrate the degree of variability that one might expect within a landfill. The COD concentrations at various locations within the refuse are very high.

These findings affirm that with lower moisture content, the decomposition rates and flushing decreases, thereby increasing the contaminating life span of the landfill.

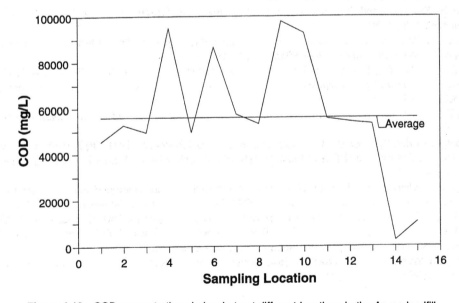

Figure 1.12   COD concentrations in leachate at different locations in the Azusa landfill.

1.4 IMPLICATIONS TO DESIGN OF CONSTRUCTED WETLANDS

The processes involved in the decomposition in landfills are relatively well understood. The reasons for the temporal variabilities in the quantity and quality of leachate in response to a number of factors are also clear. However, the heterogeneity of landfills and the variabilities in factors that influence leachate composition result in uncertainties in predictions of concentrations. Consequently, constructed wetlands must incorporate flexibility in design. Even though the temporal trends of leachate characteristics can be understood by quantifying the various aspects of the refuse, flexibility in the design of constructed wetlands must still be incorporated.

REFERENCES

Bagchi, A., 1990. *Design, Construction and Monitoring of Sanitary Landfill,* John Wiley & Sons, New York.

Chian, E., and DeWalle, F., 1977. Evaluation of Leachate Treatment, Vol. 2: Biological and Physical-Chemical Processes, U.S. EPA, 600/2/-77-186b, November.

Christensen et al., 1994. Attenuation of landfill leachate pollutants in aquifers, *Critical Reviews, in Environmental Science and Technology,* CRC Press, Boca Raton, FL.

Cureton, P. M., Groenvelt, P. H., and McBride, R.A., 1991. Landfill leachate recirculation: effects on vegetation vigor and clay surface cover infiltration, *Journal of Environmental Quality* 20:17–24.

Fungaroli, A., and Steiner, A., 1979. Investigation of Sanitary Landfill Behavior, Vols. 1 and 2, U.S. EPA - 600/2-79-053a, July.

Geoscience Support Services, Inc., and Conestoga–Rovers and Associates, 1995. *Evaluation of Proposed Vertical Expansion and Assessment of Groundwater Contamination,* report to Metropolitan Water District of Southern California and Azusa Landfill Task Force, Los Angeles, CA.

Lechner, P., Lahner, T., and Binner, E., 1993. Reactor landfill experiences gained at the Breitenau Research Landfill in Austria, paper presented at 16th International Madison Waste Conference, University of Wisconsin–Madison, Sept. 22–23, 1993.

Lu, J., Eichenberger, B., and Stearns, R., 1985. *Leachate from Municipal Landfills, Production and Management,* Noyes Publications, Park Ridge, NJ.

McBean, E., Rovers, F., and Farquhar, G., 1995. *Solid Waste Landfill Engineering and Design,* Prentice-Hall, Englewood Cliffs, NJ.

Murray, G., McBean, E. A., and Sykes, J., 1995. Estimation of leakage rates through flexible membrane liners, *Ground Water Monitoring and Remediation,* Fall:148–154.

Oweis, I. and Kehra, R., 1986. Criteria for geotechnical construction of sanitary landfills, in Fang, Hsai-Yang Fang, Ed. *International Symposium on Environmental Geotechnology — Vol. I,* Envo Publishing, Cincinnati, Ohio, 205–222.

Rovers, F. and Farquhar, G., 1973. Infiltration and landfill behavior, *Journal of Environmental Engineering* 99:671–690.

Schroeder, P., Dozier, T., Zappi, P., McEnroe, B., Sjostrom, J., and Peyton, R., 1994. The Hydrologic Evaluation of Landfill Performance (HELP) Model, EPA/600/R-94/168b, Risk Reduction Engineering Laboratory, Cincinnati, OH.

Walsh, J. and Kinman, R., 1979. Leachate and gas production under controlled moisture conditions, in *Municipal Solid Waste: Land Disposal,* EPA-600/9-79-023a, U.S. EPA, Cincinnati, OH, 41–57.

Walsh, J. and Kinman, 1981. Leachate and gas from municipal solid waste landfill simulators, in Schultz, D.W., Ed. *Land Disposal: Municipal Solid Waste,* EPA-600/9-81-002a, U.S. EPA, Cincinnati, OH, 67–93.

Wigh, R., 1979. Boone County field site interim report, EPA-600/2-79-058.

Zeiss, C. and Major, W., 1993. Moisture flow through municipal solid waste: patterns and characteristics, *Journal of Environmental Systems* 22:211–232.

CHAPTER 2

Constructed Wetlands
for Treating Landfill Leachate

Robert H. Kadlec

CONTENTS

ABSTRACT: Old and new landfills produce leachates, typically formed from infiltrating waters and the products of solid waste decomposition. Those contaminated leachate waters are a potential threat to surface and subsurface receiving waters. A very wide spectrum of pollutants is possible, reflecting the character of the materials contained in the stack. The principal categories of undesirable substances are (1) volatile organic compounds (VOCs), (2) nutrients, notably nitrogen, (3) heavy metals, and (4) priority toxic organic compounds. Wetland treatment of the leachate is one option for water quality improvement. The volumetric flow of leachate is often fairly small, compared with other wastewaters such as domestic, animal, and industrial or urban or agricultural runoff. Thus, long detention times are possible for small wetland/landfill area ratios.

Wetlands offer a wide spectrum of natural processes that may serve to reduce leachate contaminants. VOCs are air-stripped from the surface of the wetland waters and biodegraded by consortia of wetland microbes. Ammonium nitrogen may also volatilize and undergo nitrification and denitrification. The wetland carbon cycle provides the energy source for nitrate reduction. Nutrients are seasonally utilized by wetland biota, and residuals accrete as new wetland sediments and soils. Metals are sequestered in tissues of growing plants, ion-exchanged onto wetland sediments, and precipitated as sulfides and oxyhydroxide coprecipitates. Many toxic organics are biodegraded, either in the oxic surface zones, or in the anoxic or anaerobic sediments. The anoxic sediments are the site of sulfate reduction, which provides sulfides for metal removal.

Treatment wetlands may be designed to operate in a passive mode, and only minimal and infrequent maintenance activities are necessary. These features are exceptionally attractive for treatment in perpetuity, because of low present worth compared with other technologies. However, it is difficult, if not impossible, to design for removal down to detection limits, as may be required by regulation. The stochastic component of wetland performance is larger than for other technologies. Design data are lacking for many of the leachate constituents. Some leachate concentrations may be toxic to wetland biota, either the vegetation that supports wetland functions or the animals that may be attracted to the system. In the former case, pretreatment may be necessary. In the latter case, the more costly subsurface flow variants of the technology may be required.

2.1 INTRODUCTION

2.1.1 Treatment Wetlands in General

About 1000 wetland systems exist in North America, and a comparable number in Europe. These treat a variety of wastewaters, including municipal, mine drainage, urban and agricultural storm water, sludges, leachates, and various industrial effluents. Wetland treatment can take place in natural or (much more often) constructed wetland sites. Principal categories of constructed wetland systems include densely vegetated overland flows, subsurface systems, pond and island systems, and channels with floating plants.

Wetland treatment technology had its origins in Europe in the 1960s, with reed beds for reduction of organics in industrial wastewaters (Seidel, 1966; 1976). North American experience commenced in 1975, with projects at Houghton Lake, MI (Kadlec et al., 1979), as well as in Florida and Wisconsin. There now exist more than two dozen books, technology assessments, and design manuals on the subject. In addition, there is a large scientific literature that focuses on individual mechanisms of water purification.

A constructed surface-flow (SF) treatment wetland is a shallow pond (about 1 ft deep) that contains emergent wetland plants, such as cattails, bulrushes, or common reeds. Water is introduced at one end and proceeds across the wetland to the discharge point. If the basin is filled with pea gravel or other substrate, flow is subsurface (SSF wetland), in and around the roots of the wetland plants. Flow through the gravel may also be vertically downward.

Several component wetland processes combine to provide the observed overall treatments. Sedimentation and filtration remove solids. Chemical precipitation, ion exchange, and plant uptake remove metals. Nutrients are utilized by plants and algae and cycled to newly formed sediments. Volatile substances are gasified. Many materials undergo microbial transformations. These processes all lead to the transformation and transfer of a "removed" pollutant either to the atmosphere or to the wetland sediments and soils. The vegetation is extremely important for nutrient transformations and transfers, because it plays a key role in the cycling and temporary storage of many substances.

Removals proceed over the time water is held in the wetland. Therefore, detention time (or the equivalent hydraulic loading rate) becomes the key design variable. The longer the water is held in the wetland, the better the treatment. However, wetlands possess irreducible background concentrations of some substances: about 5 mg/L of biochemical oxygen demand (BOD) and total suspended solids (TSS), for instance. Typical detention times range from 1 to 10 days for existing treatment wetlands.

This technology requires land instead of mechanical devices to accomplish treatment. If the necessary land is available, it typically offers modest capital savings over competitive processes. However, it typically offers a very large advantage in operating costs, because operation is simple and maintenance is very low.

Other types of natural ecosystem may be used in series with wetlands. Lagoons, cascades, sand filters, intermittent vertical flow wetlands, and overland flow meadows all offer different facets of natural pollutant removal and processing. Such integrated natural systems offer capabilities outside the range of any individual component (Mæhlum, 1995).

2.1.2 Treatment Wetlands for Landfill Leachate

Wetland treatment of landfill leachates has been successfully tested at several locations. A facility at Ithaca, NY has been operating since 1989 (Staubitz et al., 1989; Surface et al., 1993), which has utilized SSF wetlands. SF wetlands have been operating successfully in Escambia County, FL since 1990 (Martin et al., 1993; Martin and Moshiri, 1994; Martin and Johnson, 1995). Cold climate systems are functioning properly in Norway (Mæhlum, 1994), as well as at several locations in Canada: Sarnia, ONT., Richmond, BC; Sackville, NS (Birkbeck et al., 1990; Pries, 1995). Reed beds are used to treat leachate in the U.K. (Robinson, 1990), Slovenia (Urbanc-Bercic, 1994; Bulc et al., 1997), and Poland (Agopsowicz, 1991).

Based on the current understanding of the effectiveness of wetland treatment of leachates, several U.S. projects are in planning and design stages. There has been project planning for Berrien County, MI project (New and Kadlec, 1993) and for projects in Elkhart Co., IN and Jackson Co., MI. In addition to the Isanti-Chisago leachate treatment wetland in Minnesota (MPCA, 1995), there are about a half-dozen other projects under development in various locations, such as Mississippi, Indiana, Pennsylvania, and West Virginia (Martin, 1994; Ogden, 1994; Knight, 1995). Wetlands have been proposed for control of storm water runoff from capped landfills (Mackie and Murphy, 1992).

Leachate water quality is quite variable from site to site, depending upon the contents of the stack and its hydrology. It is typically high in ammonium nitrogen and chemical oxygen demand (COD), with moderate quantities of volatile organics and metals (Table 2.1). However, the leachate flows are small compared with municipal wastewater flows, with a typical range of 40 to 400 m³/day (10 to 100,000 gal/day).

2.2 TREATMENT EFFICIENCIES AND LOADINGS

Wetland performance may be characterized by concentration reduction, by mass reduction, or by areal load reduction. With reference to Figure 2.1, these are defined by

$$\% \text{ Concentration Reduction} = 100(C_i - C_e)/C_i \qquad (2.1)$$

$$\% \text{ Mass Reduction} = 100(Q_iC_i - Q_eC_e)/Q_iC_i \qquad (2.2)$$

$$\text{Inlet Hydraulic Loading (HLR)} = q_i = Q_i/A \qquad (2.3)$$

$$\text{Inlet Areal Loading} = (Q_iC_i)/A \qquad (2.4)$$

$$\text{Load Removed} = (Q_iC_i - Q_eC_e)/A \qquad (2.5)$$

where $\quad A \;=\;$ wetland surface area, m^2
$\qquad C_e \;=\;$ outlet concentration, mg/L
$\qquad C_i \;=\;$ inlet concentration, mg/L
$\qquad q_i \;=\;$ inlet hydraulic loading, m/day
$\qquad Q_e \;=\;$ outlet flow, m^3/day
$\qquad Q_i \;=\;$ inlet flow, m^3/day

Concentration reduction may be the more important measure for potentially toxic materials, whereas mass reduction may be more important to the receiving ecosystem. Areal loading, or the areal load removed, is often the best measure of the stress placed upon the treatment wetland, or the mean storages occurring there.

Table 2.1 Selected Constituents in Leachates from Five Midwestern Landfills

		Isanti–Chisago, MN	Fulton Co., IN	Sarnia, Ont.	City Sand, MI	Saginaw, MI
BOD	mg/L	—	390	407	312	729
COD	mg/L	—	1540	1036	3203	—
TSS	mg/L	—	7840	—	241	—
N and P						
NH$_4$–N	mg/L	—	284	254	2074	322
NO$_3$–N	mg/L	—	3	<0.3	0	0
TKN	mg/L	—	284	328	—	670
TP	mg/L	—	0.92	—	—	—
VOCs						
BTX	µg/L	10	287	235	316	211
Light chlorinates	µg/L	33	0	10	5	78
Total	µg/L	342	1047	—	—	349
Metals						
Arsenic	µg/L	33	25	<100	—	6
Cadmium	µg/L	2	15	<3.4	2	30
Chromium	µg/L	6	164	22	216	99
Copper	µg/L	12	269	<35	23	54
Lead	µg/L	2	220	<75	13	5
Mercury	µg/L	1	<1	—	—	—
Nickel	µg/L	58	246	73	202	177
Iron	mg/L	18.0	178.0	17.6	—	22
Manganese	mg/L	2.2	2.7	0.8	—	0.3
Zinc	mg/L	0.30	1.82	0.36	0.81	0.19

Note: BOD, biochemical oxygen demand; COD, chemical oxygen demand; TSS, total suspended solids; TKN, total Kjeldahl nitrogen; TP total phosphorus; VOCs, volatile organic compounds; BTX, benzene, tolunene, xylenes.

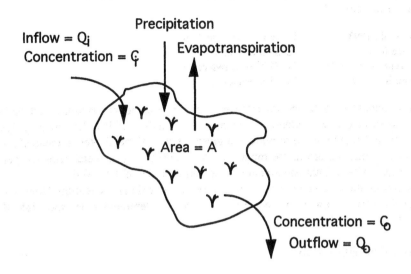

Figure 2.1 Treatment wetland notation.

2.3 PERFORMANCE

Detailed methodologies for treatment wetland design are available for many classes of pollutants (Kadlec and Knight, 1996). Some results for relevant substances are briefly illustrated below.

2.3.1 Toxic Organics

The number of organics that can potentially create environmental problems is very large. Not all of these have been studied with respect to wetland technology. Those that have been investigated show removals due to volatilization and biological degradation in the wetland environment. Wetlands can provide both aerobic and anaerobic zones in close proximity, which foster the different microbial processes along the redox gradient.

A shallow wetland water body provides the opportunity for air stripping of volatiles. The efficiency is not as great as for mechanical devices, but the difference is more than compensated by the long detention times in the wetland. Half-lives (time to volatilize one half the substance) range from 2 to 4 days for insoluble compounds such as benzene, toluene, and naphthalene (Mackay and Leinonen, 1975; Shugai et al., 1994), and project to be less for more volatile compounds, such as vinyl chloride and chloromethane.

More soluble, less volatile organics are less likely to volatilize, but degradation then has an opportunity to occur. As a result, half-lives for substances such as phenol, tetrahydrofuran, and methanol are also fairly short; these have been measured to be in the range of 10 to 40 h in wetlands (Srinivasan and Kadlec, 1995).

2.3.2 Metals

Metals are removed by cation exchange to wetland sediments, precipitation as sulfides and other insoluble salts, and plant uptake. Plant storage is typically predominantly to the roots (Surface et al., 1993). The anaerobic sediments provide sulfate reduction to sulfide and facilitate chemical precipitation. As a result, good removals of metals are reported for operating wetland facilities. For example, for zinc removal:

- Eger et al., 1993   90–96% mass reduction
  Surface flow   22–34 h detention
- Sinicrope et al., 1992   71–79% mass reduction
  Subsurface flow   24–31 h detention

Similar removals are obtained for other metals; for instance, chromium is reduced by 70% in about 70 h in surface flow wetlands (Srinivasan and Kadlec, 1995). The areal design approach selects the wetland area required to remove a given amount of metal. Recommended areas for iron and manganese reduction are in the range of 100 to 500 m^2/kg of metal removed per day (U.S. Bureau of Mines, 1991). This corresponds to detention times of 1 to 10 days.

Less is known about some elements, such as boron, arsenic and selenium. However, enough is known that the correct conditions for removal may be intentionally designed into the wetland (Masscheleyn et al., 1991).

2.3.3 Organics: BOD, COD

Nonvolatile, nontoxic organics are often lumped as COD or BOD, because they can create oxygen deficiencies in receiving waters. A very large body of knowledge has been generated from wetlands treating domestic wastewaters of medium strength and treating food processing and animal wastewaters of high strength. Approximately 90% of the incoming COD load can be eliminated by 7 days detention in an SF wetland (Kadlec and Knight, 1996). Mechanisms include oxidation and anaerobic digestion.

2.3.4 Nutrients

Reduction of nutrients — phosphorus and nitrogen compounds — requires the longest detention of any of the pollutants. Approximately 90% of the incoming nutrient load can be eliminated by 14 days detention in an SF wetland (Kadlec and Knight, 1996). These substances are required, at low loading rates, to sustain a healthy wetland. Nitrogen removal may require active or passive reaeration to promote nitrification. Phosphorus is stored in new wetland sediments. Nitrogen is also stored, but in larger measure is sequentially transformed, ultimately leaving the wetland as dinitrogen or ammonia.

2.3.5 Unexpected Substances

It is not possible to forecast the types of substances that will be contained in leachates over the entire history of anticipated discharge. Treatment wetlands have capabilities for removing a wide variety of substances, and hence operate as "broad-spectrum" treatment technology. They have the further property of pulse averaging, because of long detention times. A brief "spike" of a given substance will, at a minimum, be diluted by the relatively large volume of water in the wetland. Averaging is approximately over the detention time of the wetland.

2.4 WETLAND VEGETATION

A number of robust wetland plants are used in treatment wetlands, including cattails (*Typha* spp.), bulrushes (*Scirpus* spp.), and common reeds (*Phragmites australis*). These plants, as well as others, have a wide range of acceptable water quality, but do have limits outside which they cannot survive (Kylefors et al., 1994). Therefore, pretreatment may sometimes be necessary to ensure vegetation survival. Options include dilution with groundwater or recycled treated leachate.

Vegetation plays many important roles in wetland performance (Brix, 1997). Several processes are envisioned as being effective in pollutant reduction: phytoextraction, phytostabilization, transpiration stripping, and rhizofiltration. In particular, for the leachate context, vegetation fosters and provides several storage and reduction mechanisms.

Phytoextraction refers to plant uptake of toxicants, which is known to occur and has been studied in the storm water and mine water treatment wetland context. However, in many cases the contaminant is selectively bound up in belowground tissues, roots and rhizomes, and is not readily harvested. Metals are taken up by plants, and in many cases stored preferentially in the roots and rhizomes (Sinicrope et al., 1992). Phytostabilization refers to the use of plants as a physical means of holding soils and treated matrices in place. This process is also one of the chief underpinnings of treatment wetlands, as it relates to sediment trapping and erosion prevention in those systems. Wetland plants possess the ability to transfer significant quantities of gases to and from their root zone and the atmosphere (Brix, 1997). This ability is part of their adaptation required for survival in flooded environments. Stems and leaves of wetland plants contain airways (aerenchyma) that transport oxygen to the roots and vent water vapor, methane, and carbon dioxide to the atmosphere (Sorrel and Boon, 1994). There may also be transport of other gaseous constituents, such as dinitrogen and nitrogen oxides, and volatile hydrocarbons. The dominant gas outflow is water vapor, creating a transpiration flux upward through the plant. Rhizofiltration refers to a set of processes that occur in the root zone, resulting in the transformation and immobilization of some contaminants. Plants help create the vertical and microrhizal redox gradients that foster degrading organisms. Detritus provides both a food source and habitat for those microorganisms.

2.5 TYPE A AND TYPE B WETLANDS

There are both passive (type A) and active (type B) treatment wetlands. In the totally passive mode, the wetland conversions and storages proceed with no routine management, and only long-term sustainable processes are considered in design. Practically, this means that any potential "turnaround" activities would occur at intervals of 30 years or more. An example might be the increase in levee height to restore freeboard lost to sediment accretion. Type A wetlands can be either free water surface (FWS) or SSF, but type B wetlands would usually be SSF.

It is possible to utilize short-term, nonsustainable wetland functions. For instance, the sorption and cation exchange capacity of the original wetland substrate for metal removal will become exhausted after some period of time. Eger et al. (1993) provide insights into the capacity of some substrates, and the amounts of metals that may be stored on peat substrates. The design concept is a shallow basin filled with the selected substrate, arranged with suitable hydraulic conductivity to allow contacting of the entire bed of solids. Vegetation is established, and water flow initiated through the system. After some period, the sorption sites will be filled, and the metal removal potential will revert to a lower level, characteristic of the long-term sustainable mechanisms.

The life of the bed determines whether or not this mode of operation is feasible. At the end of that lifetime, the spent bed material must be excavated and disposed. The vegetative cover is lost. Fresh bed material is installed in the basin and revegetated. If the bed becomes saturated within a year or two, the system will be in a disrupted state virtually all the time, and it is likely that the economics will be unfavorable. If the life cycle is 5 to 10 years or longer, it becomes more probable that the physical and economic considerations will become more favorable. Nevertheless, the advantage of low operations and maintenance costs associated with type A wetlands is lost.

2.6 DESIGN MODELS

Design often contemplates a stable period of operation, over which input and outputs are averaged. The rate and temperature equations are of the form:

$$J = k \, (C - C^*) \tag{2.6}$$

$$R = k_V \, (C - C^*) \tag{2.7}$$

$$k = k_{20} \theta^{(T-20)} \tag{2.8}$$

$$k_V = k_{V20} \theta^{(T-20)} \tag{2.9}$$

where
C = concentration, mg/L
C^* = background concentration, mg/L
k_V = volumetric removal rate constant at e T °C, L/day
k_{V20} = volumetric removal rate constant at 20°C, L/day
J = areal removal rate, g/m²/year
R = volumetric removal rate, g/m³/day
k = areal removal rate constant at T °C, m/year
T = temperature, °C
k_{20} = areal removal rate constant at 20°C, m/year
θ = temperature coefficient, —
ε = porosity, —

The two alternate rate constants are related by the free-water depth, with $k = (\varepsilon h)k_V$, usually together with a change in timescale from years to days. These rates are used with the water mass balance to obtain pollutant concentration profiles along the flow direction. If flow is plug flow, with constant $(P - ET)$, exponential profiles are predicted (reaching a plateau of $C = C^*$):

$$\frac{C - C^*}{C_i - C^*} = \exp\left[-ky/q\right] \tag{2.10}$$

$$= \exp\left[-k_V \tau y\right] \tag{2.11}$$

where
q = hydraulic loading rate, m/year
τ = nominal detention time, day
y = fractional distance through wetland, = x/L

However, if flow varies due to $\alpha = P - ET \neq 0$, then power law profiles are predicted:

$$\frac{C - C'}{C_i - C'} = \left(1 + \frac{\alpha y}{q}\right)^{-\left(1 + \frac{k}{\alpha}\right)} \tag{2.12}$$

$$C' = C^*\left[\frac{k}{k + \alpha}\right] \tag{2.13}$$

Evapotranspiration (rain) has two effects: lengthening (shortening) of detention time and concentration (dilution) of dissolved constituents. The use of Equations 2.10 and 2.11 with an average flow rate compensates for altered detention time, but not for dilution or concentration. The fractional error due to flow averaging is approximately equal to α/q, for $\alpha/q > -0.5$. Thus, if 25% of the inflow evaporates, use of Equations 2.10 and 2.11 with average flow predicts concentrations 25% lower than required by the mass balance. If rain adds 25% to the flow, use of Equations 2.10 and 2.11 predicts concentrations 25% higher.

Kadlec and Knight (1996) surveyed information from many FWS and SSF wetlands treating domestic wastewater, and determined the central tendencies for areal rate constants, background concentrations, and temperature factors (Table 2.2). In the case of nitrogen species, the rate constants are for the sequential conversion processes, and not for system inputs and outputs. Considerable intersystem variability was found, as might be anticipated from the large variability in wetland characteristics. The designer should modify values in Table 2.2 according to the specifics of the intended wetland.

Table 2.2 Rate Constants, Background Concentrations, and θ Factors

	Surface Flow			Subsurface Flow		
	k_{20}, m/yr	C^*, mg/L	θ	k_{20}, m/yr	C^*, mg/L	θ
Based on Extensive Data						
BOD	34	3+	1.000	180	3+	1.000
TSS	1000	5+	1.000	3000	7+	1.000
TN	22	1.5	1.050	27	1.5	1.050
TP	12	0.02	1.000	12	0.02	1.000
Based on Scant Data						
VOCs	20–200	≈0.0				
Cadmium	25–35	≈0.0				
Chromium	25–35	≈0.0				
Copper	20–50	≈0.0				
Nickel	15–18	≈0.0				
Zinc	20–30	≈0.0				

Note: BOD, biochemical oxygen demand; TSS, total suspended solids; TN, total nitrogen; TP, total phosphorus; VOCs, volatile organic compounds.

Adapted from Kadlec and Knight, 1996.

Stochastic variability occurs in treatment wetland performance. Causes include short-term dynamics; variations in input flow and concentration; meteorological events of rain, drought, and heat waves; and biological influences due to algae, insects, fish, birds, and other animals. The result is a large degree of "chatter" about the mean performance (Figure 2.2). The output does not always track the input, and measurements occupy a wide band about the mean.

In addition to the mean behavior described by the equations given above, measures of the variance about this mean are required. The frequency distributions of inlet and outlet concentrations provide this additional description (Figure 2.3). Current regulatory requirements in the U.S. often dictate a maximum monthly value; other countries place a maximum on a given percentile, typically the 80th or 90th. Design must acknowledge regulatory requirements in most cases, and so must account for stochastic as well as deterministic effects. Where seasonal patterns are known to be significant, the design equations may be applied on that seasonal basis, but random variability still remains. Design can include this chatter, if the design target is adjusted downward by approximately a factor of 2.0 to meet a monthly maximum cap. Wetland sizes for different first-order removal rate constants are fairly small for low leachate flows (Table 2.3).

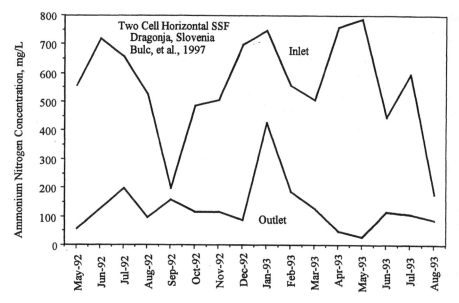

Figure 2.2   Time sequence of inlet and outlet ammonium nitrogen for a constructed wetland treating leachate. (From Bulc, T. et al., *Water Sci. Technol.*, 35, 301–306, 1997. With permission.)

2.7 BIOHAZARDS AND ACCUMULATION

Leachates often contain metals and organics in concentrations that constitute potential hazards to wildlife and other biota. It is therefore necessary to remain cognizant of the levels that may accrete in a treatment wetland and to take precautions to prevent deleterious contacts. The hazard may take the form of either acute or chronic toxicity, or more subtle effects caused by increased body burdens for the exposed organisms. For some constituents, such as mercury, the contaminant

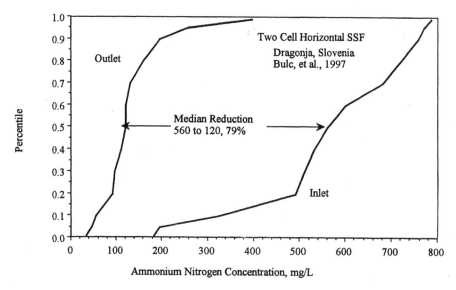

Figure 2.3   Frequency distribution of inlet and outlet ammonium nitrogen for a constructed wetland treating leachate. (Based on Bulc, T. et al., *Water Sci. Technol.*, 35, 301–306, 1997.)

Table 2.3 Wetland Areas (ha) Required for a Flow of 40 m³/day (10,000 gal/day)

% Reduction to Background	k = 5 m/yr	k = 15 m/yr	k = 50 m/yr
99	2.69 (202)	0.90 (67)	0.27 (20)
95	1.75 (131)	0.58 (44)	0.17 (13)
90	1.34 (101)	0.45 (34)	0.13 (10)

Note: A chatter factor of 2.0 has been applied to meet max monthly limit. The detention time at a free water depth of 30 cm is in parenthesis. Rain and evapotranspiration are assumed to balance.

may be stored in selected organs and build up over time, which is the process of bioaccumulation. There is also sometimes the potential for the stored substances to pass on to the next level consumers, which then concentrate the lower-level accumulations. This is the process of biomagnification.

Three methods for controlling contact are dilution, deep pond accretion, and use of SSF wetlands. Dilution of incoming leachate with either another wastewater or recycle from a downstream section of the system can reduce the concentration in the water, but not the loading to the system. However, the greatest amounts of contaminants are often in the wetland sediments, which creates a hazard for sediment-grazing organisms. It is therefore necessary to consider the concentration of the contaminant in the sediments. For instance, metals are removed by a number of mechanisms that all end up with storage in the upper soil sediment horizon. There is also dilution in the sediment layer by newly formed solids, which contain the recalcitrant residuals formed by the biogeochemical cycle. Wetlands have a strong carbon cycle, created by growth, death, litter fall, and decomposition of the microbes, invertebrates, and vegetation. Residuals cause "peat" accretion, at rates up to 1000 g/m² year. Metal accretion is then buried in this newly forming matrix, to yield an overall new sediment contaminant level.

The budgets for solids and contaminants in solids in the wetland are therefore linked to the budget for contaminants in the water in the wetland (Figure 2.4). Removal from the water to the sediments transfers the pollutant, and its potential hazards, to a new location. As a consequence, consideration must be given to both sediment and surface water quality standards. An example of a wetland hydraulically loaded at 5 cm/day is given in Table 2.4. The detention time for this wetland

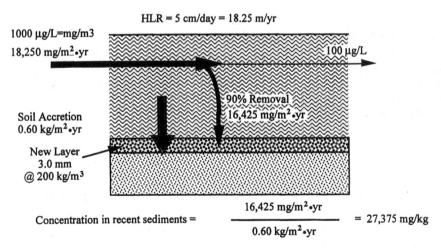

Figure 2.4 Both metals and new soils accrete in surface layers that are foraged by organism that may then bioaccumulate the metal.

Table 2.4 Sustainable Wetland Metals Criteria

| | If Sediment Standards[a] Are | | Then Allowable Surface Water Concentrations Are | | Compare to Surface Water Standards | |
	Ontario Severe Effect, mg/kg	USDA Proposed, mg/kg	Ontario Severe Effect, µg/L	U.S. EPA Proposed, µg/L	Michigan, µg/L	USDA µg/L
Cadmium	10	18	0.3	0.6	3	0.66
Chromium	110	2000	3.6	65.8	11	50
Copper	110	1200	3.6	39.5	30	30
Lead	250	300	8.2	9.9	20	30
Mercury	2	15	0.07	0.49	ND	0.2
Nickel	75	500	2.5	16.4	21	88
Zinc	820	2700	27.0	88.8	100	59

Note: New sediment accretion dilutes removed metals. Allowable surface water concentrations meet sediment standards and are comparable with the separately determined surface water quality standards. See Figure 2.4.

[a] HLR = 5 cm/day, % removal = 100%, and accretion = 0.6 kg/m²yr.

would be on the order of a week, for 35 cm free water depth. It can be seen in this example that the two sets of standards are compatible for the treatment wetland context, because the water concentrations that do not threaten the sediments (severe effects) compare favorably with the surface water standards. Cross-budget calculations vary with removal rates, hydraulic loading, and carbon cycling and, consequently, must be repeated for the specific site circumstances.

The wetland sediments may be isolated from contact with higher organisms by ensuring that they accrete in isolation. In turn, that may be accomplished by using a SSF wetland, or a floating vegetative mat "wetland." The gravel bed hydroponic system (SSF) system, when operated properly, places the water and aquatic sediments belowground and out of reach of sediment foragers. The deep pond achieves the same effect by accreting residuals at depth sufficient to be out of reach of those sediment foragers.

2.8 AIR POLLUTION

Constructed wetlands, especially FWS wetlands, serve as low-efficiency air-strippers, or water weathering systems. Volatile materials leave the water and enter the atmosphere, thus transforming a water contaminant to an air contaminant. Due consideration must therefore be given to the potential air quality degradation that results from this process. For trace organic contaminants, this air pollution load is often not a source of concern. For ammonia volatilization, there may be a local airborne "fertilizer" concern. Air quality regulations may, in the future, require design constraints based on volatile emissions.

2.9 ADVANTAGES AND DISADVANTAGES

Constructed wetlands have the advantage of offering long-term, sustainable treatment with very low operations and maintenance costs for type A systems. This is especially important for leachate control, which often requires indefinite treatment lifetimes. It is also often important to build projects with guaranteed long-term stewardship. Passive constructed wetlands offer very long lifetimes, with little or no equipment replacement. Type B wetlands offer the same advantages for the online period, but do require periodic refurbishment and spent substrate disposal.

In contrast to chemical and physical processing alternatives, wetlands provide insurance against unanticipated new pollutants. For instance, conventional air-stripping can be effectively used to reduce concentrations of ammonia nitrogen and other volatiles. But that technology has no capacity to deal with metals. Wetlands have capacity to deal with both. If in the lifetime of the leachate source it becomes a metals source, the wetland will have some capacity to treat this new pollutant.

In general, treatment wetlands are a land-intensive technology, and can be precluded on the basis of land availability. This is a serious disadvantage for treating large flows, but is less of a constraint for leachate flows of low volume. Candidate sites may include the landfill cap or the buffer zone around it.

Perhaps the most important limitation of wetlands technology is the difficulty of producing very high purity waters. Two wetland features create this difficulty: the unavoidable presence of small preferential channels from inlet to outlet and the stochastic nature of the biological system. Tiny bypasses do not preclude reductions up to 90%–95%, but do have a strong influence if removals are to be in the range of 99 and higher. Stochastic effects include a variety of biological factors, such as insect attack and seasonal variability, as well as environmental factors, such as drought and high rainfall. High evapotranspiration can drive up concentrations via water loss, despite high load reductions. High rainfall can create temporary flushing spikes.

The potential for bioaccumulation and resultant biohazards can place constraints on the implementation of wetland technology. While these may be avoided by the methods outlined above, the result may be a more costly, less passive alternative.

2.10 ECONOMICS

Capital costs for constructed wetlands vary, depending upon factors such as liners, pumps, and land costs. A median cost for SF wetlands is $50,000/ha ($20,000/acre), based on information in the U.S. EPA Treatment Wetland Database (Knight et al., 1993). The median cost for SSF wetlands, from the same source, is $360,000/ha ($145,000/acre). It is quite possible to achieve significantly lower costs for specific projects. Liners and land acquisition, pumps and piping all add to the basic costs of excavation and vegetation establishment.

Type A wetlands require very little operation and maintenance (O&M). The U.S. EPA Treatment Wetland Database (Knight et al., 1993) shows a median O&M cost of $400/acre/year, but that result is for large systems. Large components of O&M are contributed by pumping energy and sample acquisition and analysis.

2.11 CONCLUSIONS

Treatment wetlands can contribute to the improvement of leachate water quality. It is probable that they would be part of an integrated natural system, comprised of aquatic and terrestrial components together with wetlands. The spectrum of possibilities is bracketed by completely passive FWS wetlands at one extreme, and periodically refurbished SSF wetlands at the other extreme. Early experiences with the application of this technology have been positive.

It is necessary to understand thoroughly the fate of the leachate contaminants, to be able to assess the potential for wildlife exposure and other biohazards fully. Appropriate design methods may be able to deal with issues of accumulation and contaminant transfer, but pretreatment or dilution may be necessary.

On the positive side of the ledger, wetlands may offer a passive, sustainable remedy for a large number of chemical constituents. They are a low-cost alternative. Because leachate flows are relatively modest, the land-intensive attribute of treatment wetlands does not cause severe siting limitations. On the negative side, removals may be limited to something less than complete elimination, and performance may vary stochastically. The list of leachate pollutants is impressively long, and design information on these numerous compounds is not available at this time. The database on performance is growing rapidly, and more specific advantages and disadvantages will be understood in the near future.

REFERENCES

Agopsowicz, M., 1991. Contamination of groundwater near landfills in Poland and possibilities for protecting it using plants, in Etnier, C., and Guterstam, B., Eds. *Ecological Engineering for Wastewater Treatment*, Bokskogen, Gothenburg, Sweden.

Birkbeck, A. E., Reil, D., and Hunter, R., 1990. Application of natural and engineered wetlands for treatment of low-strength leachate, in Cooper, P. F., and Findlater, B. C., Eds. *Constructed Wetlands in Water Pollution Control*, Pergamon Press, Oxford, U.K., 411–418.

Brix, H., 1997. Do macrophytes play a role in constructed treatment wetlands? *Water Science and Technology* 35:11–17.

Bulc, T., Vrhovsek, D., and Kukanja, V., 1997. The use of constructed wetlands for landfill leachate treatment, *Water Science and Technology* 35:301–306.

Eger, P., Melchert, G., Antonson, D., and Wagner, J., 1993. The use of wetland treatment to remove trace metals from mine drainage, in Moshiri, G.A., Ed. *Constructed Wetlands for Water Quality Improvement*, Lewis Publishers, Boca Raton, FL, 171–178.

Kadlec, R. H. and Knight, R. L., 1996. *Treatment Wetlands*, CRC Press, Boca Raton, FL, 893 pp.

Kadlec, R. H., Tilton, D. L, and Schwegler, B. R., 1979. Three-year Summary of Pilot Scale Operations at Houghton Lake, Report to the National Science Foundation, February, NTIS PB 295965.

Knight, R. L., 1995. Personal communication on projects underway.

Knight, R. L., Ruble, R. W., Kadlec, R. H., and Reed, S. C., 1993. Wetlands for wastewater treatment performance database, in Moshiri, G. A., Ed. *Constructed Wetlands for Water Quality Improvement*, Lewis Publishers, Boca Raton, FL, 35–58.

Kylefors, K., Grennberg, K., and Lagerkvist, A., 1994. Local treatment of landfill leachates, in *Proceedings of the Fourth International Conference on Wetland Systems for Water Pollution Control*, Guangzhou, China, 539–548.

Mackay, D. and Leinonen, P. J., 1975. Rate of evaporation of low-solubility contaminants from water bodies to the atmosphere, *Environmental Science and Technology* 9:1178–1180.

Mackie, D. H. and Murphy, D. A., 1992. Using wetlands to control stormwater runoff from a municipal landfill: one facet of a comprehensive environmental strategy, in *Proceedings of the Water Environment Federation 65th Annual Conference*, IX: 139–148.

Mæhlum, T., 1995. Treatment of landfill leachate in on-site lagoons and constructed wetlands, *Water Science and Technology* 32:129–135.

Martin, C. D., 1994. Personal communication on projects underway.

Martin, C. D. and Johnson, K. D., 1995. The use of extended aeration and in-series surface-flow wetlands for landfill leachate treatment, *Water Science and Technology* 32:119–128.

Martin, C. D. and Moshiri, G. A., 1994. Nutrient reduction in an in-series constructed wetland system treating landfill leachate, *Water Science and Technology* 29:267–272.

Martin, C. D., Moshiri, G. A., and Miller, C. C., 1993. Mitigation of landfill leachate incorporating in-series constructed wetlands, in Moshiri, G. A., Ed. *Constructed Wetlands for Water Quality Improvement*, Lewis Publishers, Boca Raton, FL, 462–473.

Masscheleyn, P. H., Delaune, R. D., and Patrick, W. H., Jr., 1991. Arsenic and selenium chemistry as affected by sediment redox potential and pH, *Journal of Enviornmental Quality* 20:522–527.

Minnesota Pollution Control Agency (MPCA), 1995. Isanti-Chisago Landfill Groundwater Treatment, Project Description Brochure, MPCA Public Information Office, 520 Lafayette Road, St. Paul.

New, J. F. and Kadlec, R. H., 1993. Report on the Feasibility of Treating Contaminated Groundwater at the Southeast Berrien County Landfill Using Constructed Wetlands, Report Southeast Berrien County Landfill Authority, Buchanan, MI.

Ogden, M. H., 1994. Personal communication on projects underway Southwest Wetlands Group.

Pries, J. H., 1994. Wastewater and Stormwater Applications of Wetlands in Canada, North American Wetlands Conservation Council, Ottawa, Ontario.

Robinson, H., 1990. Leachate treatment to surface water standards using reed bed polishing, in *The Use of Macrophytes in Water Pollution Control Report,* IAWPRC Newsletter, Vol. 3, p. 32, IAWQ, London.

Seidel, K., 1966. Reinigung von Gewässern durch höhere Pflanzen, *Naturwissenschaften* 53:289–297.

Seidel, K., 1976. Macrophytes and water purification, Chapter 14, in Tourbier, J. and Pierson, R.W., Jr., Eds. *Biological Control of Water Pollution,* University of Pennsylvania Press, Philadelphia, 340 pp.

Shugai, D. et al., 1994. Removal of priority organic pollutants in stabilization ponds, *Water Research* 28:681–685.

Sinicrope, T. L., Langis, R., Gersberg, R. M., Busnardo, M. J., and Zedler, J. B., 1992. Metal removal by wetland mesocosms subjected to different hydroperiods, *Ecological Engineering* 1:309–322.

Sorrell, B. K. and Boon, P. I., 1994. Convective gas flow in *Eleocharis sphacelata* R. BR.: methane transport and release from wetlands, *Aquatic Botany* 47:197–212.

Srinivasan, K. and Kadlec, R. H., 1995. Wetland Treatment of Oil and Gas Well Wastewaters, Report to U.S. Department of Energy, Contract DE-AC22-92MT92010.

Staubitz, W. W., Surface, J. M., Steenhuis, T. S., Peverly, J. H., Lavine, M. L., Weeks, N. C., Sanford W. E., and. Kopka, R. J., 1989. Potential use of constructed wetlands to treat landfill leachate, in Hammer, D. A., Ed. *Constructed Wetlands for Wastewater Treatment,* Lewis Publishers, Chelsea, MI, 735–742.

Surface, J. M., Peverly, J. H., Steenhuis, T. S., and Sanford, W. E., 1993. Effect of season, substrate composition, and plant growth on landfill leachate treatment in constructed wetlands, in Moshiri, G. A., Ed. *Constructed Wetlands for Water Quality Improvement,* Lewis Publishers, Boca Raton, FL, 461–472.

Trautman, N. M., Martin, J. H., Jr., Porter, K. S., and Hawk, K. C., Jr., 1989. Use of artificial wetlands for treatment of municipal solid waste landfill leachate, in Hammer, D. A., Ed. *Constructed Wetlands for Wastewater Treatment,* Lewis Publishers, Chelsea, MI, 245–251.

Urbanc-Bercic, O., 1994. Investigation into the use of constructed reedbeds for municipal waste dump leachate treatment, *Water Science and Technology* 29:289–294.

U.S. Bureau of Mines, 1991. Technology Transfer Announcement. U.S. GPO: 1991-511-508.

CHAPTER 3

Wetlands for Treatment of Landfill Leachates in Cold Climates

Trond Mæhlum

CONTENTS

ABSTRACT: The best prospects for wetland treatment systems are in the warm regions; however, the number of wetland systems treating different kinds of wastewater, including landfill leachate, is steadily increasing in cold temperate regions. The paper reviews how cold weather conditions affect wetland processes and treatment results, and how these impediments can be overcome by proper design and operation. Three major engineering concerns in the use of constructed wetlands (CWs) in cold climates are ice formation, hydrology, and temperature effects on biologically or microbiologically mediated treatment processes. Filter clogging by metal oxides requires high removal of biodegradable matter and suspended solids before treatment of leachate in horizontal or vertical subsurface-flow wetlands. On-site, "high-tech" leachate treatment systems are often avoided in Scandinavia because of expensive construction and operation. The paper presents examples of wastewater wetland treatment and leachate treatment systems in cold climatic regions. Low-cost extended aeration lagoons and sequenced batch reactors (SBR) have often been successful in removing organic matter (chemical oxygen demand, COD, and biochemical oxygen demand, BOD) and NH_4–N prior to CWs. Use of constructed and natural wetlands (peat irrigation) as secondary or tertiary treatments are still in the process of development. Experience shows that integrated systems can meet effluent criteria for most water quality parameters. In general, subsurface-flow CWs provide greater thermal protection than free-water systems, because of the insulation effect of the unsaturated surface layer. Vertical subsurface-flow systems, with subsurface intermittent loading, may be superior to horizontal-flow systems, because of higher availability of oxygen in vertical filters. In locations with air temperatures less than –15°C, seasonal storage of the winter load may be necessary.

3.1 INTRODUCTION

Infiltration of precipitation and transport of water through municipal sanitary waste (MSW) landfills produces leachates containing varying quantities of undesirable and even toxic organic and inorganic substances. Unlike municipal wastewater, leachate composition and flow rates vary widely, depending on the type and age of wastes, landfill design, and climate.

On-site, "high-tech" leachate treatment systems are often avoided because of high costs of construction and operation. An alternative is cost-efficient natural treatment systems, such as constructed wetlands (CWs) for secondary or tertiary treatment (Sandford et al., 1990; Surface et al., 1992; Robinson, 1993; Martin and Moshiri, 1994; Bulc et al., 1997). CWs can be designed as free-water systems (FWS) and subsurface vertical-flow (SVF) or horizontal-flow (SHF) systems. The advantages of using CWs to treat leachate include, among others, large adsorptive surfaces, aerobic–anaerobic interfaces, and diverse microbial populations.

Climate influences all stages of the planning, design, construction, and maintenance of CWs. It is a major determinant in the success of this ecotechnology because thermal conditions affect removal processes. In northern environments, for example, ice formation can interrupt physical, chemical, and biological activities. Although the best prospects for successful wetland treatment are in the warm regions, studies in Scandinavia (Schierup et al., 1990b; Sundblad and Wittgren, 1991; Wittgren and Tobiason, 1995; Mæhlum et al., 1995; Wittgren and Mæhlum, 1997) and North America (Herskowitz, 1986; Lakshman, 1992; Reed et al., 1995; Lemon et al., 1996; Pries, 1996; Kadlec and Knight, 1996) have shown that wetland treatment is a feasible technology in cold regions. These documented studies suggest that wetland treatment is supplementing and occasionally supplanting conventional treatment methods. However, regulatory agencies remain skeptical about the use of CWs in cold climatic regions.

In this paper the term *cold climate* refers to environments where the coldest and warmest months have mean temperatures below –3°C and above 10°C, respectively (Köppen–Geiger–Pohl classification, described by Strahler and Strahler, 1992). Generally speaking, these environments are snow-covered for at least 1 month annually, and the forests are primary boreal. This classification includes most of Canada, Alaska and northern U.S., Scandinavia, eastern Europe, Russia, and northeastern China.

This paper reviews how cold weather conditions affect wetland processes, and how CW design and operation can mitigate the adverse effects of ice. In terms of removal and cycling processes of chemical constituents in leachates, this review is limited to carbon (C), nitrogen (N), phosphorus (P), and iron (Fe). The review is mainly based on experiences with municipal wastewaters and landfill leachate from Scandinavian research projects (Schierup et al., 1990b, Jenssen et al., 1996, Wittgren and Mæhlum, 1997; Mæhlum and Jenssen, 1998); recent books by Reed et al. (1995) and Kadlec and Knight (1996), papers presented at the Symposium on Constructed Wetlands in Cold Climates held in Niagara-on-the-Lake, Ontario, in June, 1996; and papers presented at the "5th International Conference on Wetland Systems for Water Pollution Control" in Vienna, in September, 1996 (Haberl et al., 1997).

3.2 GEOGRAPHICAL DISTRIBUTION OF CONSTRUCTED WETLANDS IN COLD CLIMATIC REGIONS

Several efforts have been made during the last few years to compile databases about treatment wetlands. Since the number of treatment wetlands has been increasing steadily, and not all are reported in the databases, the numbers reported in this paper are conservative estimates. As far as we know, there is no database specially devoted to CWs for landfill leachate treatment. The status of wetland treatment in Europe will be reported in a forthcoming book, *Constructed Wetlands for Wastewater Treatment in Europe* (Vymazel et al., 1998).

3.2.1   Canada and the U.S.

A national survey identified 67 wetlands treating wastewater and storm water in Canada (Pries, 1994; 1996). Of these, 67% were full-scale operating systems. These systems have successfully met effluent criteria across Canada, and as far north as the Yukon and the Northwest Territories. At several locations, wastewater is stored during the winter and then discharged to the wetland during the spring, summer, and autumn. The advantage of this approach is the use of design for warm weather conditions, but the disadvantage is the requirements for storage lagoons. Research on winter performance of subsurface vertical flow CWs receiving lagoon effluent is in progress at Niagara-on-the-Lake, Ontario (Lemon et al., 1996).

From the North American Treatment System Database, Kadlec and Knight (1996) reported 176 wetland treatment sites in Canada and the U.S. Approximately 60 of these are located in the cold climatic regions, including more than 40 in South Dakota. Almost all of the cold climate wetlands were FWS, and almost all (90%) treated municipal wastewater. CWs treating landfill leachate have been established in New York (Surface et al., 1992). Other contributions in this book present new experiences of design and operation of leachate treatment in CWs in both Canada and the U.S.

3.2.2   Scandinavia

In Scandinavia landfill leachate is mainly discharged directly to receiving waters or municipal sewers, on-site treatment lagoons, or aquifers. Spray irrigation of leachate onto grassland, woodland, and peat slopes has been successful in both Sweden and Norway, but these systems do not include CWs. Denmark has pioneered the promotion of SHF wetlands (root-zone method) for municipal wastewater treatment. The Danish database contains 109 systems constructed between 1983 and 1990, most treating municipal wastewater (Schierup et al., 1990a). Designs and operations from 71 were summarized and have been reviewed by Schierup et al. (1990b) and by Kadlec and Knight (1996). In Sweden there are at least six FWS wetlands and eight SHF wetlands treating municipal or domestic wastewater. In addition, there are at least 20 FWS wetlands treating agricultural runoff or storm water. In Norway, as in Sweden, treatment wetlands have not yet gained final approval from the regulatory authorities; therefore, few full-scale systems are commercially operating. Fifteen full-scale SHF systems and two FWS wetlands have been built in Norway (Mæhlum and Jenssen, 1998) for household wastewater, package treatment plant effluent, gray water, and landfill leachate. The majority of the Norwegian CWs are experimental and monitored under different research programs. Preliminary results of leachate monitoring and treatment efficiency of two Norwegian leachate treatment systems including extended aeration lagoons and CWs are presented by Mæhlum et al. (Chapter 9).

3.2.3   Eastern Europe

The use of wastewater treatment wetlands in Eastern Europe is most extensive in the Czech Republic, where 28 systems have been built, or are under construction, since 1989. In addition, 54 systems are in the design stage (Vymazal, 1996). All systems are of the SHF type, mostly for mechanically pretreated domestic or municipal effluent. In Slovenia, a SHF CW system treating landfill leachate has been in operation since 1992 (Bulc et al., 1997). In Estonia, there are at least 12 CWs (SHF, SVF, or FWS). These systems are small, 90 to 2400 m², but several larger FWS systems have been designed (Mander and Mauring, 1997). In Poland, there are several facilities of SHF, SVF, and FWS CWs treating wastewater in the Gdansk region, northern Poland (Obarska-Pempkowiak, 1996). In several former Soviet republics, now constituting the CIS (Community of Independent States), wetlands have been constructed since the mid-1980s to treat various waste-waters (Magmedov, 1996). By 1996, more than 30 systems had been constructed throughout the CIS. Several of them are SVF systems designed to have a free-water surface.

3.3   CLIMATE-DEPENDENT PROCESSES

Three major concerns in cold climates are (1) ice formation, (2) hydrology and hydraulics, and (3) the thermal effects on biologically or microbiologically mediated treatment processes.

3.3.1   Ice Formation and Insulation

Although stored energy from the underlying soil during the warm season retards ice formation (Kadlec, 1996), ice begins to form on a water surface when the bulk water temperature reaches 3°C because of density differences and convective losses (Reed et al., 1995). Leachate temperature is normally >10°C due to year-round biological processes in the landfill. However, the leachate temperature entering the wetland unit will be modified by pretreatment; for example, lagoons with surface aerators can decrease leachate temperature to near 0°C during cold periods. The water in natural swamps and marshes often does not freeze in winter, due to an insulating layer of snow, often enhanced by standing dead vegetation, which collects drifting and falling snow. Snow depth is thus often greater than the accumulated snowfall. If snow accumulates before a significant ice layer is formed, subsequent freezing is greatly reduced. In addition, wetland vegetation can reduce wind-induced heat loss (Reed and Calkins, 1996).

The presence of some ice can be a benefit providing insulation, thereby slowing the cooling of the underlying water. If, however, the ice is held in place by the vegetation, the water flow will be reduced as the ice layer thickens. The constriction of flow beneath the ice layer leads to subsequent flooding, freezing, and hydraulic failure. A Canadian FWS CW operated successfully for 4 years by raising the water level at freezing time, thereby providing an ice cap to trap insulating air as the water level dropped (Herskowitz, 1986).

To predict ice formation and thickness requires calculating energy balances and water temperatures. Major factors to consider are the wetland dimensions, the temperature and loading rate of wastewater, the ambient air temperature and wind speed, and the depth and thermal conductivity of different layers (snow, ice, water, plant litter and soil, or other porous media). Techniques for calculating energy balance and predicting ice thickness in FWS wetlands are detailed by Reed et al. (1995), Reddy and Burgoon (1996), Kadlec and Knight (1996), and Kadlec (1996). Reed and Calkins (1996) recommend a thermal analysis for any cold climate project to ensure that the wetland will function during the winter. If the thermal models for either SF or FWS wetlands predict sustained internal water temperature $< 1°C$, a wetland may not function in winter months. Most of the parameters in the thermal models described by Reed et al. (1995) are easily defined and measured, the one exception being the effective heat transfer coefficient U, because specific information related to CWs is limited.

The thermal regime of a subsurface flow wetland is complex because of the various layers in the system: saturated media, unsaturated media in overlain layers, plant litter, snow, and ice. Available background data of thermal conductivity (Kadlec and Knight, 1996; Jenssen et al., 1996), in combination with site-specific weather and wastewater data, make reliable predictions possible for several idealized situations. In real life, however, the heterogeneity of full-scale wetlands makes it necessary to apply the calculation methods conservatively. To model thermal performance accurately, more data are needed from existing cold climate treatment systems. Reed et al. (1995) suggested that latent heat of the water, energy transfer to and from the ground (likely a net gain during the winter), and solar radiation should be excluded in the calculations.

As an example of modeling, Jenssen et al. (1996) established a mathematical thermal model based on a ten-person equivalent wastewater SHF CW design. The model indicated that proper insulation could avoid hydraulic failure due to freezing. The calculations suggest that 10 cm of extruded polystyrene (XPS) insulation on top, and extending 1 m horizontally outside the perimeter, is sufficient to avoid freezing when the temperature is $-10°C$ for several weeks without any insulating snow layer. The modeling also showed that 1-m-depth insulation extending along the perimeter of the wetland is more efficient in reducing frost than insulation extending horizontally 1 m to the sides along the perimeter. The model, however, has not been calibrated for full-scale SHF CWs.

3.3.2 Hydrological and Hydraulic Conditions

Snow and ice influence the wetland water balance as a result of low evapotranspiration during dry winter periods and snow melting and flooding in spring. The highly variable leachate flow is mainly based on the rate of precipitation percolating through the landfill. Cold climatic regions often have higher precipitation than warmer regions, and in such areas landfill leachates seem to be more diluted than in temperate climate regions (Mæhlum and Haarstad, 1997). A severe winter with high snow accumulation and little leachate will increase leachate production if the snow melt percolates through the waste. Increased leachate volume and snow melting in the wetlands will decrease the hydraulic residence time (HRT) in the treatment system and the concentration of pollutants will be diluted. The magnitude of the influence of snow melting will largely depend on the size of the landfill area and the wetland-to-catchment ratio. If this ratio is small, HRT may decrease dramatically during snowmelt. Hydraulic overloading of SHF systems will lead to short-circuiting, and in such situations, it is advisable to provide for storage in a buffering lagoon prior to treatment in CWs.

Deposits of suspended solids within SHF systems clog the system, especially near the inlet. Degradation products from the volatile portion of the total suspended solids, TSS (processes that slow down during winter) can accumulate in the filter together with the mineral fraction of TSS, especially Fe, Mn, and Ca precipitates. These deposits can block the pores and decrease the hydraulic conductivity. In cold periods, nutrient uptake, oxygen transport to the roots, and micro-biological rhizosphere activity cease or slow down. Lack of such processes in the rhizosphere may result in the accumulation of TSS (organic material), which will also decrease the hydraulic conductivity of the system. Soil microorganisms are surrounded by polysaccharides containing structures (glycocalyx) of bacterial origin. The microorganisms and bacterial glycocalyx together form a highly organized matrix (biofilm), where most of the degradation and transformation processes occur. Production of glycocalyx can increase as a result of environmental stress, for example, during temperature changes. Increased thickness of the biofilm due to a fall in temperature and decreased oxygen availability in winter may therefore lead to decreased hydraulic conductivity in SHF treatment wetlands. However, the results whether the biofilm will increase or decrease during low temperatures are contradictory (e.g., Kristiansen, 1981).

3.3.3 Biogeochemical Processes

Nutrient uptake by plants and microbial transformations of wastewater components and plant litter are affected by climatic conditions. The direct effect is on plant physiology, governed by solar radiation and temperature, and on microbial processes, governed by temperature. The indirect influence is the dependence of biological and biochemical processes on physical conditions, e.g., HRT, oxygen availability, and freezing/thawing of soil.

The biological reactions responsible for the decomposition of organic matter (BOD), nitrifica-tion, denitrification, and removal of pathogens are generally known to be temperature dependent in all wastewater treatment processes, including CWs (Reed et al., 1995). Table 3.1 shows the influence of low temperature on removal processes. Although Table 3.1 suggests that treatment performance diminishes during the winter, the temperature effect in many full-scale CW systems is insignificant, which may be attributed to other conditions masking the effect of temperature.

Reddy and Burgoon (1996) discuss the influence on biogeochemical processes (C, N, and P cycling) in CWs. Many biogeochemical reactions proceed at a faster rate as the temperature of the medium is increased. The enzymatic reactions mediated by microbes behave similarly up to a certain temperature. Each microbial species and each strain has its own minimum, optimum, and maximum temperature for microbial activity. Optimal temperature for maximum microbial activity may vary depending on the interactive factors such as hydraulic loading, effluent quality, vegetation, and soil/substrate within the wetland. Reddy and Burgoon (1996) discuss the relationship between the reaction rate constant, activation energy, and temperature given by van't Hoff–Arrhenius equa-tion. Three types of relationships between reaction rate constant and temperature are discussed in their paper. The temperature effect on reaction rate is expressed as Q_{10}, which is the factor by which rate of a given reaction increases by a 10°C increase in temperature. This approximation is based on the assumption that the reaction rate follows a simple exponential function of temperature. In general, many biological reaction rates double ($Q_{10} = 2$) with a 10°C increase in temperature. However, Q_{10} values for CWs range from 1 to 20 (Reddy and Burgoon, 1996).

The temperature dependence of reaction rate constants is commonly expressed as

$$k_{T1} = k_{T2} * \Theta^{(T1 - T2)} \tag{3.1}$$

where k_{T1} and k_{T2} are first-order rate constants at temperatures $T1$ and $T2$, respectively, and Θ is the temperature coefficient. Equation 3.1 is a simplification of the van't Hoff–Arrhenius expression, and used since the van't Hoff–Arrhenius expression contains factors that are difficult to determine experimentally. Small-scale and well-controlled experiments often show a clear temperature depen-dence for microbiologically mediated processes, such as BOD removal and N transformations. In full-scale treatment wetlands, the temperature dependence is often not as strong as indicated by the temperature coefficients (Θ) shown in Table 3.2. Kadlec and Knight (1996) noticed that BOD

Table 3.1 Selected Physical, Chemical, and Biological Processes Regulating Nutrient Removal in CWs and Expected Influence of Low Temperature on the Process

Process	Effluent Parameters	Influence of Low Temp.
Physical		
Flocculation Sedimentation Filtration	Suspended solids, particulate organic C, N, and P	*
Chemical		
Sorption	Dissolved organic compounds, anions (PO_4^{3-}) and cations (NH_4–N, metals)	*
Precipitation	Inorganic P, sulfides, and metals	*
Volatilization	NH_4–N and volatile organic compounds	**
Microbiological		
Respiration	Biochemical oxygen demand, O_2, NO_3–N, SO_4^{2-}, HCO_3^-, and volatile fatty acids	***
Nitrification	NH_4–N	**
Denitrification	NO_3–N and NO_2–N	**
Mineralization	Organic N and P	*
Assimilation	Nutrients	**
Competition, dying	Pathogenic organism	**
Biological (Plants)		
Growth and uptake	Nutrients	***
Gas transport	O_2 and related reactions	***

Note: * Little, ** moderate, and *** large influence of low temperatures <5°C (meaning low reaction rate due to low temperatures).

removal is rarely affected by temperature, i.e., observed Θ values are close to 1. In FWS CWs, ice formation and snow layer can disrupt regular diel fluctuations of aerobic/anaerobic interfaces in sediment redox gradient, due to increased O_2 production during the day and depletion of O_2 at night. Snow and ice therefore slow nitrification–denitrification reactions, precipitation of P, and oxidation of CH_4, sulfides, and other reduced species (Reddy and Burgoon, 1996). For the FWS system in Listowel, Ontario, which received secondary effluent, the summer performance was somewhat lower than the winter performance for both BOD and SS, because of algal blooms in the summer. The reduction in removal efficiency during periods with low temperature was significant for N, slight for P, and nonexistent for BOD and TSS (Herskowitz, 1986).

Table 3.2 Temperature Coefficients for Biochemical Oxygen Demand (BOD) Removal and Nitrogen Transformations

Process	Temperature Interval (°C)	Temperature Coefficient (Θ)
BOD removal	4–20	1.13
	>20	1.06
Nitrogen mineralization	5–15	1.07–1.13
	15–25	1.07–1.16
Nitrification	5–10	1.11–1.37
	10–15	1.07–1.16
	15–20	1.06–1.12
Denitrification	5–10	1.52–1.61
	10–15	1.14–1.16
	15–25	1.07–1.08

Literature data from Reddy and Burgoon (1996).

3.3.3.1 Organic Matter

Degradation of organic matter involves hydrolysis and catabolic activity by heterotrophic microorganisms, which are sensitive to seasonal temperature changes in the system. According to Reddy and Burgoon (1996), microbial respiration is frequently limited by electron acceptor availability, rather than electron donor supply (C availability). Reduced oxygen availability and reduced nitrification will therefore influence C degradation. Temperature does not affect treatment as much as the type of organic matter. Reed et al. (1995) proposed a relationship to define the temperature dependence of BOD removal in a first-order plug flow model.

There is no consensus on BOD removal in cold climate wetlands. Although performance data from some FWS wetlands support the temperature dependency of BOD removal (e.g., Reed et al., 1995; Wittgren and Mæhlum, 1997), other cases show no obvious temperature dependence (Kadlec and Knight, 1996; Lemon et al., 1996). According to Reddy and Burgoon (1996), BOD removal is more dependent on physical settling than microbial breakdown. This is evidenced by frequent high removal rates close to the inflow of CWs. Another explanation is the long hydraulic residence time, provided by many CW systems, which tend to compensate for the lower reaction rates during the winter months.

3.3.3.2 Nitrogen

Wastewater CW systems remove less N during the winter than during the summer because microbially-mediated N transformations (mineralization, nitrification, and denitrification) are temperature sensitive. Aboveground biomass dies during winter and accumulates as detritus; consequently, a significant portion of N is released or translocated to the roots. According to Reddy and Burgoon (1996), N release during decomposition of detritus is higher in summer than in winter months. The temperature effect varied between different species, suggesting detritus quality was masking the influence of temperature on decomposition. Mineralization was shown to decrease with decreasing temperature and essentially cease when soil is frozen. The Q_{10} values are higher than 2 at temperatures less than 15°C. Ammonia volatilization decreases at low temperatures, with Q_{10} values in the range of 1.3 to 3.5, for temperatures in the range of 0 to 30°C. Nitrification is sensitive to temperature, with Q_{10} values in the range of 3 to 23 at <10°C. Denitrifying organisms have optimum temperatures between 35 and 45°C and a Q_{10} value of 21 was observed between 5 and 15°C. Reddy and Burgoon (1996) suggest that longer HRT are needed at low temperature for effective removal of NO_3–N.

Vegetation plays a significant role by assimilating N into plant tissue and providing an environment for nitrification–denitrification in the root zone. The annual plant uptake of nutrients and, thus, the potential for harvest of nutrients are higher in warm regions (2000 to 6000 kg N/ha/year) than cold temperate regions (20 to 70 kg N/ha/year) (Wittgren and Mæhlum, 1997). As far as treatment wetlands are concerned, it is unlikely that harvesting of plants is feasible in cold climates.

Jenssen et al. (1991) and Lemon et al. (1996) suggested ammonia reduction in CW due to cation exchange in filter media in fall and winter as one possible N sink during operation in cold climatic regions. Sorbed ammonia is oxidized when the plants start transporting oxygen to the rhizosphere in the spring, and during the summer the bed is renewed for the next cool session.

Freezing and thawing of soil induces transient pulses of high biological activity, due to physical disruption of the soil structure. The frequency of freeze/thaw cycles is likely to show large year-to-year variation, and may possibly cause large variation in N losses from soil–plant systems (Bakken, 1995).

Because the ammonia concentrations in most leachates are high (often >500 mg/L), it is not possible to achieve high tot-N removal in cold climate CWs without adding supplemental oxygen for nitrification and a C source for denitrification.

3.3.3.3 *Phosphorus*

For most MSW landfills, P is not regarded as an important pollutant. However, with very low concentrations, there may be insufficient P to support biological processes because of the high N:P ratio, often >500. Optimization of biological leachate treatment requires regular addition of supplemental P. If significant amounts of supplemental P is added in a pretreatment unit, for example, in an extended aeration lagoon or a sequenced batch reactor, P removal in the CW unit increases. Both the physical (sedimentation) and chemical (sorption) P removal processes are less directly sensitive to temperature than oxygen availability, a result of redox sensitive sorption to ferrous/ferric oxides. According to Reddy and Burgoon (1996), few wastewater wetland studies have shown decreased P sorption where water temperature was lowered from 20 to 5°C. However, with high concentrations of both Fe and P in the leachate and anaerobic conditions in CWs, it can be expected that both Fe–P precipitates and sorbed Fe and P will be released from the wetlands during the winter.

3.4 TREATMENT PERFORMANCE IN SCANDINAVIA

Wittgren and Mæhlum (1997) compiled data in Scandinavian wastewater treatment wetlands, which show that the temperature dependence is not as strong as indicated by the temperature coefficients (Θ) in Table 3.2. Lack of a strong temperature dependence has been demonstrated in the efficiencies of COD removal in the Haugstein and Tveter SHF systems in Norway (10 PE, Q_{dim} = 2 m³/day, area 100 m²) (Mæhlum et al., 1995) and for BOD removal (Θ = 1.01) in the Oxelösund FWS wetland in Sweden (12500 PE, Q = 6000 m³/day, area 22 ha) (Wittgren and Tobiason, 1995; Sundblad and Wittgren, 1996). Neither tot-P nor tot-N removal were temperature dependent in the Norwegian wetlands. In Oxelösund, however, removal of tot-N showed significant temperature dependence, with Θ = 1.09. Organic matter may be in particulate form, and P may be associated with particles, which are partly removed by settling. Another, and perhaps more important, explanation for lack of temperature dependence in the Scandinavian systems is low loading rates (long HRT). Measurements of the influent and effluent will not detect the temperature dependence in such systems.

Ammonium was the dominating N fraction in the Norwegian and Swedish wastewater treatment wetlands, making up on an average 87% and 80% of the tot-N, respectively (Mæhlum et al., 1995; Sundblad and Wittgren, 1996). Nitrogen removal by cation exchange might thus be an explanation for more efficient removal in SHF systems than in FWS systems during the cold season, since there is better contact between water and soil/filter media in SHF systems. Efficient insulation and heat transfer from the surrounding ground reduces water temperature fluctuations in SHF wetlands, and is another explanation for less temperature dependence in removal efficiency. Finally, the adaptations of the microbial community to cold climate may explain why wetlands in cold climates show less temperature dependence compared with laboratory-scale studies. If laboratory studies are performed with soils from a warm climatic region, the microbial processes might be more sensitive to low temperatures than soils from cold regions. This kind of microbial climate adaptation has been demonstrated for denitrification by Powlson et al. (1988).

The performance of CWs in Norway were put to test during the severe winter of 1995–96. In Norway (at 60° north latitude), the coldest average monthly temperature is seldom below –10°C, and average temperatures below –20°C for more than 2 weeks is unusual, especially in the autumn and before snowfall. According to the Norwegian Meteorological Institute, the 1995–96 winter experienced one of the worst cases of frost penetration in southern Norway during this century. The most critical frost period is before snow cover. From December temperatures dropped below –10°C for several weeks and snow did not arrive until late January. It is important to note that none of the 12 CW systems failed, despite the lack of supplementary insulation (Mæhlum and Jenssen, 1998).

Results from two Norwegian integrated landfill leachate systems (Chapter 9) showed that extended aeration lagoons can be efficient pretreatment units prior to the use of CWs. However, lagoon effluents with low water temperatures (0 to 5°C) had significant effect on the effluent quality of the SHF wetlands, with reduced oxidation of COD, BOD, TOC, NH_4–N, and Fe_{II}. During periods of flooding, treatment performance was also low. Removal of tot-N was found to be between 20 and 40% during the summer (HRT > 5 days) in the SHF CWs.

The Rebneskogen landfill leachate system in Norway (Lyche, 1994) uses spray irrigation on a peat slope during the warmest 6 months to polish stored, pretreated leachate from an extended aeration lagoon and biological degradation/storage pond. This treatment system has been operating in a cold climate region since 1988. Treatment efficiency for NH_4–N and tot-N in the 2.5-ha peat area has been 70 to 80%, and 25% for TOC and COD.

Schierup et al. (1990a) have investigated 109 wastewater CW systems in Denmark. They conclude that there are only small seasonal variations in BOD and SS removal. The Swedish plant at Snogeröd (Gumbricht, 1991) treating secondary wastewater effluent with moderate hydraulic loading (0.1 m³/m²/day) had about 30% N-removal efficiency during winter. Other studies indicate N-removal capabilities of about 40% at air temperatures around zero (Gumbricht, 1993).

3.5 DESIGN AND OPERATION

At all landfill sites, the design of treatment systems should take advantage of natural features to reduce treatment costs. Removal of organic matter, ammonia, and tot-N in wetland systems are temperature dependent and can be expected to vary on a seasonal basis. The effects of temperature on treatment efficiency can be offset by increasing the treatment area or by increasing HRT. The size of a CW in cold climate should be based on low-temperature conditions using thermal models. These models can estimate water temperature and potential ice depth. Experience with full-scale wastewater SHF wetland systems indicates that by simple design measures potential hydraulic problems during winter can be avoided. Snow cover or ice on the surface of the wetland can provide significant thermal protection; but for a conservative design, it is prudent to assume that snow cover may not be available under the assumed winter conditions.

3.5.1 Wetland Type

SHF wetlands can provide greater thermal protection than FWS systems because of the insulation effect of the unsaturated surface layer. The SHF wetland has a more complex thermal regime because the water surface is not exposed to the atmosphere and is below a layer of dry filter media, with the accumulated plant litter on the top. Subsurface vertical flow CWs with subsurface loading, as reported by Lemon et al. (1996), might be superior to SHF systems, as a result of higher oxygen availability in vertical filters with intermittent loading.

The presence of the standing vegetation and litter makes FWS more thermally stable than an open pond of similar size and depth. The vegetation shades the water surface from direct solar radiation and significantly reduces the effect of wind action. As a result, there is a minimal turbulence in the water and wind-induced heat losses tend to be minor. Use of FWS wetlands, separate or in combination with subsurface flow wetlands, appear better adapted as second- or third-level treatment to minimize clogging problems.

3.5.2 Filter Media

Natural sand, gravel, clay/silt soil, peat, and manufactured lightweight aggregates have been used as filter material in both SHF and SVF systems in cold climatic regions. Media with high hydraulic conductivity (e.g., washed gravel or lightweight aggregates) may prevent effluent flow reaching the surface. Lightweight aggregates have high insulation value. The recommended width:length ratio of SHF wetlands treating leachate is >1, which will assist in preventing potential clogging problems in the inlet zone.

3.5.3 Pretreatment

Pretreatment of the leachates is an important design feature for leachate treatment wetlands both in cold and warm climates. An aerobic pretreatment and sedimentation unit before the CWs can enhance nitrification and decrease the loading of organic matter and reduced metals like Fe_{II} and Mn_{II}. Integrated systems with extended aeration lagoons (see Chapter 9) or SBRs (see Chapter 6) prior to CWs or other natural systems (peat or sand/gravel filtration) appear to be alternatives for many landfill sites in cold climatic regions. Some examples of possible combinations between conventional and natural landfill leachate treatment systems are shown in Figure 3.1.

3.5.4 Heat Loss and Insulation

Heat loss, prior to discharge to the wetland, can be minimized if the supply piping system and the pretreatment units are insulated. Plant material will provide insulation, but this may not be sufficient, especially in the first years of operation when the layer of plant litter is thin. Additional insulation may therefore be necessary unless a snow cover is formed in early fall. Although snow cover is often present in colder climates, as indicated earlier it is prudent for design purposes to assume that snow will not be present. SHF surfaces can be covered with a porous media of low thermal conductivity, such as expanded clay aggregates, which should be kept unsaturated during the winter (Jenssen et al., 1996). Guidelines often recommend the depth of SHF systems not to exceed 60 cm. Another recommendation is to provide >80 to 90 cm depth for SHF wetlands so that the upper 10 to 30 cm can freeze, while the lower part of the system still has hydraulic capacity. The obvious drawback with increased depth will be the flow of water below the root zone reducing treatment efficiency. For FWS wetlands, it may be beneficial to raise the water level prior to freezing and provide a gravel channel in the bottom layer to maintain hydraulic conductivity during periods with ice. When a sufficient ice cover has developed, the level can be lowered and an insulating air layer formed.

3.5.5 Winter Storage or Bypass

For wetlands with small wetland-to-catchment ratios, short-term storage is recommended. In locations that experience extended periods of less than –10°C air temperatures, seasonal storage of winter load may be necessary. A number of wastewater CW systems in South Dakota and northwest Canada operate this way. For polishing leachate treatment wetlands, it might be feasible to bypass the wetland during the winter, if the compliance discharge criteria are set on an annual basis.

ACKNOWLEDGMENTS

The preparation of this paper was supported by the Natural Systems Technology for Wastewater Treatment (NAT) research program and the Centre for Soil and Environmental Research (Jordforsk).

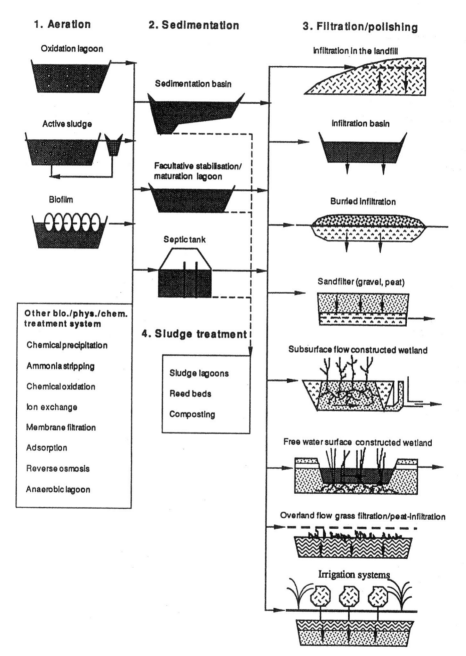

Figure 3.1   Integrated landfill leachate treatment systems: possible combinations between conventional and natural systems. (From Mæhlum 1995, *Water Science and Technology*, 32(3): 133. With permission.)

REFERENCES

Bakken, L., 1995. Nitrogen mineralization at low temperatures, in *Nitrogen Supply and Nitrogen Fixation of Crops for Cool and Wet Climates*, Report European Commission: EUR 16757 EN, Luxembourg.

Bulc, T., Vrhovsek, D., and Kukanja, V., 1997. The use of constructed wetland for landfill leachate treatment, *Water Science and Technology* 35:301–306.

Gumbricht, T., 1993. Nutrient removal capacity in submersed macrophyte pond systems in a temperate climate, *Ecological Engineering* 2:49–61.

Haberl, R., R. Perfier, J. Laber, and P. Cooper, (Eds.), 1997. Wetland systems for water pollution control, 1996. *Water Science and Technology* 35(5). (Selected proceedings of the 5th Int. Conf. on Wetland Systems for Water Pollution Control, Vienna, September 1996).

Herskowitz, J., 1986. Town of Listowel artificial marsh project, Project Report 128 RR. Ontario Ministry of Environment, Toronto, Ontario.

Jenssen, P. D., Krogstad, T., and Mæhlum, T., 1991. Wastewater treatment by constructed wetlands in the Norwegian climate — pretreatment and optimal design, in Etnier, C. and Guterstam, B., Eds., *Ecological Engineering for Wastewater Treatment*, Bokskogen, Göteborg, Sweden, 227–238.

Jenssen, P. D., Mæhlum, T., and Krogstad, T., 1993. Potential use of constructed wetlands for wastewater treatment in northern environments, *Water Science and Technology* 28:149–157.

Jenssen, P. D., Mæhlum, T., and Zhu, T., 1996. Construction and performance of subsurface flow constructed wetlands in Norway, paper presented at the *Symposium on Constructed Wetlands in Cold Climates*, June 4–5, Niagara-on-the-Lake, Ontario.

Kadlec, R. H., 1996. Physical processes in constructed wetlands, in *Proceedings of the Symposium on Constructed Wetlands in Cold Climate*, June 4–5, Niagara-on-the-Lake, Ontario.

Kadlec, R. H. and Knight, R. L., 1996. *Treatment Wetlands*, Lewis Publishers, Boca Raton, FL, 893 pp.

Kristiansen, R., 1981. Sand-filter trenches for purification of septic tank effluent: I. The clogging mechanism and soil physical environment, *Journal of Environmental Quality* 3:353–357.

Kuusemets, V., Mauring, T., and Koppel, A., 1996. Domestic wastewater purification in energy forest — case study of a willow plantation, Estonia, in preprints from the *5th International Conference on Wetland Systems for Water Pollution Control*, September 15–19, Vienna, Austria.

Lakshman, G., 1992. Design and operational limitations of engineered wetlands in cold climates — Canadian experience, in Mitsch, W. J., Ed., *Global Wetlands — Old World and New*, Elsevier Science, Amsterdam, The Netherlands, 399–409.

Lemon, E., Bis, G., Rozema, L., and Smith, I., 1996. SWAMP pilot scale wetlands — design and performance, Niagara-on-the-Lake, Ontario, Paper presented at the *Symposium on Constructed Wetlands in Cold Climates*, June 4–5, Niagara-on-the-Lake, Ontario.

Lyche, C., 1994. Sigevannskontroll. Vurdering av utslipp til vassdrag fra Rebneskogen avfallsfylling 1991–1993. Hjellnes COWI-report (in Norwegian), April 1994.

Magmedov, V. G., 1996. Combined systems of constructed wetlands in CIS: the first ten years experience, in preprints from the *5th International Conference on Wetland Systems for Water Pollution Control*, September 15–19, Vienna, Austria, pp. IX/10, 1–8.

Mander, U. and Mauring, T., 1997. Constructed wetlands for wastewater treatment in Estonia, *Water Science and Technology* 35:323–330.

Martin, C. D. and Moshiri, G. A., 1994. Nutrient reduction in an in-series constructed wetland system treating landfill leachate, *Water Science and Technology* 29:267–272.

Mæhlum, T., 1995. Treatment of landfill leachate in on-site lagoons and constructed wetlands. *Water Science and Technology*, 32(3):95–101.

Mæhlum, T., 1998. Cold-climate constructed wetlands: aerobic pretreatment and subsurface flow wetlands for domestic sewage and landfill leachate purification. Doctor Scientarum Thesis. 1998:9. Agricultural University of Norway.

Mæhlum, T. and P. D. Jenssen, 1998. Use of constructed wetlands in Norway, in Vymazal, J., H. Brix, P. F. Cooper, M. B. Green, and R. Haberl, Eds., *Constructed Wetlands for Wastewater Treatment in Europe*. Backhuys Publishers, Leiden, The Netherlands, ISBN 90-73348-72-2, pp. 207–216.

Mæhlum, T. and Haarstad, K., 1997. Lokal behandling av sigevann fra fyllplasser. En oversikt over naturbaserte og prosesstekniske rensemetoder. Jordforsk report 18/96 (In Norwegian). Ås, Norway.

Mæhlum, T., Jenssen, P. D., and Warner, W. S., 1995. Cold climate constructed wetlands, *Water Science and Technology* 32:129–135.

Obarska-Pempkowiak, H., 1996. Recent experience in operation of constructed wetlands in the northern Poland, in preprints from the *5th International Conference on Wetland Systems for Water Pollution Control*, September 15–19, Vienna, Austria, pp. IX/8 1–10.

Powlson, D. S., Saffigna, P. G., and Kragt-Cottaar, M., 1988. Denitrification at sub-optimal temperatures in soils from different climatic zones, *Soil Biology and Biochemistry* 20:719–723.

Pries, J., 1994. *Wastewater and Stormwater Applications of Wetlands in Canada*, Sustaining Wetlands Issues Paper No. 1994-1, North American Wetlands Conservation Council, Ottawa, Ontario.

Pries, J., 1996. Constructed treatment wetlands in Canada, in *Proceedings Symposium on Constructed Wetlands in Cold Climate*, June 4–5, The Friends of Fort George, Niagara-on-the-Lake, Ontario.

Reddy, K. R. and Burgoon, P. S., 1996. Influence of temperature on biogeochemical processes in constructed wetlands — implications to wastewater treatment, paper presented at the *Symposium on Constructed Wetlands in Cold Climates*, June 4–5, Niagara-on-the-Lake, Ontario.

Reed, S. C. and Calkins, D., 1996. Thermal aspects of constructed wetland system design, paper presented at the *Symposium on Constructed Wetlands in Cold Climates*, June 4–5, Niagara-on-the-Lake, Ontario.

Reed, S. C., Middlebrooks, E. J., and Crites, R. W., 1995. *Natural Systems for Waste Management and Treatment*, 2nd ed. McGraw-Hill, New York.

Robinson, H. D. 1993. The treatment of landfill leachates using reed bed systems, in Christensen, T. H., Cossu, R., and Stegmann, R., Eds., *Proceeding of Sardinia 1993 Fourth International Landfill Symposium*, Cagliari, Italy, 907–922.

Sanford, W. E., Steenhuis, T. S., Kopka, R. J., Peverly, J. H., Surface, J. M., Campbell, J., and Lavine, M. J., 1990. Rock-reed filters for treating landfill leachate, American Society of Agricultural Engineers, Paper No. 90-2537.

Schierup, H.-H., Brix, H., and Lorenzen, B., 1990a. Spildevandsrensning i rodzoneanlæg. *Spildevandsforskning fra Miljøstyrelsen*, 8, National Agency of Environmental Protection, Copenhagen, Denmark (In Danish).

Schierup, H.-H., Brix, H., and Lorenzen, B., 1990b. Wastewater treatment in constructed reed beds in Denmark — state of the art, in Cooper, P. F. and Findlater, B. C., Eds., *Constructed Wetlands in Water Pollution Control*, Pergamon Press, Oxford, 495–504.

Strahler, A. H. and Strahler, A. N., 1992. *Modern Physical Geography*, 4th ed., John Wiley & Sons, New York, 156–157.

Sundblad, K. and Wittgren, H. B., 1991. Wastewater nutrient removal and recovery in an infiltration wetland, in Etnier, C. and Guterstam, B., Eds., *Ecological Engineering for Wastewater Treatment*, Bokskogen, Göteborg, Sweden, 190–198.

Sundblad, K. and Wittgren, H. B. 1996. Nitrogen removal in relation to speciation in cold climate surface-flow wetlands, in preprints from the *Fifth International Conference on Wetland Systems for Water Pollution Control*, September 15–19, Vienna, Austria, pp. I/19 1–7.

Surface, J. M., Peverly, J. H., Steenhuis, T. S., and Sandford, W. E., 1992. Constructed wetlands for landfill leachate treatment, in *Proceedings on Constructed Wetlands for Water Quality Improvement, International Symposium*. Pensacola, FL.

Vymazel, J., 1996. The use of sub-surface-flow constructed wetlands for wastewater treatment in the Czech Republic, *Ecological Engineering* 7:1–14.

Vymazel, J., H. Brix, P. F. Cooper, M. B. Green, and R. Haberl, Eds., 1998. *Constructed Wetlands for Wastewater Treatment in Europe,* Backhuys Publishers, Leiden, The Netherlands.

Wittgren, H. B. and Mæhlum, T., 1997. Wastewater treatment wetlands in cold climate, *Water Science and Technology* 35: 45–53.

Wittgren, H. B. and Tobiason, S., 1995. Nitrogen removal from pre-treated wastewater in created wetlands, *Water Science and Technology* 32:69–78.

CHAPTER 4

Substrate Type, Flow Characteristics, and Detention Times Related to Landfill Leachate Treatment Efficiency in Constructed Wetlands

William E. Sanford

CONTENTS

ABSTRACT: In 1989, a 3-year cooperative study was started at a municipal solid waste landfill near Ithaca, NY to test the efficiency of treatment of landfill leachate by constructed wetlands. Four parallel wetland plots were constructed 3 m in width, 30 m in length, 0.6 m in depth, and lined with 1.5-mm, high-density polyethylene. Each plot was constructed with a bottom slope of 0.5% and with a horizontal upper surface. Three types of substrate material were used: coarse gravel (2–4 cm), pea gravel (0.5 cm), and two plots filled with a sand-and-gravel mixture. Plots of each substrate were planted with the reeds *Phragmites australis* with the second sand-and-gravel plot left unplanted as a control.

The hydraulic conductivities of the substrates were measured at five intervals over the first 26 months of operation. The results indicate that the sand-and-gravel substrates became almost completely clogged and that the presence of reeds did not maintain or increase the conductivity. The hydraulic conductivity of the two gravel beds also decreased during this time period but to a much lesser degree. As a result of the clogging, overland flow of leachate was observed in the two sand-and-gravel plots and tracer studies showed that there was little or no flow through the substrates. Combined results from tracer studies and specific conductance measurements indicate that preferential flow through the gravel substrates occurred as a result of the more dense leachate flowing beneath less dense rainwater. The effects of the density stratification was to reduce the detention time by as much as 50%.

Numerical simulations confirmed the effects of density differences on the development of stratification within the substrate. Density stratification has the potential for reduction of treatment efficiency by significantly reducing the residence time of the leachate within the constructed wetland and by causing the denser leachate to flow beneath the rhizosphere and bypass the most effective treatment zone.

4.1 INTRODUCTION

The current design of landfills is such that the amount of precipitation infiltrating into the body of waste is minimized and the leachate that is formed by the interaction of precipitation and groundwater with the waste is collected and treated in some manner. For many older landfills put into operation before strict environmental legislation was in place, there was no liner to prevent the formation and migration of leachate. Present regulations require leachate collection systems to be retrofitted to the old design, resulting in the collection, treatment, and disposal of relatively large quantities of wastewater. At newer landfills, less leachate is produced and collected but is often of a much higher concentration. For either case, leachate may need to be collected and treated for decades, adding long-term costs to an already expensive process. One of the methods for treating the leachate is the use of constructed wetlands, which are built to replicate and exploit some of the biological and chemical processes of natural wetlands that can remove contaminants from the wastewater or transform them into products that have minimal impact on human health or the environment. Some of the major advantages of the use of constructed wetlands for treating landfill leachate are that, once built, significant treatment of the leachate can be achieved continuously with minimal operation and maintenance costs.

A common design of constructed wetland uses the subsurface flow of wastewater through a matrix of gravel in which is planted wetland plants, such as reeds or cattails (Hammer, 1989; Moshiri, 1993; Kadlec and Knight, 1996). It is within the root zone and rhizosphere of the wetland system that the desired chemical and biochemical reactions occur. One of the main concerns with the use of constructed wetlands for treating landfill leachate is to maintain contact of the inflowing wastewater with the treatment zone within the rhizosphere. A decrease in the efficiency of treatment can occur when there is preferential flow of leachate through the system, reducing the amount of leachate coming into contact with the treatment zone or decreasing the contact time.

In this paper, the hydraulic performance and flow characteristics of experimental constructed wetlands for treating landfill leachate are summarized. The results are combined with some numerical models of wastewater flow through the system to illustrate the direct link between the hydraulic characteristics of the constructed wetland and treatment efficiency.

4.2 METHODS

4.2.1 Constructed Wetland Design

In 1989, a 3-year cooperative study involving the U.S. Geological Survey, Cornell University, and the Tompkins County, NY, Departments of Planning and Solid Waste, was initiated at a municipal solid-waste landfill near Ithaca, NY to test the efficiency of treatment of landfill leachate by constructed wetlands. Four parallel wetland plots were constructed 3 m in width, 30 m in length, 0.6 m in depth, and lined with 1.5-mm, high-density polyethylene. Each plot was constructed with a bottom slope of 0.5% and with a horizontal upper surface (Figure 4.1). Three types of substrate

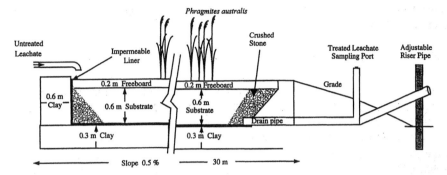

Figure 4.1 Schematic cross section of a constructed wetland used in the study.

material were used: coarse gravel (2–4 cm), pea gravel (0.5 cm), and two plots filled with a sand-and-gravel mixture. Plots of each substrate were planted with the reeds *Phragmites australis* with the second sand-and-gravel plot left unplanted as a control. Leachate was added to each plot by gravity flow from a feed tank at the head of the plots (Figure 4.2). The design detention time of the leachate of 15 days in each plot was controlled by valves on the inflow pipe, and the water level in the plots was controlled by risers located at the outlet end (Figure 4.1). Totalizing volumetric flowmeters were used to record the total inflow and outflow of leachate, and a rain gauge was located on-site to record the amount of water added to the system from precipitation. Five 5-cm-diameter slotted wells were installed at roughly 6-m intervals along the length of each bed (Figure 4.2) for collection of pore water samples and water-level measurements. For a more detailed description of the physical layout and operation of the experiment and for more information on

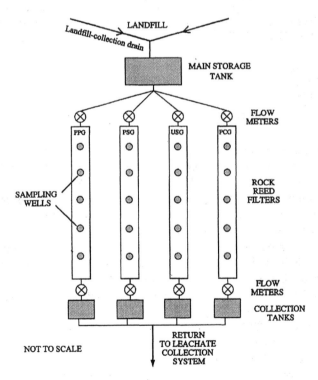

Figure 4.2 Plan view of the constructed wetland research plots. PCG is planted coarse gravel, USG is unplanted sand-and-gravel, PSG is planted sand-and-gravel, and PPG is planted pea gravel.

the treatment efficiencies, see Surface et al. (1993), Surface (1993), Peverly et al. (1994), and Sanford et al. (1995a; b).

4.2.2 Hydraulic Assessment

During the course of this project, two studies were performed on the investigation into how the hydraulics of the constructed wetlands can influence the efficiency of treatment. These two studies (Sanford et al., 1995a and b) are summarized in this section.

The first study involved the measurement of saturated hydraulic conductivity of the substrates used in the experiments at various times during the development of the wetland systems (Sanford et al., 1995a). The hydraulic conductivities of the substrates were measured five times during the first 26 months of operation by using drainage experiments in which the plots were completely drained. The resulting cumulative outflow volume vs. time curves were fitted using an equation developed by Sanford et al. (1993). The results, summarized in Table 4.1, indicate that there was little or no change in hydraulic conductivity of the coarse gravel bed and minimal decrease for the pea gravel. However, both sand-and-gravel plots experienced large reductions in hydraulic conductivities over the study period, with both beds becoming nearly completely clogged. It was also observed that the presence of plants growing in the substrate had little or no effect on the maintenance of increased hydraulic conductivity (Sanford et al., 1995a). The result of these severe reductions of hydraulic conductivity was that at the designed flow rate of about 1 m³/day the incoming leachate flowed across the surface of the substrate, bypassing the treatment zone within the rhizosphere.

Table 4.1 Hydraulic Conductivity Values of the Substrates Used in the Constructed Wetlands

Season	Planted Pea Gravel (cm/s)	Unplanted Sand and Gravel (cm/s)	Planted Sand and Gravel (cm/s)
Summer 1989	6.0	0.22	0.12
Fall 1989	6.0	0.18	0.17
Spring 1990	4.7	0.22	0.12
Spring 1991	6.3	0.03	—
Fall 1991	4.0	Clogged	0.02

Data are from Sanford et al. (1995b).

The second investigation into the hydraulic characteristics of flow through the constructed wetlands involved the use of a tracer test combined with the measurement of the specific conductance of the leachate within the substrate (Sanford et al., 1995b). The tracer test was performed to examine if the detention time of the tracer was the same as the designed detention time. Ideal tracers of water flow are those that are not absorbed by the substrate material and do not react with the leachate in any way. The leachate contained a relatively high concentration of chloride (around 1000 mg/L), and the tracer test was performed by adding freshwater with no chloride and measuring the removal of chloride from the plots (Sanford et al., 1995b). To provide additional information on how the wastewater flowed through the substrates, the specific conductance of the leachate was measured at 7.6-cm-depth intervals within the substrate in each of the five in-plot wells both prior to and during the tracer experiment (Figures 4.3 and 4.4). The detailed results from this study are presented in Sanford et al. (1995b).

The data from the tracer test confirmed that the two plots with sand-and-gravel substrates were nearly completely clogged. The specific conductance measurements showed that there was little or no freshwater from the tracer test flowing through the substrates (Figures 4.3 and 4.4), which allows the inference that leachate was not traveling through the substrate. Most of the incoming water into these two plots was flowing over the surface to the outlet.

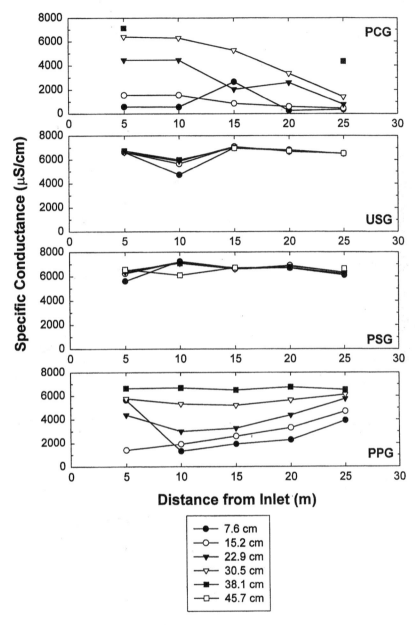

Figure 4.3 Specific conductance measurements taken in each plot prior to the tracer experiment. PCG is planted coarse gravel, USG is unplanted sand-and-gravel, PSG is planted sand-and-gravel, and PPG is planted pea gravel. (From Sanford W. E. et al., *Ecol. Eng.*, 5, 37–50, 1995. With permission.)

Perhaps the most interesting finding of the tracer test was the discovery of the stratification of the specific conductance within the two gravel plots (Sanford et al., 1995b). The measurements of specific conductance from the pea gravel plot before and during the tracer test are shown in Figures 4.3 and 4.4. What was distinctly evident in the pretracer test measurements was the fact that the highest specific conductance occurred near the bottom of the substrate and the lowest was near the top (Figure 4.3). Sanford et al. (1995b) postulated that this was the result of a density difference between the leachate coming into the constructed wetland and the rainwater that fell upon the plot, with the leachate having the greater density and higher specific conductance.

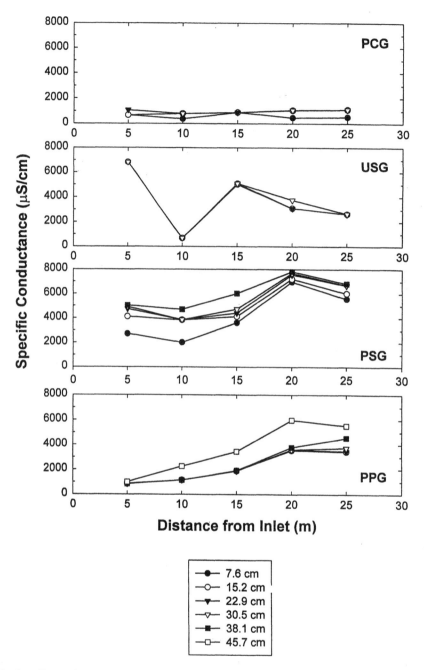

Figure 4.4   Specific conductance measurements taken in each plot on day 11 of the tracer test. PCG is planted coarse gravel, USG is unplanted sand-and-gravel, PSG is planted sand-and-gravel, and PPG is planted pea gravel. (From Sanford, W. E. et al., *Ecol. Eng.*, 5, 37–50, 1995. With permission.).

The specific conductance profiles for day 11 of the tracer test are shown for the pea gravel plot in Figure 4.4, where it can be seen that there was still some stratification evident toward the outlet end but that some complete displacement occurred in the upper portions of the substrate near the inflow end. This can be interpreted as being a result of the less dense tracer water flowing over the top of the more dense leachate.

These two previous investigations lead to the following conclusions: (1) the best substrates for maintaining high hydraulic conductivity and preventing overland flow are fine gravels and (2) density stratification caused by density differences between the leachate and rainwater can cause preferential flow of leachate through the system. The preferential flow caused by density can have two effects. One is the reduction of detention time, and the other is that the leachate could flow beneath the rhizosphere (or treatment zone) within the constructed wetland.

4.2.3 Modeling of Density Flow

To confirm the postulates of Sanford et al. (1995b) that density differences could cause a stratification of flow through a constructed wetland, numerical simulations were performed using a three-dimensional finite element model, which can be used for simulating density-dependent flow and transport through porous media (FEMWATER; Lin et al., 1996). Variables used in the modeling are similar to those reported by Sanford et al. (1995a; b).

Before discussing the models, it is helpful to discuss the physical system in some detail. The flow of water through a porous medium is described by Darcy's law

$$Q = A \, K_s \, I$$

where Q = volumetric flow rate [L^3/T]
 A = cross-sectional area through which flow takes place [L^2]
 K_s = saturated hydraulic conductivity of the medium [L/T]
 I = hydraulic gradient [L/L].

For the pea gravel constructed wetlands described above, Q = 1 m³/day, A = 1.8 m², and K_s = 0.04 m/s, which results in a hydraulic gradient of I = 0.0002, which would result in only a 0.005-m drop in water level over the 30-m length of the plot. Little head drop is to be expected due to hydraulic resistance at these flow rates in constructed wetlands using gravel substrates. Much of the head drop that is measured in these beds results from loss of water out of the plots due to evapotranspiration, which can result in apparent hydraulic gradients that are greater than those actually due to hydraulic resistance.

As a result of the low gradients resulting from the flow rate used in the field experiments, the simulations were performed using a smaller length scale. This was because the resulting gradient would be too small to create any appreciable flow in the center of the simulated region with reasonable grid spacings. The parameters used in the simulations were length = 2.0 m; width = 0.4 m; depth = 0.4 m; K_s = 0.00028 m/s (equivalent to a coarse sand); Q = 0.004 m³/h; porosity = 0.30; and a concentration-dependent density that at full concentration was equal to 1.3 g/cm³ and was equal to 1.0 g/cm³ with zero concentration. Water was added to the system at one end between 0.2 and 0.25 m height and exited from the other end at a depth of 0 to 0.05 m. The simulations were run to simulate a total of 48 h or the equivalent of 2 pore volumes of fluid run through the porous medium.

Shown in Figure 4.5 is a simulation in which a substrate initially filled with water with no concentration (density = 1.0 g/cm³) is displaced with a fluid containing a conservative tracer of constant concentration of 25 g/L but with no density difference. As can be seen, with no density contrast, the flow of tracer through the system is relatively uniform. In contrast, Figure 4.6 shows the displacement of fresh water by a fluid that has a concentration-dependent density. It can be seen that significant stratification developed with time, with the greater density being at the bottom. This is analogous to the situation observed by Sanford et al. (1995b) before the tracer experiment described above (Figure 4.3).

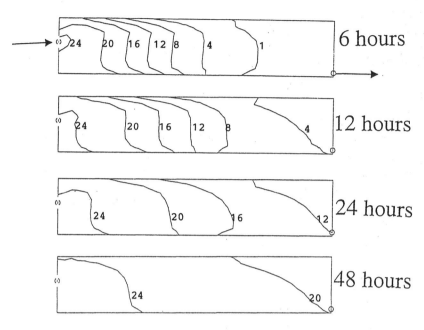

Figure 4.5 Numerical simulation of flow of a nonreactive tracer with no density contrast through a theoretical constructed wetland. Contours are for concentration in g/L.

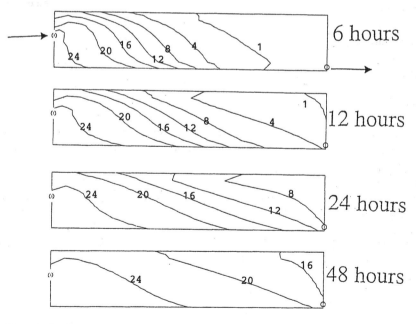

Figure 4.6 Numerical simulation of flow of a nonreactive tracer with a greater density displacing a less dense fluid in the theoretical constructed wetland. Contours are for concentration in g/L.

Another simulation was performed in which the fluid initially in the porous medium was at a concentration in which the density was 1.3 g/cm³ and was displaced by a fluid with no concentration (less dense). The results of the simulation are presented in Figure 4.7, where it is clearly seen that the less dense water rides over the top of the more dense water originally in the medium. This also resulted in the development of stratification and is analogous to what Sanford et al. (1995b) observed in the pea gravel substrate as the field-scale tracer test proceeded with time (Figure 4.4).

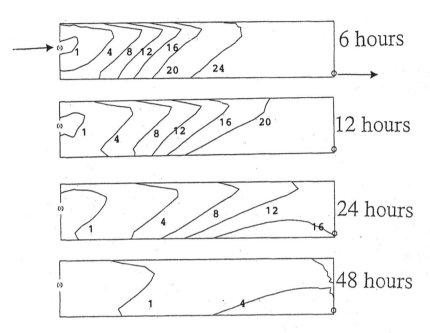

Figure 4.7   Numerical simulation of flow of a nonreactive tracer with lower density displacing a more dense fluid in the theoretical constructed wetland. Contours are for concentration in g/L.

The results from the numerical studies are similar to the results from a series of laboratory studies by Rothstein et al. (1993). In those studies, an artificial leachate with a density of 1.052 g/cm³ was created using blue dye and $CaCl_2$. The artificial leachate was added to a flume with similar dimensions as used in the numerical model filled with a pea gravel substrate. Rothstein et al. (1993) did find that the leachate flowed through the system in only the bottom one half of the substrate.

4.3 DISCUSSION

Numerical models incorporating density dependent flow and transport through constructed wetlands confirm the observations of Sanford et al. (1995b) that the density contrast between landfill leachate and rainwater can cause stratification to occur within the substrate of the wetland. The effects of this stratification can be twofold. The first is that the detention time of the leachate in the constructed wetland can be significantly decreased as a result of the leachate flowing through less of the cross-sectional area of the substrate. The second is that in constructed wetlands, where root and rhizospheres have not fully developed, the leachate could actually flow beneath the treatment zone, thereby reducing the treatment efficiency.

One possible remedy to minimize of the effects of density stratification is to incorporate low-permeability zones within the constructed wetland. These would cause vertical flow to occur as leachate flows over them, resulting in the mixing of the leachate with the less-concentrated rainwater. This design has been incorporated in experimental wetlands for landfill leachate constructed in Monroe County, NY (see Chapter 13). Rothstein et al. (1993) also found during laboratory studies that raising the outlet drain off the bottom of the wetland cell allowed for more mixing of the leachate with rainwater.

ACKNOWLEDGMENTS

Partial funding for this project was provided by the New York State Energy Research and Development Authority and the U.S. Geological Survey.

REFERENCES

Hammer, D. A., Ed. 1989. *Constructed Wetlands for Wastewater Treatment: Municipal, Industrial and Agricultural*, Lewis Publishers, Chelsea, MI, 831 pp.

Kadlec, R. H., and Knight, R. L., 1996. *Treatment Wetlands*, Lewis Publishers, Boca Raton, FL, 893 pp.

Lin, H. C., Richards, D. R., Yeh, G. T., Cheng, J. R., Chang, H. P., and Jones, N. L., 1996. *FEMWATER: A Three-Dimensional Finite Element Computer Model for Simulating Density Dependent Flow and Transport*, U.S. Army Engineer Waterways Experiment Station Technical Report, 129 pp.

Moshiri, G. A., Ed. 1993. *Constructed Wetlands for Water Quality Improvement*, Lewis Publishers, Boca Raton, FL.

Peverly, J., Sanford, W. E., Steenhuis, T. S., and Surface, J. M., 1994. *Constructed Wetlands for Municipal Solid Waste Landfill Leachate Treatment*, New York State Energy Research and Development Authority Report 94-1, Albany, NY.

Rothstein, E., Janaushek, T., Bodnar, M., Steenhuis, T. S., and Sanford, W. E., 1993. Density stratification in rock-reed filters (abstract). *EOS Transactions of the American Geophysical Union 1993 Spring Meeting*, 74 (16), 146.

Sanford, W. E., Parlange, J. Y., and Steenhuis, T. S., 1993. Hillslope drainage with sudden drawdown closed from solution and laboratory experiments, *Water Resources Research*, 29:2313–2321.

Sanford, W. E., Steenhuis, T. S., Parlange, J. Y., Surface, J. M., and Peverly, J. H., 1995a. Hydraulic conductivity of gravel and sand as substrates in rock-reed filters, *Ecological Engineering*, 4:321–336.

Sanford, W. E., Steenhuis, T. S., Surface, J. M., and Peverly, J. H., 1995b. Flow characteristics of rock-reed filters for treatment of landfill leachate, *Ecological Engineering*, 5:37–50.

Surface, J. M., 1993. Constructed wetlands for municipal solid-waste landfill leachate treatment, M.Sc. Thesis, Cornell University, Ithaca, NY, 172 pp.

Surface, J. M., Peverly, J. H., Steenhuis, T. S., and Sanford, W. E., 1993. Effect of season, substrate composition, and plant growth on landfill leachate treatment in a constructed wetland, in Moshiri, G. A., Ed. *Constructed Wetlands for Water Quality Improvement*, Lewis Publishers, Boca Raton, FL, 461–472.

CHAPTER 5

Performance of a Constructed Wetland Leachate Treatment System at the Chunchula Landfill, Mobile County, Alabama

Keith D. Johnson, Craig D. Martin, Gerald A. Moshiri, and William C. McCrory

CONTENTS

ABSTRACT: Leachate poses a number of environmental problems. This is due primarily to the extreme variability of sources of this material and, therefore, the heterogeneity of its composition. Operating and closed landfills generate leachate whose quality and quantity depend on how the landfill was constructed, operated, and ultimately closed. Currently, a number of options exist for the containment and treatment of this type of wastewater. The option presented here uses a surface-flow constructed wetland treatment system installed at the Chunchula, AL closed landfill to provide the required leachate treatment for discharge into state and federal waters in conformity with the National Pollution Discharge Elimination System (NPDES) guidelines.

The construction of this full-scale operating wetland treatment system was started during the summer of 1992 and became fully functional in October 1993. The treatment system operates under permits issued by the Alabama Department of Environmental Management (ADEM) and remains in compliance with permit conditions as established by state and federal environmental agencies.

The treatment system encompasses approximately 1.29 ha of surface-flow wetlands located within the property boundaries of the old landfill. The wetlands receive contaminated groundwater, which is pumped from a confining layer located beneath the unlined but capped landfill. The treated effluent discharging from the wetlands is subsequently mixed with untreated storm water and eventually discharged into a receiving waterway. Effluent compliance with established state and federal water quality standards is assured through monthly analyses for BOD_5, COD, pH, TOC, NH_3, and heavy metals including Ar, Cd, Cr, Cu, Pb, Hg, Ni, Ag, and Zn. This system was originally designed to treat both the generated leachate and the leachate-contaminated groundwater at a hydraulic loading rate of 5×10^2 g/day. However, in spite of the fact that the original design loading volumes have been frequently exceeded by over 20 times, there has been no noticeable change in the quality of the treated leachate. Constructed wetland design, permitting, treatment efficiencies, and removal processes are discussed.

5.1 INTRODUCTION

The design of leachate treatment systems is usually made more complex by variations in quality and quantity of leachate as affected by time and characteristics of the solid waste received by the system. Site-specific leachate variability is due to variations in the composition of the refuse and its depth and permeability, as well as the climatologic conditions of the region.

Age-related variation in the composition of sanitary landfill leachate, as it influences the effectiveness of certain treatment processes, was first described by Chian and DeWalle (1976). They concluded that the leachate generated by young landfills is highly influenced by the acid fermentation stage of anaerobic decomposition resulting in the presence of free volatile fatty acids. Wastewater thus generated is deemed to be well suited for biological treatment. In contrast, leachates from old landfills are influenced by the methane fermentation stages of anaerobic decomposition yielding primarily recalcitrant humic and fulvic compounds which are more amenable to physical/chemical treatment modalities. Table 5.1 lists observed concentration ranges for commonly measured leachate constituents.

Table 5.1 Observed Ranges of Constituent Concentrations in Leachate from Municipal Landfills

Parameter	Concentration Range (mg/L)	Parameter	Concentration Range (mg/L)
COD	50–90,000	Cd	0–0.375
BOD_5	5–75,000	Cr	0.02–18
TS	50–45,000	Hardness (as $CaCO_3$)	0.1–36,000
TDS	1–75,000	Total phosphorus	0.1–150
TSS	1–75,000	Organic phosphorus	0.4–100
Total volatile solids (TVS)	90–50,000	Phosphate (inorganic)	0.4–150
Total volatile acids (TVA)	70–27,700	Ammonia nitrogen (NH_3–N)	0.1–2,000
Volatile suspended solids (VSS)	20–750	Organic nitrogen	0.1–1,000
Fixed Solids (FS)	800–50,000	Total Kjeldhal nitrogen	7–1,970
Ma	20–7,600	Nitrate nitrogen	0.1–45
Fe	200–5,500	Acidity	2,700–6,000
Zn	0.6–220	Turbidity (jackson units)	30–450
Cu	0.1–9	pH (standard units)	3.5–8.5
Ni	0.2–79	Specific conductance (µmho/cm)	960–16,300
Mn	0.6–41	Alkalinity	0.1–20,000
Pb	0.001–1.44	Chlorides	30–5,000
Mg	3–15,600	Sulfate	25–500
K	35–2,300	Total coliforms (CFU/100 ml)	$0–10^5$
Hg	0–0.16	Fecal coliforms (CFU/100 ml)	$0–10^5$
Se	0–2.7		

Source: U.S. Environmental Protection Agency, EPA/530-510-86-054, Office of Solid Waste and Emergency Response, Washington, D.C., 1986.

As a result of this inherent variability in composition, no two landfills produce the same quality of leachate. This variability presents landfill managers with the problem of providing cost-effective, reliable, flexible, on-site technologies for pretreatment and on-site disposal, or off-site disposal such as direct discharge of treated effluent into surface waters. Typically, leachate treatment and disposal options consist of seven categories, with some solid waste facilities incorporating several technologies simultaneously (Blanchard and Martin, 1995). An unpublished survey completed by the Florida Department of Environmental Protection provides the eight categories of leachate treatment in Florida (Table 5.2).

The type of leachate management to be implemented at a particular solid waste facility is controlled by several factors that include location, leachate volume and characteristics, available disposal options, and economic factors. Many of these methods tend to have high initial capital costs and high yearly operational and maintenance expenses. One low-cost method now widely used throughout the world is the on-site treatment and disposal using constructed wetlands (Sanford et al., 1990; Robinson, 1993; Surface et al., 1993; Martin and Johnson, 1995; Mæhlum, 1995).

5.2 BACKGROUND

The Chunchula solid waste facility became operational during 1975 and received 200 yd^3 (153 m^3) of municipal waste a day from the city of Mobile, Alabama. The disposal site reached its design capacity in 1989, and was subsequently closed. This particular municipal landfill, like many others that were operating during this period, was constructed before the implementation of the Environmental Protection Agency (EPA) Subtitle D regulations which were implemented to provide national standards for landfill design, construction, and closure procedures. At the time of the operation of this facility, little consideration was given to potential hazards, such as generation of leachate and associated contaminant migration into potable water supplies. The solid waste disposal practices at the time of operation involved the excavation of a pit into which refuse was disposed. As expected, the leachate generated by such operations has come in contact with both storm water and a perched water table that exists beneath the landfill.

Borden and Yanoshak (1989) report that violations of groundwater quality standards for organic and/or inorganic pollutants were detected at 53% of North Carolina landfills examined based on adequate monitoring data. This suggests that most unregulated landfills do contaminate groundwater and that existing monitoring programs are reasonably effective in detecting contaminant trends.

During 1990–91, Mobile County officials and their consultants attempted to delineate the extent of the contaminated leachate plume originating from the Chunchula landfill. This effort entailed the drilling of 26 soil test borings and installation of 40 monitoring wells (BA-1 through BA-40) in and around the landfill area (Figure 5.1).

These borings revealed the existence of a massive clay stratum that underlies the landfill site. Although the complete thickness of this confining layer is not known, it occurs within 9 to 12 m (30 to 40 ft) of the boring termination in BA-4. This hydrogeologic evaluation concluded that the layer extends throughout the subsurface of the landfill property. Fortunately, this formation has contained the leachate originating from the landfill and has prevented the contamination of the underlying main aquifer. However, this clay layer also supports a seasonal perched water table that has become severely contaminated by leachate. Due to the complexity of the hydrodynamics of the interactions of surface and subsurface waters with refuse at this landfill, it became evident that some form of remedial action was needed to prevent further contaminant migration. A majority of the monitoring wells exhibited pollutant concentrations exceeding established drinking water standards. The water quality parameters analyzed during this groundwater characterization included the following:

Table 5.2 Leachate Treatment Categories in Florida

FDEP District	Piped to WWTP	Trucked to WWTP	Aeration & Evaporation	Pretreatment	Sprayfield/ Percolation	Wetlands Treatment	Deepwell Injection	Recirculation in Lined Cell Slurry Wall 60 mil HDPE
Northwest	1	4	4	—	—	1	—	—
Northeast	1	10	—	2	—	—	—	5
Southeast	7	2	—	2	5	—	2	2
Southwest	2	3	5	2	—	—	—	3
South	—	1	4	1	—	—	1	2
Central	3	3	3	—	—	2	—	3
Total	**14**	**23**	**16**	**7**	**5**	**3**	**3**	**15**

Note: FDEP, Florida Department of Environmental Protection; WWTP, wastewater treatment plant; HDPE, high density polyethylene.

Source: Blanchard, R. A. and Martin, C. D., in Proceedings from the 6th Annual Southeastern Regional Solid Waste Symposium, Solid Waste Association of America, Publication GR-G0155, 1995, 55–67. With permission.

COD (chemical oxygen demand)   Cl (chloride)   Ba (barium)
TOC (total organic carbon)   Ca (calcium)   Cr (chromium)
SC (specific conductance)   Mg (magnesium)   Cd (cadmium)
TSS (total suspended solids)   Fe (iron)   K (potassium)
TDS (total dissolved solids)   Zn (zinc)   Hg (mercury)
SO_4^- (sulfate)   Pb (lead)   Na (sodium)
pH   Ni (nickel)

Groundwater monitoring data were collected only from wells in the general vicinity of the landfill and are summarized in Table 5.3. The units are the same as in Table 5.1.

Specific conductance values were used as an indicator of the extent and pattern of leachate migration occurring beneath the unlined landfill. This parameter was chosen as a result of real time data retrieval, the relatively inexpensive nature of the measurement, and the ability of the parameter to indicate the extent and severity of the leachate plume. Using the data collected, a map of the resultant contamination plume was constructed. Specific conductance values ranging from 0 to 50 μmhos/cm were considered the outside limits of the contaminant plume. Monitoring well data revealed that conductivity values ranged from 3940 μmhos/cm at BA-9 on the south side of the landfill to 21 μmhos/cm at BA-35 southeast of the landfill. The extent of the leachate plume, as defined by the 50 μmhos/cm contour is shown in Figure 5.2.

Preliminary landfill closure measures consisted of the implementation of certain remedial strategies to ensure proper closure of the landfill, reduce leachate generation, and diminish potentials for the contamination of the surrounding landscape. These measures included regrading, covering

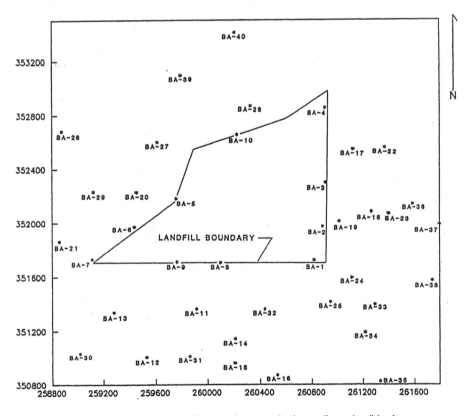

Figure 5.1 Location of shallow leachate monitoring wells and soil borings.

Table 5.3 Summary of Monitoring Well Data from the Chunchula Landfill

	BA-1	BA-2	BA-3	BA-5	BA-6	BA-7	BA-8	BA-9	Max	Min
pH	7.05	7.02	6.7	6.06	5.52	5.33	6.46	6.45	7.05	5.33
COD	675	850	265	640	300	280	250	390	850	250
TOC	180	230	27	150	51	39	23	330	330	23
SC	2700	2600	925	960	245	295	2000	1750	2700	245
TSS	2250	2360	1080	1120	530	150	320	250	2360	150
TDS	2390	1600	500	730	270	270	1360	1500	2390	270
Fe	47	100	160	340	320	130	60	72	340	47
Na	210	280	120	83	21	22	220	180	280	21
K	150	150	13	3.9	2.6	2.9	12	15	150	2.6
Chloride	12	380	99	2.6	18	17	470	8.4	470.0	2.6
Sulfate	59	15	17	11	18	5.8	33	4.9	59	4.9
Ca	38	15	22	52	12	10	160	150	160	10
Mg	10	20	7.8	25	10	9.2	14	24	25	7.8
Mn	0.78	0.52	0.4	8.1	5.6	4.9	7.4	6.2	8.1	0.4
Zn	0.062	0.092	0.062	0.052	0.08	0.049	0.19	0.049	0.190	0.049
Cu	0.049	0.049	0.049	0.055	0.049	0.049	0.049	0.049	0.055	0.049
Pb	0.019	0.110	0.021	0.019	0.190	0.019	0.230	0.019	0.230	0.019
Ni	0.085	0.049	0.078	0.080	0.080	0.100	0.080	0.100	0.100	0.049
Ba	0.56	1.52	0.52	0.70	0.44	0.40	2.20	1.36	2.20	0.40
Cr	0.075	0.180	0.180	0.100	0.210	0.049	0.060	0.049	0.210	0.049
Hg	0.0019	0.0019	0.0019	0.0019	0.0019	0.0019	0.0019	0.0019	0.002	0.0019
Ca	0.009	0.009	0.009	0.009	0.009	0.009	0.009	0.009	0.009	0.009

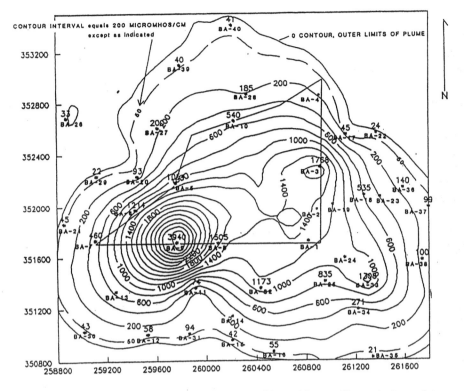

Figure 5.2 Extent and direction of the leachate plume as evidenced by specific conductance data generated from monitoring well samples sited at the Chunchula landfill.

the entire waste disposal area with a synthetic cap, placing 18 in. of soil on top of the synthetic liner, establishing vegetative cover over the entire cap, and constructing an adequate storm water drainage system and associated sedimentation ponds. These measures were implemented to minimize infiltration and maximize storm water runoff.

There are only a few studies reported in the literature that quantify the benefits of landfill caps and covers. One study, by Emrich and Beck (1981), indicates that significant groundwater improvement occurred within 1 year after construction of the cover was completed.

During planning for the installation of the stated closure measures, it was expected that these strategies would control contact of leachate with the storm water and thus only uncontaminated storm water would be discharged from the site. Based on this expectation, Mobile County applied and received an NPDES permit for storm water discharge.

During implementation of the improvements associated with the closure process, a spring of leachate-contaminated perched water was encountered in the area adjacent to the lower sedimentation pond. With the approval of ADEM, an underdrain piping was installed beneath the leachate spring to intercept the contaminated water and shunt it to the outfall structure of the pond. This did not alter the existing conditions at the site, but merely rerouted the flow of contaminated water to the outfall structure. The action was not intended to be a solution to the problem, but a measure to allow correction of related problems. These measures resulted in the discharge of storm water commingled with leachate, a condition outside the limit of the existing NPDES permit.

5.3 WETLAND DESIGN

A constructed wetland system was chosen because of its lower installation and maintenance costs, habitat and aesthetic values, and demonstrated ability to restore the quality of many types of wastewater through various physical, chemical, and biotic processes.

During the permitting process, ADEM requested and received information concerning the viability of the proposed wetland treatment technology. The case cited for this purpose was the highly successful and effective constructed wetland installed and operated by the authors 60 miles to the east at the Perdido Solid Waste Facility, Escambia County, FL (Martin and Johnson, 1995). Subsequently, regulatory approval was granted to commence design and install a full-scale constructed wetland system at Chunchula. In conjunction with the leachate treatment design, a supplementary underdrain system was installed to intercept the contaminated groundwater and direct the effluent to a wet well from which it would be pumped to the constructed wetland cells.

A total of seven in-series free-water-surface (FWS) constructed wetland treatment cells were installed with approximately 1.29 ha of actual water surface to accommodate emergent vascular macrophytes (Figure 5.3).

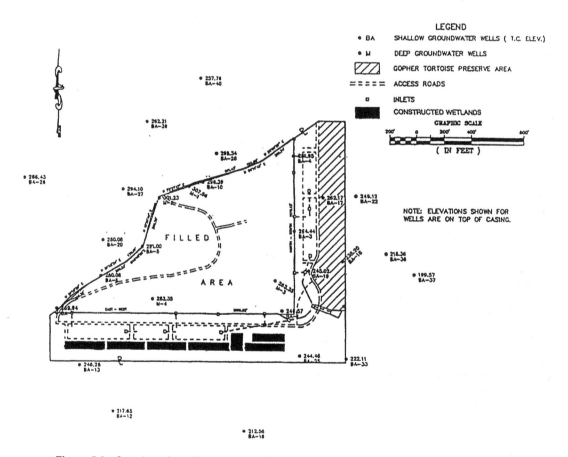

Figure 5.3 Overview of the Chunchula landfill facility and constructed wetland treatment system.

The first four emergent wetland treatment cells were constructed with equal dimensions of 100 × 17 m (0.68 ha). Cells 5, 6, and 7 total 0.59 ha and were designed somewhat differently because of limitations in the availability of land. The treatment system consists of a series of wetlands with a relatively high length-to-width aspect ratio (6:1). Individual cells are separated by 3.6-m-wide dikes to allow access to the ponds by maintenance equipment.

The treatment system has been designed to prevent storm water runoff from entering the wetland cells. Each cell contains inflow and outflow structures designed to provide even distribution of water and reduce potential for channelization. These distribution structures control water levels by head pressure, while at the same time providing a mechanism for bypassing of individual wetland cells for maintenance purposes. The inflow pipes and their associated weirs in cells 1 and 2 have been designed to provide aeration while distributing approximately 0.23 m³/min of leachate evenly into the receiving wetlands. The bottom of each cell was designed with a 2% slope, resulting in depths of 30 to 46 cm of standing water at the influent and effluent ends of each cell, respectively. This water-level design was instituted to overcome friction problems at the sediment–water interface, to ensure adequate water movement and to facilitate maintenance operations (Martin et al., 1993). Individual treatment cells contain a minimum 46 cm of freeboard sufficient to contain precipitation events associated with a 100-year storm. All side slopes of the wetland cells were constructed on a 3:1 ratio and vegetated to reduce erosional and depositional processes. Plant species used on the slopes and the freeboard areas of the ponds were selected to be compatible with those installed within the wetland cells. Exterior slopes and roadways (dikes) were stabilized through the planting of grass seeds.

Flow between the wetland cells is controlled primarily by head pressure, with hydraulic design goals aimed at ensuring a minimum of 15 days of retention time for wastewater renovation. The treated effluent from the constructed wetlands is directed back to the permitted outfall for discharge through a flowmeter and then to a surface water discharge site. Immediately before this juncture, the treated leachate is commingled with storm water runoff before the final discharge point.

The EPA Hydraulic Estimation for Leachate Production (HELP) model was used to estimate potential leachate generation from the Chunchula landfill. The values generated from the HELP model estimated leachate flow to be approximately 113 m³/day. Because over a decade of accumulated leachate was contained within the underlying confining layer, it was determined that wetland design should consider an average flow of 378 m³/day. This approach was based on these hydraulic data primarily because of the availability of a limited budget for water quality analysis. Incorporated in this design was also the consideration of 162 cm of rainfall annually. The treatment wetlands were constructed outside the 100-year flood zone and above the existing perched water table.

In order to maintain the necessary hydrologic conditions and prevent leakage, a leachate-resistant, 40-mm very low density polyethylene (VLDPE) liner was installed underneath the entire constructed wetland treatment system and extended up the sides to 15 cm above the anticipated water level. To protect the liner from puncture and degradation, it was covered with a minimum of 30 cm of sandy soil without coarse debris. To facilitate plant establishment and growth, a 30-cm layer of muck was placed on top of the sand/clay layer in the bottom of each cell (Figure 5.4). Once the liner placement was complete, berms were installed at designed intervals to separate individual constructed wetland cells.

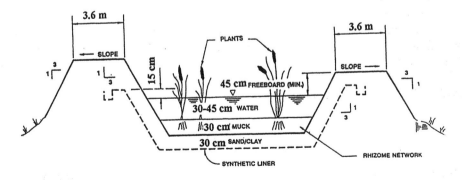

Figure 5.4 Cross-sectional detail of a typical constructed wetland cell.

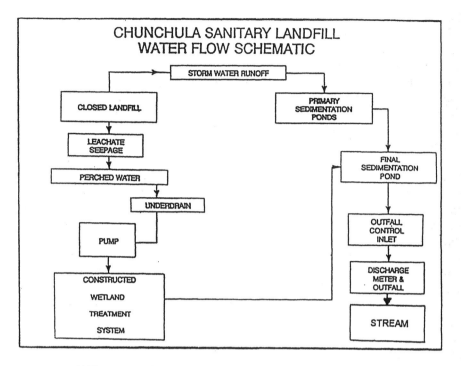

Figure 5.5   Chunchula landfill facility and treatment system.

As already stated, effluent from the final constructed wetland treatment system is mixed with storm water runoff and discharged through a flowmeter and into a small stream (Figure 5.5). Since this stream is considered waters of the U.S., an NPDES permit was required.

5.4 WETLAND VEGETATION

In constructed wetland ecosystems, the rhizosphere (root mass area) is considered the primary site where reduction of contaminants actually occur through the processes of sequestration and transformation. Emergent vascular plants have evolved special adaptations that make possible the transport of oxygen to the roots that penetrate the anaerobic sediments (Brix and Schierup, 1990). Therefore, these oxygenated plant root surfaces provide a thin aerobic interface within which numerous essential chemical and biological processes occur.

The first four cells (0.68 ha) were planted with cattails (*Typha latifolia*), because of the ability of this species to form dense populations in nutrient rich environments in a short period of time. The species also has the capability to translocate and incorporate certain heavy metals into biomass (Shutes et al., 1993). Other species, such as the arrowhead (*Sagittaria latifolia*) and the soft rush (*Juncus effusus*), were installed around the periphery at the land–water interface of each wetland cell. The remaining three treatment cells (0.60 ha) were planted with the common reed (*Phragmities communis*) and giant bulrush (*Scirpus californicus*). These two species attain considerable vertical stature (3 to 4 m) and can thus aid in increasing rates of evapotranspiration.

5.5 NPDES PERFORMANCE

The NPDES permitting program was authorized under section 402 of the Clean Water Act and requires a permit for the discharge of pollutants from any point source into waters of the United

States. The provisions of the Clean Water Act mandate that administration of the NPDES permits be delegated to those states whose program has been approved by the EPA. With the exception of the State of Florida, all of the states within EPAs Region IV have been delegated primary responsibility for administering the NPDES permit program. Although the EPA may require adherence to certain NPDES permit conditions, the states are not precluded from adopting more-stringent permit criteria (Laitos and Tomain, 1992).

States with an approved NPDES permit program are responsible for determining their own effluent limitations based primarily on the total maximum daily loads (TMDLs) established by environmental agencies of each state. The daily waste load limitations provide assurance that the quality of receiving bodies of water will meet water quality standards set by the states. Natural background contaminant levels, as well as other point-source discharges occurring within the same water body, are taken into account when determining TMDLs.

A compliance-monitoring program for the Chunchula treatment wetland system was established by ADEM. This includes monthly sampling of the effluent stream for flows and specific chemical constituents, as stated in the permit (Table 5.4) This monitoring program also regulates daily maximum and monthly average concentrations for certain parameters in the treated effluent. If a parameter exceeds the daily maximum as presented within the conditions of the NPDES permit, resampling will be required. The resultant values of the chemical constituent will be used in determining the monthly average. As noted in Table 5.4, some of the contaminant maxima listed exceed specified permit conditions. However, subsequent resampling and associated monthly average calculations have always fallen below the monthly maxima as specified in the permit. Precipitation events that generate stormwater–related contaminants (i.e., TSS and TDS) within the comingled waste stream are believed to be among the factors involved in these water quality perturbations. Until a needed remote recording meteorological system is installed on the site, this cause/effect relation is only speculative.

Table 5.5 presents concentrations for certain metals, ammonia, suspended and dissolved solids, and COD in the influent and effluent waters. Influent values were generated during the groundwater characterization study, whereas the effluent values were amassed from the monthly effluent sampling data. The monthly monitoring compliance data show that the effluent from the system is considerably less contaminated than earlier influent values indicated, demonstrating that even a heavily loaded FWS treatment system has the ability to perform satisfactorily in providing treated leachate quality that meets the mandated permit discharge limits.

5.6 CONCLUSIONS

The landfill cap has been in place since December 1991. The purpose of the cap was to decrease the volume of water that is recharging the perched water table and to eliminate further leachate generation. It is anticipated that when conditions beneath the landfill have stabilized, contamination of the perched water table will be controlled. However, landfilled waste has a great capacity to store water; therefore, the full extent of saturation of the waste cannot be determined, and the length of time needed for conditions to stabilize is unpredictable.

Precipitation-related dilution effects and the resulting unpredictable recharge of the contaminated perched water are expected to affect the quality of influent entering treatment wetlands. Conversely, drier periods may tend to concentrate the contaminant levels in the influent. At Chunchula, the constructed wetlands project was implemented as a cost-efficient, self-maintaining, fully operational system. For these reasons, limited funds were budgeted for the acquisition of related analytical data other than what is specified within the NPDES permit conditions. It is strongly recommended that a budget for the execution of a detailed groundwater and surface water quality sampling and the installation of an automated meteorological data logger be established so that efforts can be made to correlate the potential dilution/concentration effects on wastewater quality.

Table 5.4   Daily Maximum Effluent Concentration, Monthly Average Effluent Concentration Maxima, and Observed Effluent Concentrations Occurring at the Wetlands Treatment Systems Point of Discharge for 2 ½ Years of Operation, April 1994 to December 1996

Parameter	Flow	pH	TSS	TDS	BOD₅	COD	TOC	O&G	NH₃-N
Max	Monitor	8.5	Monitor	Monitor	Monitor	Monitor	Monitor	15	Monitor
Monthly Average	Monitor	—	35	500	50	30	—	—	20
Units	MGD	s.u.	mg/L	mg/L	mg/L	mg/L	mg/L	mg/L	mg/L
Min	0.0043	6.06	4.9	100	0.9	9	3	0.9	0.09
Max	9.1201	7.95	280	1090	37	190	67	15	35
Average	0.13	6.86	30	396.22	5.66	45.52	17.13	1.81	10.77
S.D.	0.0854	0.5493	44.6424	179.3531	5.1655	31.7797	10.1369	2.6297	9.3197

Parameter	Ar	Cd	Cr	Cu	Pb	Hg	Ni	Ag	Zn
Max	0.63	0.0076	4.7	0.024	0.072	0.0079	1.56	0.0012	0.20
Monthly Average	0.33	0.0027	0.58	0.017	0.0028	0.00004	0.174	mon	0.18
Units	mg/L	mg/L	mg/L	mg/L	mg/L	mg/L	mg/L	mg/L	mg/L
Min	0.09	0.001	0.009	0.019	0.004	0.0019	0.009	0.0009	0.019
Max	0.09	0.038	0.17	0.71	0.01	0.0019	0.04	0.0009	0.06
Average	0.09	0.001	0.09	0.02	0.004	0.0019	0.01	0.0009	0.03
S.D.	0	0.0011	0	0.0008	0.0014	0	0.0052	0	0.0095

Note: S.D., standard deviation.

Table 5.5 Removal Efficiencies for Required Monitored Chemical Constituents — Average Influent and Effluent Contaminant Concentrations Associated with the Chunchula Treatment Wetlands

	Influent (W-1)	Effluent (W-7)	% Removal
pH (s.u.)	6.32	6.86	—
TSS	1008	30	97
TDS	1078	396	63
COD	456	45	90
TOC	129	17	87
Cu	0.05	0.024	52
Pb	0.078	0.004	94
Hg	0.0019	0.0019	0
Ni	0.082	0.01	88
Zn	0.08	0.03	62

Note: Results are in mg/L except where noted.

Under the present conditions, and as long as the treatment system remains in compliance with the NPDES-specified parameters, the only analytical data that will be collected from this project will be those required by the NPDES protocol.

Further, it should be stated that the variability of leachate composition and the complexity of treatment and associated discharge options are just a few examples as to why it is impractical to extrapolate a reliable generic plan from a system such as Chunchula for potential application to other systems. Therefore, each system must be designed after reviewing all available information on the system including historical data, groundwater quality, and qualitative and quantitative data on the leachate.

Finally, in considering on-site leachate management strategies, it is important to contemplate such issues as regional climate, waste origination, daily volume of refuse received, susceptibility of wastewater generated to treatment modalities, and economic variables associated with each treatment alternative and discharge option. In the case of the Chunchula closed landfill, the treatment wetland approach has provided effective, consistently reliable leachate treatment at minimal installation and maintenance costs.

REFERENCES

Blanchard, R. A., and Martin, C. D., 1995. Removal processes and treatment performance of constructed wetlands treating landfill leachate at the Perdido Solid Waste Facility, Escambia County, Florida, in *Proceedings from the 6th Annual Southeastern Regional Solid Waste Symposium*, Solid Waste Association of America (SWANA), Publication GR-G0155, 55–67.

Borden, R. C., and Yanoschak, T. M., 1989. North Carolina Sanitary Landfills: Leachate Generation, Management and Water Quality Impacts. Water Resources Research Institute of the University of North Carolina, Report No. 243. UNC-WRRI-89-243, 43 pp.

Brix, H., and Schierup, H. H., 1990. Soil oxygenation in constructed reed beds: the role of macrophyte and soil-atmosphere interface oxygen transport, in Cooper, P. F. and Findlater, B. C., Eds., *Proceedings of the International Conference on the Use of Constructed Wetlands in Water Pollution Control*, Sept. 24–28, Cambridge, U.K., 53–66.

Chian, E. S. K. and DeWalle, F. B., 1976. Sanitary landfill leachates and their treatment, *Journal of Environmental Engineering* 102:215–239

Emrich, G. H. and Beck, W. W., Jr., 1981. Top sealing to minimize leachate generation—status report, in *Proceedings of the Annual Research Symposium (7th) On the Land Disposal of Hazardous Waste*, EPA/600/9-81/002B, Philadelphia.

Laitos, J. G. and Tomain, J. P., 1992. *Energy and Natural Resources Law*, West, St. Paul, MN.

Mæhlum, T., 1995. Treatment of landfill leachate in on-site lagoons and constructed wetlands, *Water Science and Technology* 32:129–135.

Martin, C. D., and Johnson, K. D., 1995. The use of extended aeration and in-series surface-flow wetlands for landfill leachate treatment, *Water Science and Technology* 32:119–128.

Martin, C. D., Moshiri, G. A., and Miller, C. C., 1993. Mitigation of landfill leachate incorporating in-series constructed wetlands of a closed-loop design, in Moshiri, G. A., Ed., *Constructed Wetlands for Water Quality Improvement*, Lewis Publishers, Chelsea, MI, 473–476.

Robinson, H. D., 1993. The treatment of landfill leachates using reed-bed systems, in Christensen, T. H., Cossu, R., and Stegman, R., Eds., in *Proceedings of Sardinia 1993 Fourth International Landfill Symposium*, Cagliari, Italy, 907–922.

Sanford, W. E., Steenhuis, T. S., Kopka, R. J., Peverly, J. H., Surface, J. M., Campbell, J., and Lavine, M. J., 1990. Rock-Reed Filters for Treating Landfill Leachates, American Society of Agricultural Engineers, Paper No. 90-2537.

Shutes, R. B, Ellis, J. B., Reviti, D. M., and Zhang, T. T., 1993. The use of *Typha latifolia* for heavy metal pollution contol in urban wetlands, in Moshiri, G., Ed., *Constructed Wetlands for Water Quality Improvement*, Lewis Publishers, Chelsea, MI, 473–476.

Surface, J. M., Peverly, J. H., Steenhuis, T. S., and Sandford, W. E., 1993. Effect of season, substrate composition, and plant growth on landfill leachate treatment in a constructed wetland, in Moshiri, G., Ed., *Constructed Wetlands for Water Quality Improvement*, Lewis Publishers, Chelsea, MI, 461–472.

U.S. Environmental Protection Agency, 1986. Subtitle D Study Phase I Report. EPA/530-SW-86-054, Office of Solid Waste and Emergency Response, Washington, D.C.

CHAPTER 6

The Use of an Engineered Reed Bed System to Treat Leachate at Monument Hill Landfill Site, Southern England

Howard Robinson, Gwyn Harris, Martin Carville, Mike Barr, and Steve Last

CONTENTS

ABSTRACT: The paper presents a brief review of landfill leachate composition, leachate management in the U.K. and landfill leachate treatment. The review is followed by a description of an engineered reed bed system to treat the leachate from Monument Hill Landfill Site in Southern England. Monument Hill is a closed landfill, in an infilled valley in southern England. The site received domestic wastes from the surrounding area during the 1970s and is typical of its era — unlined, with a culverted stream beneath the landfill in the valley bottom. The 10 to 15 m overburden of wastes caused failure of the culvert, resulting in some contamination of the stream. Even after the establishment of a new culvert, to direct the stream around the landfill, there was continued minor contamination of the stream. The case study is placed within a realistic context of state of the art leachate treatment in the U.K., where several dozen full-scale leachate treatment plants have been operating for periods of up to 15 years, and where reed beds are most commonly used for polishing of effluents, following aerobic biological pretreatment. Such combined leachate treatment schemes have been widely demonstrated to be capable of meeting extremely tight final effluent quality consents.

6.1 INTRODUCTION

The paper describes the process design, installation, and performance of remedial works that were commissioned during July 1996 at an old, closed, landfill site in southern England. Monument Hill Landfill Site is typical of many hundreds of similar old landfills that exist in the U.K., which contribute continuing, albeit minor, levels of contamination to watercourses, as relatively weak leachates seep from them. The solution adopted has potential to provide a low-technology, low-maintenance, and low-cost solution, in an environmentally friendly manner and may have wide-spread application at other sites.

It is valuable, initially, to review experiences over more than two decades in the U.K., relating to the composition and treatment of landfill leachates. This will allow the reed bed treatment solution adopted to be placed within a realistic perspective, for application to landfill leachates.

As a wastewater requiring treatment, landfill leachate has several unique properties. Primarily, it is extremely variable in composition — between different landfill sites, over time, and even spatially at specific landfills (see McBean and Rovers, Chapter 1). The main source of variability is the progressive decomposition of wastes that have been deposited in a modern landfill site. Rapidly, within days, oxygen in air spaces is consumed, and all subsequent decomposition processes take place under anoxic and ultimately anaerobic conditions.

Two major phases of decomposition are involved. First, acetogenic conditions rapidly establish and begin to hydrolyze complex organic molecules into simpler, soluble organic compounds, such as volatile fatty acids and sugars. During this stage, carbon dioxide is evolved, and ammoniacal-N is produced during degradation of proteins. Leachates are characterized by high chemical (COD) and biochemical oxygen demand (BOD_5) values (typically in excess of 20,000 mg/L), slightly acidic pH values (to about 5.5) and concentrations of ammoniacal-N that often approach 1000 mg/L. At this stage, high concentrations of metals such as iron and calcium may be present in solution because of the acidic conditions.

Second, after a period that may take months, or years, methanogenic conditions gradually become established and microorganisms begin to convert simple organic molecules into landfill gases, primarily methane and carbon dioxide, in a typical ratio of 3:2. Acetogenic processes continue throughout this stage, but as soon as simple organics are generated, they are consumed by metha-nogenic organisms and so do not appear in leachates to the same extent. At this time, leachates are characterized by COD values in the order of 1000 to 5000 mg/L, and BOD values that represent a reducing proportion of this COD. Ammoniacal-N concentrations remain high for an extended period and in the longer term represent the main contaminant of concern requiring to be treated. Processes of decomposition of landfilled wastes have been described in far greater detail elsewhere (Robinson, 1989; 1996b).

6.2 LEACHATE MANAGEMENT IN THE UNITED KINGDOM

During the last 4 or 5 years, the U.K. waste management industry has been undergoing a substantial reorganization, as public and private sector waste disposal companies have been repositioning themselves in the marketplace. After a period of uncertainty, and lack of investment in landfill operations, organizations are now seriously addressing their problem landfill sites and their new landfill developments. This has had an influence on the demand for leachate treatment technologies, which has expanded rapidly during the last few years.

The market for such treatment technology has been characterized recently by the entry of companies who bring to it heavily marketed specific systems, which, while derived from other effluent treatment industries, often have no track record of application to landfill leachate. Knowledgeable and independent advice for selection of appropriate technologies at specific sites has been in short supply. A common scenario has been the release of tender documents asking for "design and build" solutions, which have resulted in submission of schemes based on a whole range of systems (for example, aerobic biological, anaerobic, reverse osmosis, ammonia stripping, etc.), such that the client commissioning the scheme has been unable to make an informed choice between compatible technologies. In several instances, systems have been adopted that have not provided a complete solution.

Of vital importance to selection of leachate treatment systems is basic understanding of leachate quality at landfills. Research funded by the U.K. Department of the Environment, and by specific waste disposal companies, has added considerably to knowledge in these areas (Robinson and Lucas, 1985; Robinson, 1989). Most recently, a major review contract completed for the Department of the Environment by Aspinwall and Company has coordinated a mass of leachate quality monitoring information from U.K. landfills (over 5000 samples), and was published during 1996 by the Environment Agency (Robinson, 1996b). This will add considerably to knowledge in this area and provide much-needed assistance in improved design of leachate treatment systems, especially where these have to be specified and constructed in advance of waste disposal operations at new landfill sites.

6.3 A REVIEW OF U.K. LANDFILL LEACHATE COMPOSITION

A detailed review of leachate composition from 11 categories of landfill site listed below has been presented in detail elsewhere (Robinson and Gronow, 1993); a brief summary follows:

1. Large, high waste input rate, relatively dry
2. Large, high waste input rate, deep, wet, and "bioreactive"
3. Large, deep landfills, wet and "cold"
4. Small landfills, operated with limited water ingress
5. Small landfills, operated with some control of water ingress
6. Small, relatively wet landfills
7. Large, flat and shallow landfills
8. Landfills containing a high proportion of baled wastes
9. Landfills containing a high proportion of pulverized wastes
10. Valley landfills, with groundwater ingress suspected
11. Old landfills without engineered restoration

These categories were necessarily arbitrary, but were considered to provide a foundation upon which the review could be structured. For each category, significant effort was made to obtain time-series data for basic leachate composition at a minimum of three or four landfill sites. This was not always possible. Nevertheless, the objective was to characterize leachate quality over time in terms of COD and BOD_5 values, pH values, and concentrations of ammoniacal-N and chloride. Results from analysis of more than 4000 leachate samples have been included in this part of the review.

The database was complemented by an extensive program of leachate sampling and more detailed analysis at each site. To complete the database, 30 leachate samples were taken for very detailed analysis. The sites chosen for sampling were selected to be representative of overall landfill conditions in the U.K. and of the 11 site categories. To some extent resources were concentrated on the types of site currently being licensed and operated, while maintaining the broad overall coverage required. Table 6.1 indicates the range of parameters that has been determined in these limited numbers of samples. Such analyses are nevertheless seen as a first sweep and will lead to recommendations for further analytical studies in the future. Various interesting aspects are included within the final report of the study: for example, discussion of leachate components such as heavy metals and priority organic substances.

Table 6.1   Detailed Analytical Suite Selected

pH value	Cyanide	Chloroform
Alkalinity	Monohydric phenols	1,1,1-Trichloroethane
Conductivity	Dichlorvos	Carbon tetrachloride
BOD (20 day)	Malathion	Bromodichloromethane
BOD (5 day)	Parathion	Trichloroethylene
TOC	Parathion-methyl	Dibromochloromethane
Fatty acids		Tetrachloroethlene
		Tribromomethane
		1,2-Dichloroethane
Ammoniacal-N	Fenitrothion	
Kjeldahl-N	Fenthion	
Nitrate-N	Azinphos ethyl	AOX (total)
Nitrite-N	Azinphos methyl	AOX (solids-associated)
Sulfate	Gamma-HCH (lindane)	Phthalate esters
Phosphate	Aldrin	
Chloride	Dieldrin	
	Endrin	
Boron	p'p-DDT	*Microbiological*
Sodium	PCB (as Aroclor 1254)	Total viable count
Magnesium	Endosulfan-A	(22 and 37°C)
Aluminium	Endosulfan-B	Total and fecal
Silicon	Trifluralin	Coliforms
Potassium		Fecal streptococci
Calcium	Hexachlorobenzene	*Salmonella*
Vanadium	Hexachlorobutadiene	*Clostridium botulinum*
Chromium	1,2,3-Trichlorobenzene	*Clostridium perfringens*
Manganese	1,2,4-Trichlorobenzene	
Iron	1,3,5-Trichlorobenzene	
Nickel	Pentachlorophenol	
Copper	Individual phenols	
Zinc		Tritium
Arsenic	Atrazine	
Cadmium	Simazine	
Tin	Total organotin (as TBTO)	
Mercury		
Lead		

The leachate review is not intended to be a definitive or final report on the subject. Rather, it provides a starting point from where other research needs can be identified and further work begun. The demands of sustainable design will certainly affect the nature of wastes being disposed of into landfills and hence the composition of leachates produced from them. Other changes are already taking place: for example, the end of the marine disposal of sewage sludge has led to a large increase in the quantities of sludge and sludge ash being landfilled and this will have a significant local influence on leachate quality.

6.4 LANDFILL LEACHATE TREATMENT

Various wastewater treatment technologies have been applied to landfill leachates with varying degrees of success. These have included

- Aerobic biological treatment
    attached growth
    nonattached growth
- Anaerobic biological treatment
- Spray irrigation to grassland, peatland, or woodlands
- Reed bed treatment
- Ammonia stripping
- Reverse osmosis
- Ozonation

The majority of leachate treatment schemes that have been successfully installed on landfill sites in the U.K. and overseas have at their heart an aerobic biological treatment process. Sequencing batch reactors (SBRs) have proved a reliable and robust method of providing treatment of leachates to specified effluent consent values, often installed in combination with reed beds to provide consistently high-quality effluents that can safely be discharged into sensitive surface watercourses.

Experience at extended aeration SBR leachate treatment plants in the U.K. goes back more than a decade and a half and many papers have been published that provide detailed operating information over several years at several plants (Robinson and Grantham, 1988; Robinson, 1992; 1993a; 1996a; Last et al., 1993; Barr, 1994). For example, more than 20 full-scale pilot plants designed by Aspinwall and Company are at present operational in England, Ireland, Scotland, and Wales, of which half discharge into high-quality surface watercourses. In some circumstances, reed bed treatment systems have been used successfully to provide polishing of biologically treated effluents and achieve very high standards (Robinson, 1993).

U.K. expertise in treatment of landfill leachates has been used in other countries. Of particular interest is the application of proven aerobic biological treatment systems to leachate from Hong Kong landfill sites where Aspinwall and Company has been working since 1988. Landfills will continue to provide a major disposal route for most of the 30,000 t of waste per day, projected to be produced in Hong Kong by the year 2001. Three very large strategic landfill sites have been developed in more remote parts of the New Territories of Hong Kong, to be supplied with wastes by a new system of transfer stations. These landfill sites will be located in the West, North East, and South East New Territories. Aspinwall has reported research into characterization and treatment of Hong Kong leachates, including systems with full nitrification and denitrification of concentrations of ammoniacal-N as high as 4000 mg/L (Robinson and Luo, 1991; Robinson et al., 1995). Results obtained have been used as a basis for the detailed design of some of the largest leachate treatment plants in the world.

Technologically advanced and proven leachate treatment systems have been installed at sites in the U.K. and overseas during the last two decades. These systems have demonstrated that automated, robust, and reliable treatment schemes can be provided at reasonable cost and be operated easily by landfill site staff. It has been possible to design optimum solutions at a wide variety of sites, in a wide range of circumstances, ranging from pretreatment before disposal to sewer to full treatment to very high standards, for discharge of effluents into sensitive receiving watercourses. Elsewhere, detailed case studies have described full-scale leachate management and treatment schemes that are "state of the art", certainly from a U.K. perspective and probably internationally, too (Robinson et al., 1997).

Reed beds have been used very successfully for effluent polishing after aerobic biological pretreatment of leachates. However, they have generally been considered to be completely unsuitable as a primary treatment, for the likely strengths of leachate at most recent landfills, particularly in the light of documented poor performance for removal of even relatively low (<50 mg/L) concentrations of ammoniacal-N. Reed beds have, nevertheless, been identified as having great potential for the treatment of leachate emission from older, historic landfill sites, where leachates are in a strongly methanogenic state and much more diluted. The following case study, at Monument Hill Landfill Site describes in detail one such instance where a reed bed has been successfully designed and commissioned.

6.5 BACKGROUND

Monument Hill Landfill Site is situated in an infilled valley, some 2 km east of the town of Devizes in Wiltshire in southern England. The site received domestic wastes from the surrounding area during the 1970s and is typical of its era — unlined with a culverted stream beneath the landfill in the valley bottom. The 10 to 15 m overburden of wastes caused failure of the culvert, resulting in minor but persistent contamination of the stream. The site is remote and unstaffed and has no power supply or nearby sewer for disposal of effluent. Part of the site includes a wildlife reserve.

In 1985, to improve serious contamination of the stream by leachate, a new culvert was prepared by pipe jacking to direct the stream around the landfill. The old culvert remained in place and caused continuing minor pollution downstream of the site, primarily from iron and suspended solids, but also from ammoniacal-N levels, typically to 30 mg/L, and low concentrations of the herbicide Mecoprop. Remedial works were initiated during early 1994 and initially comprised a detailed monitoring and sampling exercise to determine leachate and stream flow rates and qualities and to compare these with seasonal and daily weather patterns.

As the site is remote, closed, and unstaffed, a low-maintenance, low-cost, vandal-resistant system was required for treatment of leachate flows, which were typically in the range 200 to 300 m^3/day. Coupled with physical constraints posed by the site — wildlife sensitivity, the only area available for construction being over infilled parts of the site — and required effluent standards indicated by the Environment Agency, an engineered reed bed scheme was developed.

6.6 MONUMENT HILL LANDFILL SITE

Figure 6.1 is a plan that shows the location of Monument Hill Landfill Site and indicates the general arrangement of the landfilled area and flows of water and leachate. Plate 1 is an aerial photograph taken during the spring of 1996 before remedial works began. The partially restored site is at present used by a recycling company, primarily for the separation of construction wastes and aggregates into usable fractions. The replacement bypass culvert is to the west of the landfill and was installed by pipe jacking during 1985, to divert the main flow of the stream (the Stert Watercourse) around the wastes, rather than through the original culvert in the base of the infilled valley.

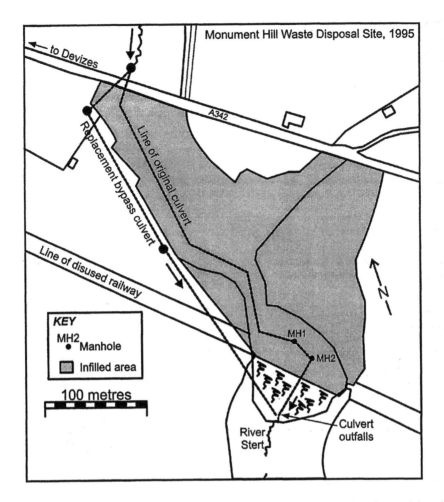

Figure 6.1 Monument Hill waste disposal site in 1995, prior to implementation of remedial works.

This stream diversion was entirely successful to the extent that leachate contamination was never detected in the Stert Watercourse, where it emerged from the "new culvert" to the south of the infilled area, at the base of the old railway embankment that formed the southernmost limit of infilling with wastes.

Nevertheless, the original culvert beneath the landfilled wastes and associated granular materials used initially to cover it, continued to collect leachate from the base of the site and to provide a route whereby this could drain into the Stert Watercourse some 10 m after it emerged from the new culvert (Plate 2). Although leachate flows were relatively weak and subject to dilution by a factor of about ten times, there remained an impact on the stream after 1985, although the stream diversion works meant that this was substantially reduced. The impact was primarily visual, as precipitation of iron on the streambed caused bright orange staining for a distance of 5 or 10 m downstream, and could be detected as far as 50 m beyond this. Growth of filamentous "sewage fungus" organisms (*Sphaerotilus natans, Leptomitus lacteus*) was never a problem, primarily because of the very low BOD values present in the leachate discharge. Low concentrations of ammoniacal-N, to a maximum of 2 or 3 mg/L, were also detectable within the Stert Watercourse, for a distance of up to 100 m downstream.

During late 1992 and early 1993, the National Rivers Authority (since 1996, the U.K. Environment Agency), began discussions with the Wiltshire County Council, to seek implementation of additional remedial works that would reduce further the impact of the old landfill on the watercourse. Aspinwall and Company was instructed by the Council during Autumn 1993 to progress toward such a solution and immediately began a structured program of site monitoring.

6.6.1 Site Monitoring

An intensive program of monitoring of the site was implemented during Autumn 1993, to complement the long period during which occasional samples of leachate had been taken by Wiltshire County Council. Monitoring included continuous measurement and recording of flows of leachate and of the flow within the Stert Watercourse (Plate 2).

Water samples were routinely obtained from the following points:

1. Leachate flowing from the *old culvert* outfall.
2. Water from the Stert Watercourse just upstream of the leachate outfall from a location referred to as the *new culvert*;
3. Water from the Stert at a monitoring point agreed with the Environment Agency, located 80 m *downstream* of the culvert outfalls.

The samples were analyzed for the parameters listed in Table 6.2, which summarizes the monitoring results.

Table 6.2   Summary of Monitoring Results for Leachate Quality in the Outfall from the Old Culvert at Monument Hill Landfill Site, 4 December 1993 to 4 October 1994 (10 months)

Determinand	Samples	Mean	Min.	Max.	Units
pH value	14	7.1	6.8	7.8	pH
COD	22	43.6	25	64	mg/L
BOD_5	21	<5	1.4	5.0	mg/L
Ammoniacal-N	21	25.5	16.7	31.0	mg/L
Chloride	20	94.7	83	108	mg/L
Suspended solids	14	57.5	5ว	70	mg/L
Alkalinity (as $CaCO_3$)	7	663	614	740	mg/L
Conductivity	7	1330	1210	1472	µS/cm
Nitrate-N	8	0.6	0.3	1.0	mg/L
Nitrite-N	8	<0.1	—	<0.1	mg/L
Sulfate (as SO_4)	6	48.3	26	86	mg/L
Phosphate (as P)	2	0.3	—	0.3	mg/L
Sodium	7	59.3	54	67	mg/L
Magnesium	7	16.8	15	20	mg/L
Potassium	12	31.8	26.0	36.4	mg/L
Calcium	12	215	196	235	mg/L
Chromium	7	<0.1	<0.01	<0.1	mg/L
Manganese	13	0.81	0.50	0.99	mg/L
Iron	20	21.2	12.0	28.0	mg/L
Nickel	7	<0.05	<0.01	<0.05	mg/L
Copper	7	<0.05	<0.01	0.03	mg/L
Zinc	18	0.08	0.05	0.11	mg/L
Cadmium	8	<0.02	<0.002	<0.01	mg/L
Lead	7	<0.05	<0.01	0.02	mg/L
Mercury	2	<1.0	<0.1	<1.0	µg/L
Arsenic	1	0.005	—	0.005	mg/L
Mecoprop	15	5.34	1.06	18.91	µg/L
1,2-Dichloroethane	14	<5	<2	12.0	µg/L

6.6.2   Contaminants of Concern

The only contaminants present in the leachate discharge that were considered to have any potential for significant adverse impact on the Stert Watercourse were iron, suspended solids, and ammoniacal-N. Iron was unlikely to be a health concern, its main impact being the iron staining that was evident for a distance of 10 m below the discharge point. This orange-colored deposit not only looked unsightly, but also carpeted the streambed, thus inhibiting the establishment of plant and animal life, and potentially upsetting the natural ecological balance within the stream (see Chapter 16). Levels of *suspended solids* in leachate associated to some extent with particulate iron were typically about 60 mg/L, and would need to be reduced. Ammoniacal-N was of concern due to its potential toxic effect on aquatic organisms. Of specific concern is the effect of the unionized fraction, present in the form of ammonia (NH_3–N) (Seager et al., 1988). Salmonid fish, such as trout, are particularly susceptible.

Mecoprop (MCPP) is a phenoxyalkanoic herbicide with the molecular formula $C_{10}H_{11}ClO_3$ that is relatively soluble in water (620 mg/L at 20°C). It is widely used in the U.K. for postemergence control of broad-leaved weeds, such as chickweed, plantain, and clover, in wheat, barley and rye crops, or in grassland and pastures for control of docks. Leachate analyses determined concentrations of up to 19 µg/L, present in the leachate discharge (values of up to 0.6 µg/L were also measured in the upstream Stert Watercourse). No other Red List substances were detected in the leachate, except occasional erratic and very low levels of 1,2 dichloroethane — only one sample exceeding the drinking water limit of 10 µg/L.

U.K. Drinking Water Quality Regulations incorporate the EC non-health-related standard of 0.1 µg/L for an *individual* pesticide (or a "surrogate zero"). Even though MCPP is of low toxicity to mammals, fish, and insects, U.K. guidance states that it should not be used near watercourses (Ivens, 1994). Clearly, in light of the above, it was considered advantageous to attempt to reduce the concentrations of MCPP entering the stream, if reasonably practicable.

Heavy metals are often stated to be of concern by regulators in dealing with discharges of landfill leachate, either for treatment in sewage works or directly into surface watercourses. Extensive study and research work has demonstrated that they are rarely, if ever, found at significant levels in any leachates (Robinson and Gronow, 1993), nor were any significant concentrations detected in samples from either culvert at Monument Hill (Table 6.3).

To confirm this further, a sample of the precipitate being deposited within the mouth of the old culvert outflow was sampled, air dried, digested, and analyzed for metals. Dried sludge was primarily iron (32% w/w) and calcium (6.8% w/w). Concentrations of toxic metals in the sludge were well below permitted values for *soils* following application of sewage sludges — values much higher than those found are permitted in sludges which are destined for land disposal.

6.6.3   Measured Dilution Values

Figure 6.2 plots concentrations of chloride at each monitoring location over time, to provide an indication of the dilution of leachate by the Stert. Chloride is probably the best conservative parameter to measure, as it is only affected minimally by chemical and biological processes in a watercourse, and so is a good measure of actual dilution effects that are taking place. Data very clearly demonstrated that a consistent degree of dilution occurred at all times, a fact that was taken into account in the design of remedial works, and in negotiations with the Environment Agency.

Table 6.3 compares mean values of all determinands in samples from the three monitoring locations, and confirmed that dilution available within the receiving watercourse was typically in the range of 6 to 25 times. Figure 6.3 is a plot of the relationship between measured daily flows in the Stert Watercourse as it emerged from the new culvert (1000 to 4000 m³/day), and daily flows of leachate from the old culvert (60 to 300 m³/day). This demonstrated that during 1994 minimal

Table 6.3 Mean Values of All Determinands in Water Samples from the Three Monitoring Points at Monument Hill Landfill Site, 4 December 1993 to 4 October 1994

Determinand	New Culvert	Old Culvert	Downstream	Units
pH value	7.7	7.1	7.7	pH
COD	<20	43.6	20.3	mg/L
BOD_5	<2	<5	<2	mg/L
Ammoniacal-N	0.11	25.5	2.5	mg/L
Chloride	21.5	94.7	29.7	mg/L
Suspended solids	18.5	57.5	22.2	mg/L
Alkalinity (as $CaCO_3$)	318	663	349	mg/L
Conductivity	646	1330	721	µS/cm
Nitrate-N	3.7	0.6	3.2	mg/L
Nitrite-N	<0.1	<0.1	<0.1	mg/L
Sulfate (as SO_4)	51.2	48.3	49	mg/L
Phosphate (as P)	0.3	0.3	0.3	mg/L
Sodium	11.4	59.3	17	mg/L
Magnesium	3.5	16.8	4.8	mg/L
Potassium	5.6	31.8	8.2	mg/L
Calcium	151	215	159	mg/L
Chromium	<0.01	<0.1	<0.02	mg/L
Manganese	0.07	0.81	0.12	mg/L
Iron	0.49	21.2	2.5	mg/L
Nickel	<0.02	<0.05	<0.03	mg/L
Copper	<0.01	<0.05	<0.02	mg/L
Zinc	0.04	0.08	0.03	mg/L
Cadmium	<0.01	<0.02	<0.01	mg/L
Lead	0.01	<0.05	0.009	mg/L
Mercury	—	<1.0	<1.0	µg/L
Arsenic	<0.001	0.005	—	mg/L
Mecoprop	0.076	5.34	0.436	µg/L
1,2-Dichloroethane	<5	<5	~8.0	µg/L

dilution available was 5:1; dilution exceeded 6:1 more than 99% of the time; and exceeded 10:1 for 70% of the time.

6.6.4 Flow Measurement Study

The flow from both culverts was continuously monitored by data loggers over a 12-month period. Rainfall records were obtained from four Meteorological Office Weather Stations located within a 5 km radius of the site. Site-specific rainfall records were also obtained by installation of a rain gauge and data logger at the site itself. The flow measurement study provided not only a record of the flow variation of both culverts, but also clearly demonstrated that the flow of leachate from the old culvert was *not* rainfall dependent. It was calculated that mean rainfall infiltration rates varied between 25 and 33 m³/day compared with flows of leachate from the old culvert, which were typically between 180 and 220 m³/day. It was concluded that these flows of leachate could not result simply from surface infiltration of rainfall into the landfilled area which would account for less than 20% of total flows being measured.

Much of the flow being discharged by the old culvert, therefore, appeared to represent ground-water inflows into the drainage system in the site base which were connected into the old culvert. A study of old maps of the area did indeed show springs emerging from the old valley sides, helping to support this conclusion. This finding had a direct bearing on the design of the remedial works eventually undertaken. Any engineering works undertaken to reduce infiltration into the landfill surface would not only be extremely expensive, but would in practical terms have minimal impact

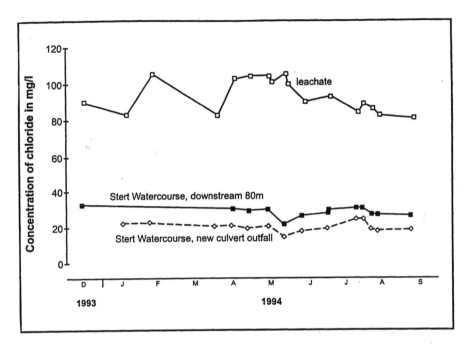

Figure 6.2   Concentrations of chloride (mg/L) in leachate, and in the upstream and downstream Stert Water-course, 1993–94.

on flows of leachate at the old culvert. Efforts were therefore concentrated on methods of treating the flow, rather than in trying to reduce it.

6.7 DESIGN OF THE LEACHATE TREATMENT SCHEME

6.7.1   Discharge Consent Values

Indicative consent conditions likely to be imposed on the treated discharge by the Environment Agency were initially as follows:

COD	60 mg/L
BOD$_5$	15 mg/L
Ammoniacal-N	5 mg/L
Iron	2 mg/L
Suspended solids	25 mg/L

6.7.2   Remedial Measures

A low-maintenance, low-cost, vandal-resistant treatment system was required, as the site was remote, closed, and unstaffed. These factors, coupled with the physical restraints of the site, such as lack of an electricity supply and that part of the site is in a wildlife trust area, all pointed toward the use of a horizontal-flow reed bed. However, reed beds are usually used for polishing effluents, or are used in conjunction with other processes. Little is published about their use for the removal of any specific pesticides, and their ability to remove ammoniacal-N is generally thought to be poor. The required size of reed bed would be determined after consideration of the process design requirements.

6.7.3   Process Design

6.7.3.1   Brief Description of a Reed Bed

The type of reed bed under consideration was a lined, gravel-filled bed, generally about 600 mm deep to be planted with the reed *Phragmites australis*. Effluent would enter at the inlet and travel slowly through the bed following a horizontal flow path, before leaving the bed via a level-control device. The reeds have the ability to transport oxygen from the air down to their rhizomes and out through the roots. Aerobic bacteria thrive in the area immediately surrounding the rhizomes — an area known as the rhizosphere. During passage through the bed, organic matter in the effluent can be oxidized by the bacteria. The gravel in which the reeds are planted also acts as a filter medium. Ammoniacal-N may be oxidized to some degree, although the process tends to be oxygen limited, especially if there is a high organic load.

6.7.3.2   Contaminant Removal Mechanisms within a Reed Bed

Iron and suspended solids are readily removed in a reed bed system, principally via physical oxidation and filtration processes. Growth of the rhizome system of the reeds within the gravel bed may contribute to improved performance, probably by enhancing the supply of oxygen, which is required to convert soluble iron to insoluble iron hydroxide.

Although reed beds have a poor record for removal of ammoniacal-N from effluents containing high levels of COD and BOD (for example, from domestic sewage), they are generally more successful in situations where concentrations of organic contaminants are much lower (as in the Monument Hill leachate) and more oxygen is therefore available to nitrifying organisms. It is these nitrifiers, principally Nitrosomonas and Nitrobacter, that convert ammoniacal-N to nitrite and then to nitrate.

No specific nitrification performance could be guaranteed, as a reed bed fails to establish completely the first 2 or 3 years. Notwithstanding this, because of the consistent and immediate dilution that had been demonstrated (Figure 6.3), the Environment Agency was able to relax its consent limit for ammoniacal-N from 5 to 23 mg/L. It is beneficial to the situation at Monument Hill that maximum rates of nitrification within a reed bed occur during the summer months, when flows in the Stert are at their lowest; the available dilution is thus least and the receiving watercourse is most sensitive.

One of the main reasons for the widespread use of mecoprop as a herbicide is that it is readily metabolized and degraded in the environment. Degradation is promoted by factors that favor aerobic microbial growth, including temperature, oxygen, and moisture. Consequently, at low rates of application, the herbicide usually disappears within 1 to 4 weeks. In addition, the phenomenon of *enhanced degradation* has been shown to occur when mecoprop is applied on several occasions to the same soil (Hassal, 1990).

It was considered that a reed bed would be the ideal environment to enhance this process of "enrichment" of certain microorganisms that are capable of rapidly metabolizing low levels of specific compounds, such as mecoprop. The correct environmental factors are present within the bed: air and bacteria in the rhizosphere; plentiful sites for adsorption prior to degradation; long retention times and slow flow velocities.

6.7.3.3   Size of Reed Bed Required

The bed was sized using the experience gained from the experimental reed bed, designed and monitored by Aspinwall for the Department of the Environment, that polished effluent from the leachate treatment plant at Compton Bassett, Wiltshire (Robinson, 1993b). Being pretreated, this effluent had a low BOD, similar to that of raw leachate at Monument Hill. Based on a population

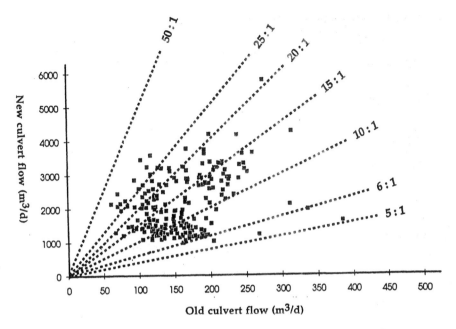

Figure 6.3 Comparison of total measured daily flows in the old and new culvert outfalls at Monument Hill Landfill Site during 1994 (results in m³/day).

equivalent for BOD removal, the Compton Bassett bed design was extrapolated to give a bed size of 1800 m² at Monument Hill.

Iron loading rates of between 3 and 20 g/m²/day have been reported (Cooper and Findlater, 1990). An 1800-m² nominal size bed resulted in a loading rate of 4 g/m²/day and so this size was considered adequate, with additional spare capacity to account for the bed possibly becoming clogged with iron deposits over time.

In order to confirm the predicted iron removal rates, simple trials were carried out on samples of leachate from the old culvert as follows:

1. A sample was collected and the completely shaken sample was analyzed;
2. A sample was filtered on site (glass fiber GF/D), and a shaken sample of the filtrate analyzed;
3. A sample was filtered after 24 h storage in a full bottle at 10°C; a shaken sample of the filtrate was analyzed.

Most iron in the leachate comes out of the culvert in solution, and then rapidly precipitates on contact with the atmosphere. In the proposed treatment system, this precipitated iron would primarily be filtered out within the reed bed.

Results are included in Table 6.4. As expected, only iron was really affected by filtration. The results demonstrated that most of the iron is in solution as it leaves the culvert (86% remaining in the sample filtered on site), but that within 24 h this had nearly all (>96%) precipitated out, leaving a residual concentration of soluble iron of less than 0.6 mg/L. This confirmed the belief that the reed bed system would very effectively reduce iron concentrations and that most of the solids in samples from the old culvert comprise precipitated iron.

6.7.4 Iron Settlement Tank

The results in Table 6.4 also demonstrate that 14% of the iron in the leachate had already been oxidized and had come out of solution before it had left the old culvert and entered the stream. In

Table 6.4 Results from Analysis of Samples Taken from the Old Culvert at Monument Hill Landfill Site, on 8 January 1996

Determinand	Total Leachate	Filtered On Site	Filtered at 24 h
COD	47	47	47
BOD$_5$	3	3	<2
Ammoniacal-N	19.2	18.9	19.4
Iron	16.6	14.3	<0.6
Zinc	0.07	0.06	0.04
Suspended solids	70	40	—
Volatile suspended solids	18	9	—
Nitrate-N	0.6	0.3	<0.3
Nitrite-N	<0.1	<0.1	<0.1

Notes: Results in mg/L; dashes equal no result.

keeping with the low maintenance requirements, and in recognition of the fact that the reed bed could progressively become clogged with iron, a settlement tank with a 2 to 3 h hydraulic retention time was included at the front end of the reed bed. This tank also contained an internal baffle system, which was devised to remove some, if not all, of the 14% of iron already oxidized. The tank can readily by cleaned out when required by a vacuum tanker. Additionally, the inlet distribution weir was made very simple — in the form of a wide, open channel — which was very easy to clean and maintain.

6.7.5 Mecoprop

As previously described, the removal mechanism for mecoprop requires air (oxygen) and time. The aim was to remove mecoprop in the reed bed, rather than have this process occurring in the stream. The size of bed chosen gave a mean hydraulic retention time of 3 to 4 days and the rhizomes provide both air and plentiful sites for bacterial degradation. Again, the immediate dilution in the receiving waters was accepted by the Environment Agency as a sufficient safeguard. [*Note:* At another Aspinwall-designed full-scale leachate treatment plant (Robinson et al., 1997), the leachate also contains mecoprop, but at concentrations of hundreds of micrograms per liter, compared with a maximum of 18 µg/L at Monument Hill. Mecoprop at that site is successfully reduced to less than detection limits by a purpose-built biological plant, followed by ozonation and final reed bed polishing stages.]

6.7.6 Shape and Design of the Reed Bed

The land available (Plate 1 and Figure 6.4), dictated the shape of the bed somewhat, but a long flowpath was maintained to discourage short-circuiting. The bed also has an irregular curved shape, which helped it to blend in with its setting in a wildlife trust area. Being constructed on waste, settlement of the bed was expected to occur, which would cause subsequent ponding. In recognition of this fact, allowance was made for extra freeboard, together with provision for controlling the level of water in the bed over a wide range.

6.8 CONSTRUCTION OF THE TREATMENT SYSTEM

The construction of the reed bed took 11 weeks, with completion and commissioning in mid-July 1996. Construction was a fairly straightforward earthworks job followed by high-density polyethylene membrane lining and filling of the bed with gravel. Care had to be taken during the earthworks stage to avoid excavating through the landfill "cap" (such as it was) to the waste below.

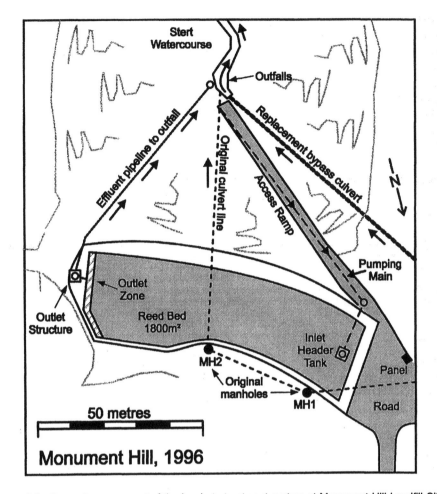

Figure 6.4 General arrangement of the leachate treatment system at Monument Hill Landfill Site.

Interception of the old culvert to install a pumping chamber near to the existing outfall produced a few surprises. Excavation demonstrated that the 300-mm-diameter culvert outfall was merely a short length of pipe, crudely concreted into the remains of the original culvert structure that was 900 mm in diameter. At least 90% of the culvert was filled by iron-rich sludge and silt that had built up over more than 30 years since installation, potentially for a distance of nearly 100 m beneath the landfill. The sludge needed to be jetted out and tankered away to a licensed sewage treatment works, quickly, before it might be washed into the stream by the flow of leachate in the culvert that continued unabated. Fortunately, as part of the water quality monitoring study described earlier, a detailed analysis of sludge scraped from the inside of the 300-mm-diameter old culvert outfall pipe had been carried out previously and the receiving sewage treatment works was able to accept the sludge, which comprised 40% dry solids, primarily iron and calcium.

6.9 PERFORMANCE OF THE TREATMENT SYSTEM

The reed bed was planted with 20,000 9-cm pot-grown plants of *Phragmites australis* during the first week of July 1996 (Plates 3 and 4). These were maintained by use of clean water until the end of the month when leachate from the old culvert was first pumped to flow through the bed.

The initial performance of the reed bed was extremely encouraging. Results obtained by analysis of samples taken 4 weeks and 8 weeks after commissioning (i.e., after a minimum of 7 and 14 bed volumes, respectively, had passed through the bed) were remarkably consistent.

Table 6.5 presents results obtained from analysis of samples taken on 5 September 1996 (Plates 5 and 6). The removal of iron can be traced through the system, with 28% being removed in the settling tank and the remainder being removed within the reed bed, resulting in the iron concentration in the final effluent discharge being reduced to below detection limits. The inlet tank had no effect on the concentration of ammoniacal-N, and was not expected to. The removal rate for ammoniacal-N within the reed bed was 40%, with subsequent dilution within the Stert Watercourse occurring at the agreed effluent discharge point. Chloride values demonstrate that the removal of iron, ammoniacal-N, and mecoprop in the reed bed is not due to dilution. The removal of mecoprop by the reed bed, from 10.5 µg/L in the influent to 2.68 µg/L in the effluent, was extremely encouraging at such an early stage in the commissioning of the scheme.

Table 6.5   Results from Analysis of Samples Taken from Different Locations at Monument Hill, 5 September 1996

Determinand	New Culvert	Old Culvert	After Settling Tank	Reed Bed Effluent	Agreed Downstream Sampling Point
pH value	8.0	6.8	6.9	7.4	7.6
BOD$_5$	<2	<2	<2	<2	<2
Ammoniacal-N	<0.3	19.4	19.6	11.8	1.8
Iron	<0.6	16.9	12.2	<0.6	0.7
Suspended solids	19	42	42	3	16
Chloride	23	78	77	76	32
Mecoprop	<0.1	9.4	10.5	2.68	0.44

Note: All results in mg/L, except mecoprop in µg/L, and pH value in pH units.

Figure 6.5 presents results for the concentration of suspended solids within the discharge to the Stert Watercourse, in samples taken from April 1994 to April 1997. An immediate improvement has been evident since operation of the reed bed began in July 1996, levels falling from typically 50 to 70 mg/L to present concentrations that rarely exceed 10 mg/L.

Figure 6.6 contains equivalent data for iron and for ammoniacal-N. Introduction of the reed bed immediately effected reliable and almost complete removal of iron, generally to background concentrations. Results for ammoniacal-N in leachate have historically shown slightly elevated concentration (typically 25 to 30 mg/L) during summer months, compared with values near to 20 mg/L during winter months. Introduction of the reed bed in July 1996 has resulted in a significant and consistent reduction of about 50% in concentrations of ammoniacal-N, to between 10 and 15 mg/L.

Figure 6.7 gives results for concentrations of mecoprop in the discharge. Introduction of the reed bed does appear to have had an immediate effect — values remaining below 3 µg/L since July 1996 — but, again, more monitoring over a longer period will be needed to see if this removal is maintained or improved in the future.

Figure 6.8 shows concentrations of chloride, and COD value, in the discharge to the Stert Watercourse from December 1993 to April 1997. Chloride results show that dilution is not a factor in treatment performance, except initially, when leachate was first fed into the reed bed that had been filled with clean water during the commissioning phase. It is not possible, at this stage, to demonstrate the extent to which reductions in COD are taking place in the reed bed, and monitoring over a longer period will again be required for this. Reductions of between 10 and 20% are expected in the longer term.

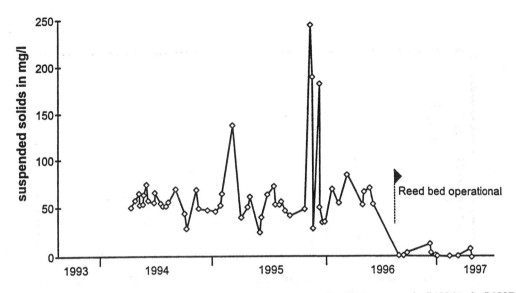

Figure 6.5 Concentrations of suspended solids in the discharge to the Stert Watercourse, April 1994 to April 1997.

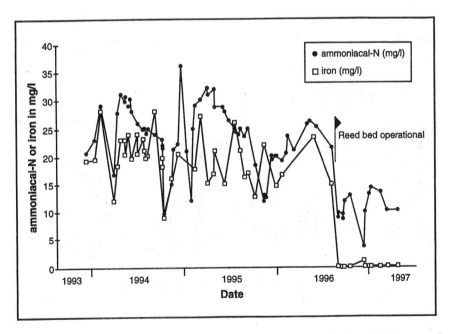

Figure 6.6 Concentrations of iron and ammoniacal-N in the discharge to the Stert Watercourse, December 1993 to April 1997.

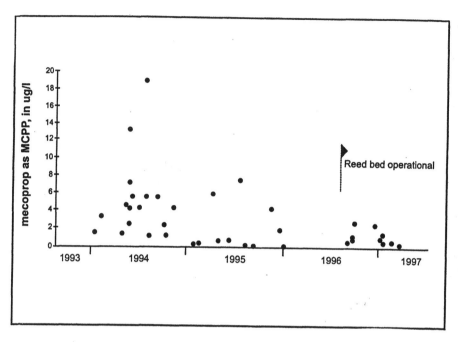

Figure 6.7 Concentrations of mecoprop, as MCPP, in the discharge to the Stert Watercourse, January 1994 to April 1997.

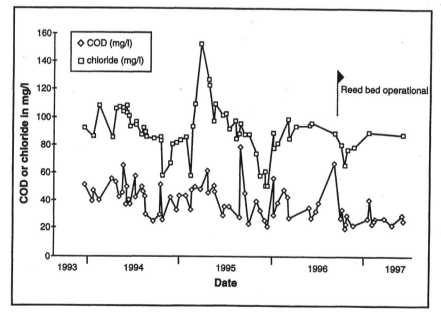

Figure 6.8 Concentrations of chloride, and COD values, in the discharge to the Stert Watercourse, December 1993 to April 1997.

6.10 FUTURE WORK

The performance of the reed bed leachate treatment scheme at Monument Hill Landfill Site is being monitored routinely by both Wiltshire County Council and by the Environment Agency (South Western Region). All operational features of the system are recorded using a data logger within the control kiosk — for example, inlet and outlet daily flow rates into and from the reed bed.

The system is fitted with a telemetry link and is able to relay alarm messages to the council in the event of any of the following:

- Failure of any pump for any reason (although duty and standby pumps are supplied),
- Mains power failure greater than 15 min,
- Restoration of mains power,
- A high level in the outlet stilling chamber, header tank, reed bed, or reed bed outlet chamber,
- Blockage of the feed pipeline (i.e., no flow in the pipeline with the pump running).

6.11 CONCLUSIONS

A leachate treatment scheme has successfully been designed, constructed, and commissioned to improve the quality of leachate from an old valley landfill site in Wiltshire, in southern England. Remedial works comprised a caisson sump to intercept leachate flows, a settlement chamber to remove precipitated iron, and a large (1800 m^2) engineered reed bed. The reed bed provides additional iron removal, degradation of residual levels of BOD, COD, and mecoprop, with some reduction in ammoniacal-N. This removal will be most effective during warmer summer months, when the stream that will receive final effluent is most sensitive.

Design details and performance data over the initial 9 months of operation have been presented and have demonstrated that, even at this early stage, the system is reliably and consistently meeting effluent quality requirements. It appears that similar schemes may have widespread application at many such closed landfill sites elsewhere in Britain.

ACKNOWLEDGMENTS

Wiltshire County Council has been responsible for the remedial works that have been put in place at Monument Hill Landfill Site, and their cooperation in the preparation of this paper is much appreciated. Particularly important has been assistance from Ian Hunt, Mike Wood, and Sandra Truscott.

The Environment Agency (formerly the National Rivers Authority), South Western Region, has also been helpful, both in fruitful and helpful discussions that have led to the solution adopted at Monument Hill, and also in making available the results from their independent sampling program for inclusion in the paper. Special thanks are due to Barry Gray and Lawrence Matthews.

Plate 1   Aerial view of Monument Hill Landfill from the south, Spring 1996.

Plate 2   View of the Stert Watercourse emerging from the bypass culvert (background) and of the old culvert and leachate release, Spring 1994.

Plate 3   Reed bed planting underway, June 1996.

Plate 4   Reed bed planting completed, July 1996.

Plate 5 Reed bed, growth by September 1996.

Plate 6 Reed bed outfall and level control, September 1996.

Plate 7   Completed reed bed treatment system, Autumn 1996.

REFERENCES

Barr, M., 1994. Leachate treatment at Scottish landfills, *Scottish Envirotech* 2:21.

Cooper, P. F. and Findlater, B. C., Eds., 1990. Constructed wetlands in water pollution control, in *Proceedings on the International Conference on the Use of Constructed Wetlands in Water Pollution Control,* Cambridge, U.K., 24–28 September, Pergamon Press, Oxford.

Hassall, K. A., 1990. *The Biochemistry and Use of Herbicides: Structure, Metabolism, Mode of Action, and Uses in Crop Protection,* 2nd ed., VCH Verlagsgesellschaft, Weinheim.

Ivens, G. W., 1994. *The UK Pesticide Guide 1994,* Guidance notes produced by the British Crop Protection Council, Centre for Agriculture and Bioscience International.

Last, S. D., Barr, M. J., and Robinson, H. D., 1993. Design and operation of Harewood Whin Landfill: an aerobic biological leachate treatment works, *Institute of Wastes Management Proceedings* 5:9–16.

Robinson, H. D., 1989. The development of methanogenic conditions within landfilled wastes, and effects on leachate quality, in *Landfill Concepts, Environmental Aspects, Lining Technology, Leachate Management, Industrial Waste and Combustion Residues Disposal, Sardinia '89, the Second International Landfill Symposium,* Porto Conte (Alghero), Sardinia, Italy, 9–13 October, paper XXIX, 9 pp.

Robinson, H. D., 1992. Leachate collection, treatment and disposal, *Journal of the Institution of Water and Environmental Management* 6:321–332.

Robinson, H. D., 1993a. Leachate technology, *Wastes Management* August:30–31.

Robinson, H. D., 1993b. The treatment of landfill leachates using reed bed systems, in *Proceedings "Sardinia '93", Fourth International Landfill Symposium,* S. Margherita di Pula, Sardinia, Italy, 11–15 October 1993, Vol. 1, 907–922.

Robinson, H. D., 1996a. The UK: leading the way in leachate treatment, *Institute of Wastes Management Proceedings* 12:4–10.

Robinson, H. D., 1996b. A review of the composition of leachates from domestic wastes in landfill sites. Report CWM/072/95, Wastes Technical Division of the Environment Agency.

Robinson, H. D. and Grantham, G., 1988. The treatment of landfill leachates in on-site aerated lagoon plants: experience in Britain and Ireland, *Water Research* 22:733–747.

Robinson, H. D., and Gronow, J. R., 1993. A review of landfill leachate composition in the UK, in *Proceedings of Sardinia 1993, the Fourth International Landfill Symposium,* S. Margherita di Pula, Cagliari, Italy, 11–15 October, 821–832.

Robinson, H. D., and Lucas, J. L., 1985. Attenuation of leachate in a designed, engineered and instrumented unsaturated zone beneath a domestic waste landfill, *Water Pollution Research Journal of Canada* 20:76–91.

Robinson, H. D., and Luo, M. W. H., 1991. Characterization and treatment of leachates from Hong Kong landfill sites; *Journal of the Institution of Water and Environmental Management* 5:326–335.

Robinson, H. D., Chen, C. K., Formby, R. W., and Carville, M. S., 1995. Treatment of leachates from Hong Kong landfills with full nitrification and denitrification, in *Proceedings of Sardinia 1995, the Fifth International Landfill Symposium,* S. Margherita di Pula, Cagliari, Italy, 2–6 October, Vol. 1, 511–534.

Robinson, H. D., Last, S. D., Raybould, A., Savory, D., and Walsh, T. C., 1997. State-of-the-art landfill leachate treatment systems in the United Kingdom, in *Proceedings of Sardinia '97, the Sixth International Landfill Symposium,* S. Margherita di Pula, Cagliari, Italy, 13–17 October.

Seager, J., Wolf, E. W., and Cooper, V. A., 1988. Proposed Environmental Quality Standards for List II Substances in Water, Ammonia WRc Report TR260.

CHAPTER 7

Leachate Wetland Treatment System in Orange County, Florida

Larry N. Schwartz, Lee P. Wiseman, and Erik L. Melear

CONTENTS

1-56670-342-5/99/$0.00+$.50
© 1999 by CRC Press LLC

ABSTRACT: Results from a research program evaluating the use of wetlands as a leachate treatment system describe water quality, soil, biological, and hydrological components of Wide Cypress Swamp. These results indicate that North Wide Cypress Swamp provides treatment of the landfill leachate mixed with storm water that is pumped from the borrow ponds to the wetland. Organic priority pollutants (OPPs) have been detected in surface water, soil, fish, and vegetation samples but at levels generally below existing standards and criteria. Therefore, at the present time, there is little measured effect from these and other detected pollutants. The overall results suggest that the landfill leachate constituents are diluted, deposited, sorbed, precipitated, degraded, or biologically assimilated within the landfill leachate treatment system. In addition, the results indicate that viable biological communities have adapted to these conditions created with the discharge from the borrow ponds. A water quality–based effluent limits (WQBEL) study established effluent limits for surface water discharge and the pollution abatement technology necessary to meet the water quality criteria in the receiving waters. The wetland treatment system should remain in operation and should be managed to provide long-term treatment to meet effluent limits for surface water discharge from the wetland treatment system to the landfill outfall canal (LOC).

7.1 INTRODUCTION

At the Orange County Landfill, leachate mixed with storm water is discharged to North Wide Cypress Swamp (the wetland) for treatment and wetland restoration. The Florida Department of Environmental Regulation (FDEP) required a research program to evaluate this wetland as a treatment system, the restoration of the hydrology, and enhancement of ecological conditions. The wetland treatment system discharges to the landfill outfall canal (LOC), which is the headwaters of the Little Econlockhatchee River. The FDEP also required that a water quality–based effluent limits (WQBEL) study be performed for the receiving waters to establish effluent limits for surface water discharge. The wetland research program and the WQBEL study are described below. A description of the site in terms of its historical development follows.

The Orange County Landfill began operation in 1971 as an EPA demonstration project for the purpose of evaluating landfill operations in high-groundwater-table areas. Perimeter drainage and collection canals were constructed to lower the groundwater table beneath the landfill. From 1971 until 1984, the canals were connected to Pond A with discharge to the LOC (Figure 7.1).

The activities to lower the groundwater table beneath the landfill also dewatered Wide Cypress Swamp, and contributed to a major burn in 1980. Since 1984, leachate mixed with storm water collected in the canals is routed to borrow ponds 1 and 2 and then pumped to North Wide Cypress Swamp for treatment and wetland restoration. The water level in the borrow ponds is maintained approximately 10 ft below water levels in this wetland. This creates groundwater gradients toward the borrow ponds, which minimizes groundwater flow from the landfill. North Wide Cypress Swamp is segregated from South Wide Cypress Swamp (the control wetland) by the LOC that transects the wetland (Figure 7.1). Surface discharge from North Wide Cypress Swamp is controlled by two sharp-crested weirs located in the south berm which flows 2.8 miles to the headwaters of the Little Econlockhatchee River. The discharge from North Wide Cypress Swamp to the LOC at this location began in 1990.

The FDEP required a research program to evaluate the wetland as a treatment system and to evaluate the ecologic and hydrologic responses of the wetland to water pumped from the borrow ponds. To assess the fate of landfill leachate constituents discharged to wetlands, a study was also performed utilizing an on-site wetland microcosm created with soil and vegetation typically found in Wide Cypress Swamp. Demonstration of the treatment of dilute leachate by the wetland treatment system is necessary to support establishment of effluent limits for surface water discharge from the wetland treatment system to the LOC. The FDEP also required that a WQBEL study be performed

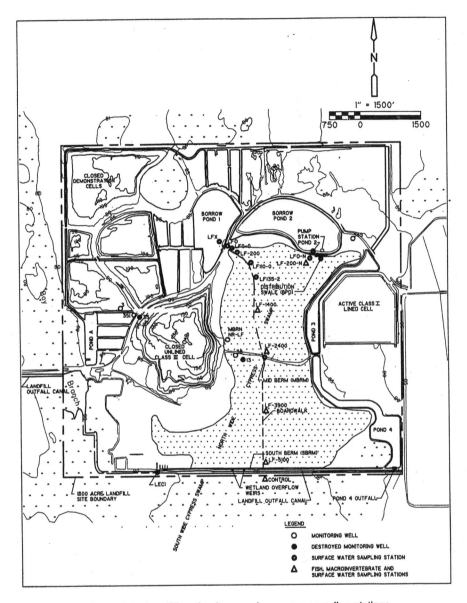

Figure 7.1 Landfill wetland research program sampling stations.

for the receiving waters to establish effluent limits for surface water discharge from the wetland treatment system to the LOC at station LEC1 (Figure 7.1).

7.2 METHODS

7.2.1 Wetland Research Program

The wetland research program consisted of the following activities. Compliance water quality data were evaluated and chemical analyses were performed on groundwater, surface water, soil, vegetation, and fish samples as summarized in Table 7.1. The physical and chemical properties of the indigenous soils of Wide Cypress Swamp were evaluated, to determine the effects of the

Table 7.1   Wetland Sampling and Analysis Matrix for the Orange County Landfill Wetland Research Program

Type of Sample	Reason for Sampling	List of Parameters	Year 1 Program (1989–1990)			Year 2 Program (1992–1993)			Year 3 Program (1993–1994)		
			Frequency	No. of Events	No. of Locations	Frequency	No. of Events	No. of Locations	Frequency	No. of Events	No. of Locations
Groundwater	Operational	IPs	SA	2	4	A	1	4	A	1	4
	Monitoring	OPPs	SA	2	4	A	1	4	A	1	4
Surface water	Operational	IPs	Q	4	9	A	1	6	SA	2	3–6
	Monitoring	OPPs	Q	4	9	A	1	6	SA	2	3–6
Soil	Operational	IPs	SA	2	25	A	1	25	A	1	25
	Monitoring	OPPs	SA	2	9	A	1	7	A	1	6
Vegetation	Operational monitoring	OPPs	A	1	5	A	1	4	A	1	6
Fish	Operational monitoring	OPPs	M	4	5	A	1	2	A	1	2

Note: SA = semiannual; Q = quarterly; A = annual; M = monthly.

discharge to these soils. A comprehensive vegetation-monitoring program was implemented to determine the effects of the discharge on plant communities in North Wide Cypress Swamp. A faunal-monitoring program was carried out to document the effects of the discharge to macroin-vertebrate and fish communities in North Wide Cypress Swamp. A water budget for North Wide Cypress Swamp was calculated in order to develop an understanding of the hydrology of the wetland. The water budget was used in conjunction with the concentration of selected water quality constituents — total nitrogen, total phosphorus, and 5-day biochemical oxygen demand (BOD$_5$) — to perform mass balance analyses to determine the treatment performance of the wetland.

7.2.1.1 Chemical Analyses

For regulatory purposes the organic chemical constituents are referred to as organic priority pollutants (OPPs), which are grouped by analytical method of determination as volatile organic compounds (VOCs) or base/neutral and acid extractables (BNAEs). The locations of the water quality, macroinvertebrate, and fish sampling stations are shown on Figure 7.1. Sampling and analyses were conducted twice during the first-year study period (1989 to 1990). Samples obtained from four groundwater wells and nine surface water stations were analyzed for indicator parameters (IPs) and OPPs. IPs in soil samples from 25 stations and OPPs in soil samples from 9 stations were determined. Also processed for determination of OPPs were 10 fish samples and 17 vegetation samples. Sampling and analyses were conducted once during each of the second- and third-year study periods (1992 to 1993 and 1993 to 1994). Samples obtained from four groundwater wells and six surface water stations were analyzed for IP and OPPs. IPs were determined in soil samples from 25 stations. Soil samples from seven stations in the second-year study period and six stations in the third-year study period were analyzed for OPPs. Analyses were compiled for OPPs in 15 fish samples and 11 vegetation samples in the second-year study period and 20 fish samples and 16 vegetation samples in the third-year study period.

7.2.1.2 Soils

The physical and chemical characteristics of the wetland soils were determined and included redox potential (Eh), bulk density, soil water content, soil reaction (pH), total phosphorus (TP), total nitrogen (TN), water soluble phosphorus (WSP), KCl-extractable ammonium nitrogen, and metals (Na, Mn, Fe, Al, K, Ca, Zn, Mg, Cd, Cr, Pb, Ni, and Cu). In the third-year study period, acid volatile sulfides (AVS) were also measured in soil samples. The location of the soil sampling stations are shown on Figure 7.2.

7.2.1.3 Vegetation

The vegetation-monitoring program included the determination of baseline conditions to which all changes can be compared. Permanent quantitative vegetation quadrats were established in North Wide Cypress Swamp and in South Wide Cypress Swamp (the control wetland) for woody vege-tation, and a modified line-intercept technique was used for herbaceous vegetation in these wetlands. Measurements of woody (canopy- and subcanopy-size) vegetation and herbaceous vegetation were taken in each of the three study periods. Viability categories were used to describe the condition of woody vegetation accurately.

7.2.1.4 Macroinvertebrates and Fish

Baseline data were collected to establish diversity indexes and pollution class distribution of macroinvertebrates and to determine fish species composition in the wetland treatment system and in the control wetland. An evaluation of macroinvertebrate communities was used to assess the

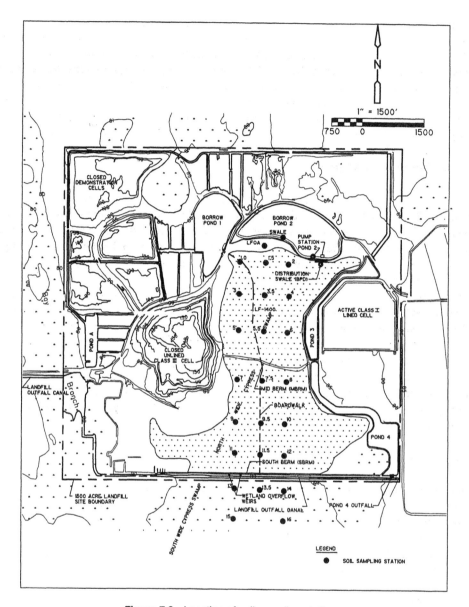

Figure 7.2 Location of soil-sampling stations.

ecological integrity of the system by reflecting both current and past water quality conditions. Fish populations were also evaluated as water quality and hydrologic conditions influence their species composition, abundance, and diversity. Both taxonomic and functional characteristics of the faunal populations were evaluated.

7.2.1.5 Hydrology

A major emphasis of the research program was to develop an understanding of the hydrology of North Wide Cypress Swamp (390 acres). Therefore, a water budget was prepared for this wetland. All components in the water budget were either measured or calculated from information collected on site and coupled with the results of previous investigations. The hydrologic cycle can be expressed in the form of a steady-state water budget model and can be written as

$$P + PUMP + SSI\ VP\ ET\ SO\ SSO = \Delta S/t$$

where inflows are:

    P = Precipitation
   VP = Vertical percolation

and outflows are:

  PUMP = Pumped inflow
    SSI = Subsurface inflow
     ET = Evapotranspiration
    SO = Surface outflow
  SSO = Subsurface outflow
  $\Delta S/t$ = Change in storage over time

7.2.1.6 Wetland Microcosms

Cattails and duckweeds were established in separate replicated wetland microcosms, which received an amended landfill leachate solution containing metals (lead and cadmium) and organic compounds (trichloroethylene and benzene) for 13 months. Sampling of the various wetland microcosm components (water, soil, flora, fauna) were conducted periodically to assess the fate of metal and organic compounds in the wetland in order to determine the fate of these constituents in the wetland systems.

7.2.2 Water Quality-Based Effluent Limits Study

Water quality sampling and flow measurement for 13 locations (Figure 7.3) were conducted during both a low-flow and a high-flow event. The samples were analyzed for the chemical constituents listed in Table 7.2. The results of the water quality sampling and flow measurements were used to construct and calibrate a water quality model of the LOC between the landfill and the Little Econlockhatchee River (See Figure 7.3). The U.S. EPA QUAL2E model (Brown and Barnwell, 1987) was used to perform the water quality simulations. The calibrated model was then used to establish water quality limits for the landfill discharges to the LOC at station LEC1 in the LOC (See Figure 7.3). Pollution abatement technology facilities may need to be constructed to meet the water quality criteria at this location.

7.2.2.1 Hydrologic Conditions for Completing Intensive Sampling Events

Available rainfall, flow, and stage data for the LOC and the Little Econlockhatchee River were reviewed to evaluate suitable conditions for conducting low-flow and high-flow intensive surveys. Based on a detailed analysis of U.S. Geological Survey (USGS) flow data for the only gauging station in the Little Econlockhatchee River and limited flow data for station LEC1, a reasonable 7Q10 low-flow (7-day minimum flow that occurs once every 10 years) for the landfill ranges from 0.45 ft³/s (cfs) to 0.53 cfs (0.013 to 0.015 m³/s). During the low-flow intensive survey, flow at LEC1 was measured at 0.49 cfs (0.014 m³/s) and there was no surface water discharge from the landfill.

Since the high-flow event is driven by rainfall, a correlation was developed between rainfall and discharge at the same USGS gauging station. A 2-in. (5.08-cm) rainfall within a 24-h period during the "wet season" was considered to be representative of average high flow conditions. A 2-in. (5.08-cm) rainfall within a 24-h period was estimated to produce a flow of approximately

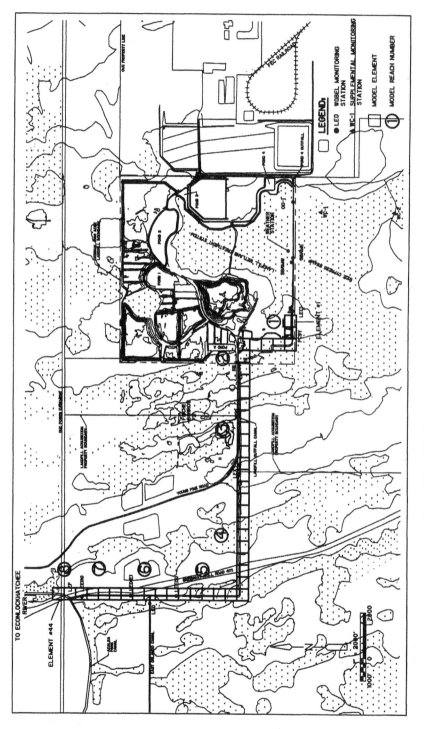

Figure 7.3 WQBEL Monitoring stations and modeling network.

Table 7.2 Surface Water Quality Parameters Measured for the WQBEL Study

Parameter	Units	EPA Method	Method Detection Limit	Surface Water MCL[a]
pH	Standard	150.1	0.01	6.0–8.5
Specific conductance	µhos/cm	120.1	0.1	1275
Dissolved oxygen	mg/L	360.1	0.1	5.0
Temperature	°C	170.1	0.5	
BOD_5, BOD_{10}, BOD_{15}, BOD_{20}	mg/L	SM 5210[b]	2.0	NS
Total Kjeldahl nitrogen	mg/L	351.4	0.04	NS
Ammonia nitrogen	mg/L	350.1	0.02	0.02
Nitrite nitrogen	mg/L	354.1	0.005	NS
Nitrate nitrogen	mg/L	353.2	0.02	10
Total phosphorus	mg/L	365.1	0.02	NS
ortho-Phosphate	mg/L	365.1	0.01	NS
Chlorophyl-a	mg/m³	SM 10200 H[b]	0.1	NS
Cadmium	mg/L	200.7	0.005	0.005
Chromium	mg/L	200.7	0.01	0.1
Copper	mg/L	200.7	10.0	Calculated
Iron	mg/L	236.1	0.01	1.0
Lead	mg/L	200.7	0.01	0.015
Zinc	mg/L	200.7	0.03	Calculated
Chlorides	mg/L	325.3	1.0	NS
Total alkalinity	mg/L as $CaCO_3$	310.1	0.2	>20
Total hardness	mg/L as $CaCO_3$	SM 2340B[b]	1.0	NS

[a] Maximum Contaminant Levels for Class III Surface Water Bodies as listed in Chapter 62-302, FAC, Surface Water Quality Standards.

[b] Standard Methods 2340B, 5210, and 10200H, as described in Standard Methods for the Examination of Water and Wastewater.

Note: NS = no standard.

107.0 cfs (3.03 m³/s) at the USGS gauging station. By using the contributing area method to estimate flow at the landfill, the corresponding flow at station LEC1 is calculated to be 3.6 cfs (0.10 m³/s). During the high-flow intensive survey, flow at station LEC1 was measured at 8.1 cfs (0.23 m³/s). Although flow conditions during the high-flow intensive survey were much higher than calculated average high flow conditions, they were acceptable to the FDEP.

7.2.2.2 Water Quality Modeling

The purpose of developing the QUAL2E model was to establish upper limits on the concentration of BOD_{ult} in the landfill discharge (as measured at station LEC1) that will not decrease the dissolved oxygen (DO) concentration in the LOC.

The model developed for this project consists of eight reaches representing a total length of 4.4 miles (7.1 km). The first reach in the model begins at station LEC1, which is the FDEP surface water discharge compliance point for the landfill discharge. A map of the site showing the model grid superimposed on the LOC is shown on Figure 7.3. The reaches are further subdivided into uniformly spaced elements each representing a length of 528 ft (160.9 m). A total of 44 elements were used in the model network. The hydraulics of the LOC and the point source inflows determine the number and length of the reaches and elements. The hydraulic gradient and flow velocities are very low, and long reach element lengths would create long residence times that make the model numerically unstable. Additionally, the hydraulics are fairly uniform within a given reach and each point-source inflow is placed near the beginning of a reach. Flow and water level data have been measured for the WQBEL stations; these stations influence the discretization and configuration of the model grid.

The model includes one headwater, four point-source inflows, and eight non-point-source inflows. Based on the flow and mass balances for the low- and high-flow intensive surveys, non-point-source inflows occur along the first seven reaches. The headwater represents the combined landfill surface water discharge, which includes storm water discharge from a retention pond in the lined portion of the landfill and the wetland treatment system discharge. The four point-source discharges are all off-site discharges. The first two off-site point source discharges are wetland discharges that occur between stations LEX and LEE(S) during high flow conditions (see Figure 7.3). The first wetland discharge, referred to as "wetland culvert," connects the wetland located between station LEX and the Greenway reach in the LOC. The second wetland point-source discharge, referred to as "wetland ditch," occurs at the 90° bend in the LOC. The third and fourth point-source discharges are the East Orlando Canal and the Azalea Park Canal, respectively.

The model was calibrated by adjusting the input parameters (reaction rate coefficients) within an acceptable range until the model was capable of predicting organic nitrogen, ammonia, nitrite, nitrate, ortho-phosphate, organic phosphorus, chlorophyll a, DO, and BOD_{ult} measured during the low-flow intensive survey. The results of the calibration indicated that deoxygenation and sediment oxygen demand were very low in the LOC and that reaeration capacity was low upstream of station LEX and increased progressively downstream of station LEX.

The model input parameters, determined from the calibration process, were used to predict the DO and BOD_{ult} measured during the high-flow intensive survey. The model was found to be sufficiently reliable to predict the water quality conditions in the LOC measured during the high-flow intensive survey.

The calibrated and confirmed water quality model was used to perform 15 predictive simulations, 8 using low-flow conditions and 7 using high-flow conditions with various discharge DO levels. A summary of the model simulations and the results of each simulation are presented in Table 7.3.

The first two low-flow simulations (Simulations 1 and 2) considered the measured low-flow discharge from the landfill of 0.01 cfs, and the remaining six low-flow simulations (Simulations 3 through 8) considered an average discharge from the landfill of 3.4 cfs. From Table 7.3, the first six simulations were used to evaluate the existing conditions in the LOC and the individual and combined effects of the landfill discharge and the non-point-source (incremental) inflows on DO levels in the LOC. The last two low-flow simulations (Simulations 7 and 8) evaluated the combined impact of an aeration system (meeting the FDEP Class III DO standard of 5.0 mg/L at station LEC1) and LOC improvements (improved reaeration) using an average discharge from the landfill of 3.4 cfs.

For all seven of the high-flow simulations, the discharge from the landfill was 10.26 cfs (the maximum conveyance at LEC1), and measured water quality was used to establish downstream LOC and point-source and non-point-source inflow water quality. The primary differences between the high-flow simulations were DO and BOD_{ult} levels at LEC1 and DO in the incremental inflows. As indicated on Table 7.3, the first four high-flow simulations (Simulations 9 through 12) were used to evaluate the existing conditions in the LOC and the individual and combined effects of the landfill discharge and the non-point-source inflows on DO levels in the LOC. Simulations 13, 14, and 15 evaluated the impact of LOC improvements on DO levels, the impact of an aeration system and LOC improvements on DO levels in the LOC, and to establish the high-flow water quality–based effluent limit for BOD_{ult} for the landfill discharge (at station LEC1) with an aeration system (at station LEC1) and LOC improvements, respectively.

7.3 RESULTS AND DISCUSSION

7.3.1 Wetland Research Program

A more comprehensive presentation of the results is provided in "Orange County Landfill Wide Cypress Swamp Wetland Research Program Final Summary Report" (Camp Dresser & McKee,

Table 7.3 Description and Results of Water Quality Modeling Simulations for the Orange County Landfill Level II WQBEL Study

Simulation No.	Flow	Description	Purpose	Results
1	Low	No discharge from OCLF & NPS inflows have existing DO and BOD_{ult} levels	Establishes baseline WQ in LOC	Low DO in the LOC without the OCLF discharge
2	Low	No discharge from OCLF and NPS inflows meet the DO STD	Cause and contribute demonstration for OCLF discharge	Gives DO concentrations throughout LOC for cause & contribute demonstration
3	Low	Existing loading to the LOC	Establishes baseline WQ in LOC	BOD loading in OCLF discharge & NPS inflows keep DO below STD
4	Low	Existing loading to the LOC, except that NPS inflows meet the DO STD	Evaluates impact of OCLF discharge on LOC WQ	Increasing DO in NPS inflows to Std cannot keep DO in LOC above SD
5	Low	Existing loading to the LOC with an aeration system	Evaluates impact of DO addition on LOC WQ	BOD loading in OCLF discharge & NPS inflows keep DO below STD
6	Low	Existing loading to LOC with aeration system & NPS inflows meet DO STD	Evaluates impact of NPS inflows & OCLF discharge on LOC WQ	Aeration system alone cannot keep DO above STD in LOC
7	Low	Existing loading to LOC with canal improvements & aeration system	Evaluates impact of DO addition & LOC improvements on LOC WQ	OCLF discharge meets DO STD & does not cause or contribute to DO violation
8	Low	WQBEL limit scenario with LOC improvements & aeration system	Establishes WQBEL at LEC1	OCLF discharge limit is 10 mg/L BOD_{ult}
9	High	Existing loading to the LOC	Establishes baseline WQ in LOC	BOD loading in OCLF discharge & NPS inflows keep DO below SD
10	High	Existing loading to the LOC, except that NPS inflows meet the DO STD	Cause and contribute demonstration for OCLF discharge	Increasing DO in NPS inflows to STD cannot keep DO in LOC above STD
11	High	Existing loading to the LOC with an aeration system	Evaluates impact of OCLF discharge on LOC WQ	BOD loading in OCLF discharge & NPS inflows keep DO below STD
12	High	Existing loading to LOC with aeration system & NPS inflows meet DO STD	Evaluates impact of NPS inflows & OCLF discharge on LOC WQ	Aeration system alone cannot keep DO above STD in LOC
13	High	Existing loading to the LOC with the canal improvements	Evaluates impact of LOC improvements on LOC WQ	LOC improvements help downstream WQ but, DO in LOC never reaches STD
14	High	Existing loading to LOC with canal improvements & aeration system	Evaluates impact of DO addition & LOC improvements on LOC WQ	OCLF discharge meets DO STD & does not cause or contribute to DO violation
15	High	WQBEL limit scenario with LOC improvements & aeration system	Establishes WQBEL at LEC1	OCLF discharge limit is 20 mg/L BOD_{ult}

Abbreviations: NPS = non-point source; LOC = landfill outfall canal; DO = dissolved oxygen; OCLF = Orange County Landfill; WQ = water quality; STD = Class III Surface Water Quality Standard for dissolved oxygen (5.0 mg/L).

1996). An overview of the results follows. A literature review indicated the types of chemical constituents found in landfill leachate. Although landfill leachate composition is waste and site specific, in general it is anoxic with high BOD and may have high concentrations of heavy metals, pesticides, chlorinated and aromatic hydrocarbons, and other toxic chemicals. The literature review also indicated that natural wetlands have been impacted by landfills and that natural and constructed wetlands have been successfully used to treat landfill leachate. The properties of wetlands that are conducive for treating landfill leachate include high plant productivity, large adsorptive surfaces on soils and plants, an aerobic–anaerobic interface, and an active microbial population (Staubitz et al. 1989). Absorption of organics and nutrients by plants, adsorption of metals on soils and plants, microbially mediated oxidation of metals, and simultaneous aerobic decomposition and anaerobic digestion of organic compounds can transform and immobilize a wide range of chemical constituents added to these systems (Staubitz et al. 1989). These processes can reduce or eliminate the impact on aquatic organisms and protect animals higher up the food chain.

7.3.1.1 Chemical Analyses

Historical and compliance monitoring data and results of this study indicate that certain groundwater wells at the landfill are affected by leachate (from old, unlined Class I cell areas, Figure 7.1) primarily vinyl chloride and aromatic compounds associated with petroleum fuels Table 7.4a. Although these compounds were present in the groundwater, it appears as though they are diluted, degraded, or deposited during transport to or within the borrow ponds, as the concentration of VOCs in the surface water in the borrow ponds were below detection limits or were very close to the detection limits (Table 7.4b and Table 7.5). The concentration of VOCs in the surface water in the wetland were also below the detection limit or were very close to the detection limit.

BNAEs were found in low concentrations or below the detection limit in certain groundwater wells. BNAEs were found in the water discharged to the wetland and in soils and surface water farther downstream in the wetland, suggesting that some accumulation of these compounds may be occurring in the wetland.

OPPs were detected at low concentrations in large predatory fish (Florida gar and bowfin) and in plant tissues. There were no apparent trends in concentration of OPPs in fish or plant tissue with distance from the points of discharge to the wetland. The most widely distributed chemicals in fish and vegetation were polycyclic aromatic hydrocarbons (PAHs) and phthalate esters (degradation products of plastics). The burning of the wetland (in 1980) may have contributed to the high concentrations of PAHs measured in biota and surface water.

Historical and compliance monitoring data and results of this study also indicate that surface water quality parameters decreased in concentration with increasing distance from the points of discharge to the wetland. These decreases in concentration may be due to a combination of factors including dilution, biological uptake, soil sorption or precipitation, or conversion (degradation) as water moves through the wetland.

A statistical analysis was performed to determine if there were significant differences between the mean concentrations of selected water quality parameters for pollutant sources and successive stages of the landfill leachate treatment process. The data were grouped by source and represent the following: source 1 — leachate-impacted groundwater at well 35I; source 2 — the dilute leachate collected in ponds 1 and 2 and discharged to the wetland (borrow pond discharge BPD); source 3 — the midpoint of the wetland treatment system (mid-berm MBRM); and source 4 — the discharge from the wetland to the LOC (south berm SBRM) (Figure 7.1). The mean concentrations for the water quality parameters (conductivity, chemical oxygen demand, total phosphorus, ammonia nitrogen, total nitrogen, total suspended solids, 5-day biochemical oxygen demand, and total organic carbon) were reduced after each succeeding stage of treatment (Figures 7.4 and 7.5). Statistically significant differences ($p = 0.05$ for all parameters except BOD_5; where $p = 0.08$) between groundwater and the discharge from the wetland indicate that treatment is provided through

Table 7.4a   Comparison of Detected Pollutants in Groundwater with Existing State and Federal Drinking Water Standards

Parameter	EPAᵃ MCL (mg/L)	FDEPᵇ Drinking Water Standards (mg/L)	Well 38 Aug-89	Well 38 Jan-90	Well 38 Aug-92	Well 38 Sep-93	Well 13ᶜ Jan-90	Aug-89	Well 35I Jan-90	Well 35I Aug-92	Well 35I Sep-93	Well 38 Aug-92	Well 38 Sep-93
	EPAᵃ MCL (mg/L)	FDEPᵇ Drinking Water Standards (mg/L)	Pollutant Concentration (mg/L)										
Vinyl chloride	2	1	—	—	4.84	7.8	—	—	—	5.06	1.72	48.41	1.2
Benzene	5	1	11.95	7.12	1.44	5.7	2.53	19.23	9.75	2.39	15.2	13.5	4
Trichloroethane	5	3	—	—	—	—	—	—	—	—	—	9.21	—
1,2 Dichloropropane	5	5	—	—	—	—	—	—	—	—	—	7.88	—

ᵃ Maximum contaminant levels per U.S. EPA, 1993. Drinking Water Regulations and Health Advisories.

ᵇ FDEP, 1993. Florida Drinking Water Standards. Chapter 62-550, Florida Administrative Code.

ᶜ Subsequently destroyed.

Table 7.4b   Comparison of Detected Pollutants in Surface Water with Existing State and Federal Drinking Water Standards

Parameter	WQCᵃ Acute/Chronic (mg/L)	FDEPᵇ Class III (mg/L)	Pond 2 Aug-92	LF-200N Aug-92	MBRM Aug-89	MBRM NR LF Jan-90–Mar-90	SBRM Aug-92
			Pollutant Concentration (mg/L)				
Total Polycylic Aromaticsᶜ	30/6.3	≤0.031	8.27	7.69	—	—	12.06
Benzo (a) anthracene	—	—	4.9	3.32	—	—	—
Benzo (a,k) fluoranthene	—	—	—	4.37	—	—	8.74
Chrysene	—	≤370	3.32	—	—	—	3.32
Fluoranthene	—	—	3.32	—	—	—	—
Phenanthrene	—	≤1	0.05	—	—	—	—
4-Chloro-3-methylphen	30/—	—	—	—	20.3	22.6	—

ᵃ U.S. EPA, 1992. Water Quality Criteria Summary Concentrations.

ᵇ FDEP, 1993. Florida Drinking Water Standards. Chapter 62-550, Florida Administrative Code.

ᶜ The sum total of the five PAHs listed below cannot exceed the stated criteria.

Note: No surface water parameter in the third-year study period (September 1993 sampling event) exceeded state and/or federal standards.

Table 7.5　Existing State and Federal Standards

Regulated Parameter	Groundwater	
	EPA[a] MCL (mg/L)	FDEP[b] Drinking Water Standards (mg/L)
VOCs		
Vinyl chloride	2	1
trans-1,2-Dichloroethane	100	100
1,2 Dichloroethane	5	3
cis-1,2-Dichloroethane	70	70
Chloroform	100	100
Benzene	5	1
Trichloroethane	5	3
1,1,1 Trichloroethane	200	200
1,2 Dichloropropane	5	5
Toluene	1,000	1,000
Chlorobenzene	100	100
Ethylbenzene	700	700
m,p-Xylene	10,000	10,000
o-Xylene	10,000	10,000
1,4 Dichlorobenzene	75	75
1,3 Dichlorobenzene	600	
1,2 Dichlorobenzene	600	600
BNAEs		
1,2 Dichlorobenzene	600	600
1,3 Dichlorobenzene	600	
Naphthalene	100[f]	
Phenol	20000[f]	
Metals		
Fe	300	300
Cu	1,000	1,000
Cd	5	5
Pb	15	15
Ni	100	100
Zn	5,000	5,000

Regulated Parameter	Surface Water	
	WQC[c] Acute/Chronic (mg/L)	FDEP[d] Class III
VOCs		
Benzene	5,300	71.28
1,2 Dichloroethane	1,120/763	
1,3 Dichlorobenzene	1,120/763	
1,4 Dichlorobenzene	1,120/763	
Toluene	17,500	
cis-1,2-Dichloroethane	11,600	
1,1,1 Trichloroethane	18,000	173,000
BNAEs		
Phenol	10,200/2,560	4,600,000
PAHs		≦0.031
Anthracene		110,000
Benzo(a) anthracene		
Benzo(a,k) fluoranthene		
Chrysene		
Fluoranthene	3,980	370
Phenanthrene	30/6.3	
Pyrene		11,000
Benzo(b+k)-fluoranthene		e
Benzo(a) anthracene		e
Metals		
Fe	1000	1000
Cu	18/12	12
Cd	3.9/1.1	1.1
Pb	82/3.2	3.2
Ni	1,400/160	160
Zn	120/110	106

Soils

Regulated Parameters	NOAA[a] Guidance Concentrations (mg/kg)
Organic Priority Pollutants	
Phenanthracene	0.225–1.38
Chrysene	0.4–2.8
Benzo(a) anthracene	0.23–1.6

[a] Maximum contaminant levels per U.S. EPA, 1993. Drinking Water Regulations and Health Advisories.

[b] FDEP, 1993. Florida Drinking Water Standards. Chapter 62-550, Florida Administrative Code.

[c] USEPA, 1992. Water Quality Criteria Summary Concentrations.

[d] FDEP, 1993. Florida Surface Water Quality Standards, Chapter 62-302, Florida Administrative Code.

[e] National Oceanic and Atmospheric Administration, 1991. Technical Memorandum NOS OMA 52- August 1991.

[f] EPA drinking water equivalent level (DWEL) is a guidance concentration based on a lifetime exposure that is protective of adverse, noncancer health effects, which considers exposure to a compound from a drinking water source. U.S. EPA, 1993.

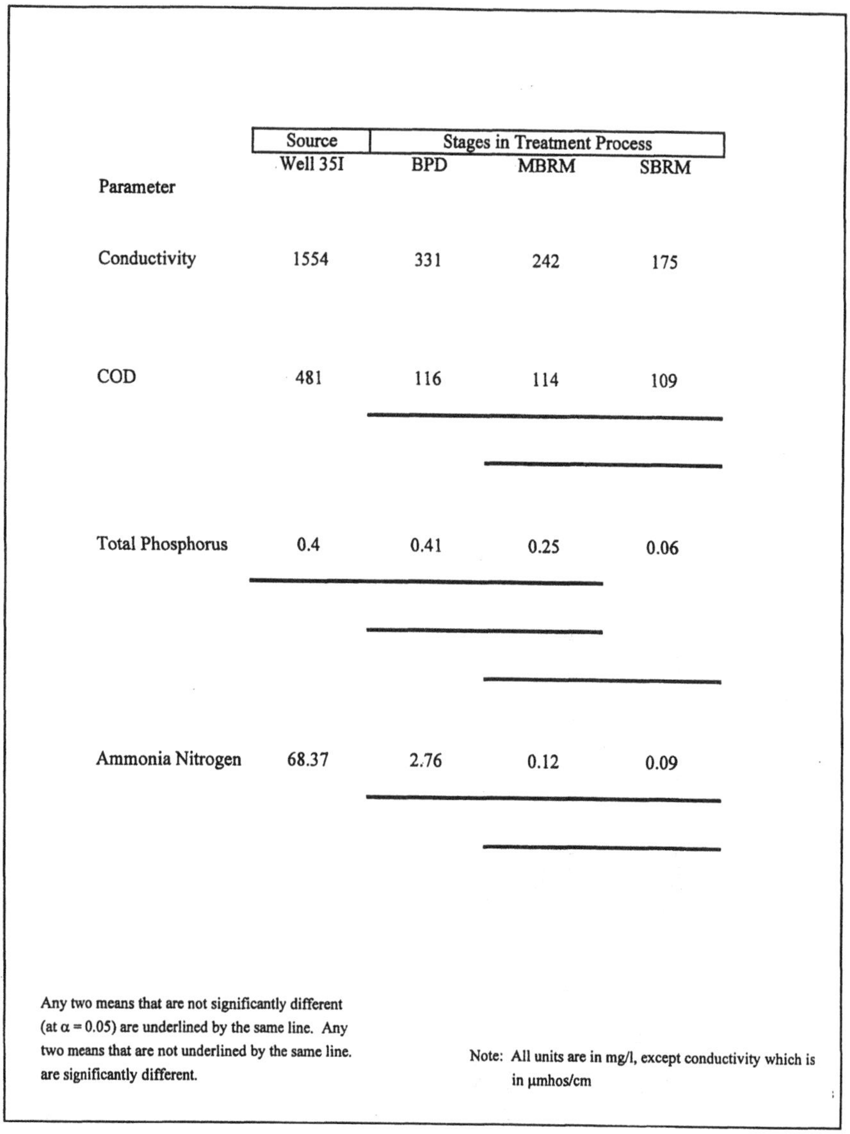

Parameter	Source Well 35I	Stages in Treatment Process		
		BPD	MBRM	SBRM
Conductivity	1554	331	242	175
COD	481	116	114	109
Total Phosphorus	0.4	0.41	0.25	0.06
Ammonia Nitrogen	68.37	2.76	0.12	0.09

Any two means that are not significantly different (at $\alpha = 0.05$) are underlined by the same line. Any two means that are not underlined by the same line. are significantly different.

Note: All units are in mg/l, except conductivity which is in μmhos/cm

Figure 7.4 Tukey HSD sample mean comparison for selected water quality parameters.

the process of collecting and diluting the leachate in the borrow ponds and applying the dilute leachate to the wetland treatment system. In addition, results from the WQBEL study indicate that treatment is occurring in the wetland treatment system since its receiving waters meet the FDEP Class III water quality criteria, except for dissolved oxygen and pH, which are naturally low in wetland surface waters.

7.3.1.2 Soils

Accumulation of heavy metals (Cr, Cu, Mn, Ni, Zn) was observed in the soils, presumably due to the discharge of dilute leachate since the bulk of the accumulation occurred in surface soils near the points of discharge to the wetland (Table 7.6). Metal concentrations in the soil may not be toxic to wetland biota because the soils have moderate levels of AVSs, which have been shown to

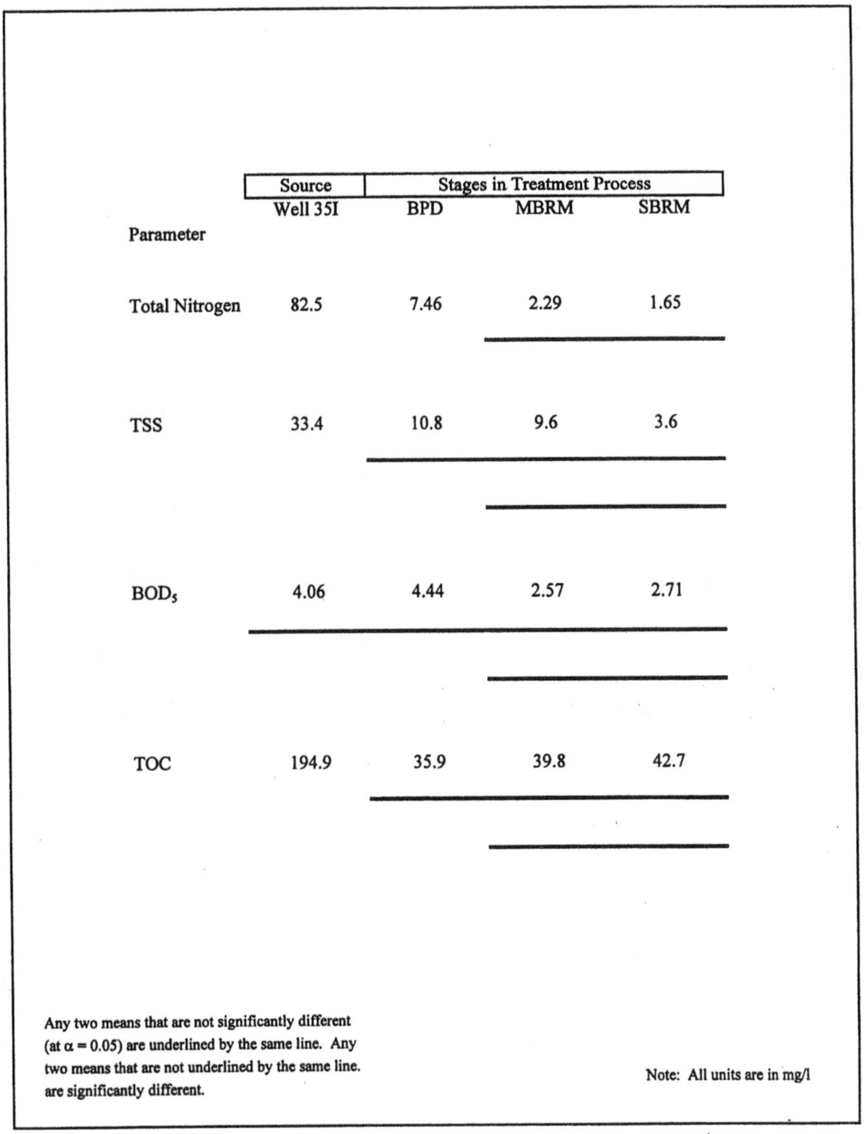

Parameter	Source	Stages in Treatment Process		
	Well 35I	BPD	MBRM	SBRM
Total Nitrogen	82.5	7.46	2.29	1.65
TSS	33.4	10.8	9.6	3.6
BOD$_5$	4.06	4.44	2.57	2.71
TOC	194.9	35.9	39.8	42.7

Any two means that are not significantly different (at $\alpha = 0.05$) are underlined by the same line. Any two means that are not underlined by the same line, are significantly different.

Note: All units are in mg/l

Figure 7.5 Tukey HSD sample mean comparison for selected water quality parameters.

immobilize these metals, and thereby reduce the risk for bioaccumulation (Di Toro et al., 1992). Biotoxicity testing would be required to confirm this assumption. Soils analyses revealed considerable temporal variability in physical and chemical parameters as well as spatial variability within the wetland. Differences in soil types within the wetland and the burning of the wetland in 1980 may have contributed to the observed variability in soil constituents.

7.3.1.3 Vegetation

Due to the fire in 1980, deep peat deposits in the wetland were burned, as well as most of the canopy-size individuals, creating a deep-water marsh dominated by floating leaved aquatic herbs, ferns, and emergent vegetation distributed on floating mats. The most dominant species displaying the highest total coverage are floating leaved aquatic herbs, which include the duckweeds

CONSTRUCTED WETLANDS FOR THE TREATMENT OF LANDFILL LEACHATES

Table 7.6 Metals Concentrations of Soil at Wide Cypress Swamp

Location[a]	Depth (cm)	Cr				Cu				Mn			Ni				Zn			
		1989	1990	1992	1993	1989	1990	1992	1993	1989	1990	1992	1989	1990	1992	1993	1989	1990	1992	1993
Swale	0–5	—	—	15.60	40.20	—	—	3.60	7.50	—	—	—	—	—	7.80	31.50	—	—	95.40	24.30
	5–20	—	—	5.60	—	—	—	2.10	—	—	—	—	—	—	2.60	—	—	—	17.80	—
	20–50	—	—	12.20	—	—	—	6.20	—	—	—	—	—	—	5.60	—	—	—	36.60	—
LFOA	0–5	35.60	38.96	11.60	11.90	42.46	37.95	10.60	10.00	60.00	83.50	11.60	24.35	12.94	8.40	50.70	70.77	140.31	17.80	108.80
	5–20	10.10	7.22	11.60	—	12.89	5.69	3.80	—	29.30	17.50	4.60	0.00	2.98	10.40	—	21.78	14.50	22.60	—
	20–50	5.84	3.57	10.00	—	5.45	3.65	0.80	—	4.80	4.10	4.60	0.00	0.00	2.00	—	6.53	12.81	4.20	—
1.0	0–5	12.79	7.63	18.00	17.80	56.15	9.58	20.60	8.00	31.60	22.10	12.60	16.35	3.91	15.40	15.40	89.20	29.77	76.60	79.50
	5–20	3.55	6.01	4.80	—	15.93	7.95	0.80	—	7.80	7.20	5.80	10.01	0.00	2.20	—	19.62	15.70	6.20	—
	20–50	1.94	2.11	1.40	—	13.59	2.88	<0.20	—	0.00	1.20	1.40	0.00	0.00	1.00	—	14.14	7.93	5.20	—
1.5	0–5	45.40	37.72	39.00	10.40	26.75	18.10	31.20	14.00	15.60	11.30	23.00	10.55	6.88	16.40	21.50	21.85	12.84	133.80	200.50
	5–20	34.65	39.51	12.60	—	25.63	22.50	7.40	—	16.40	12.60	1.50	10.92	6.80	7.40	—	25.15	17.92	27.80	—
	20–50	18.37	24.92	2.20	—	7.91	14.73	<0.20	—	10.00	14.50	1.20	6.33	7.93	0.60	—	8.88	13.29	<2	—
2.0	0–5	10.40	32.42	20.00	14.40	32.40	38.15	21.80	43.50	15.50	36.00	11.60	12.33	8.68	8.40	23.40	37.96	185.17	22.60	49.30
	5–20	3.20	3.99	6.80	—	10.57	13.16	3.20	—	0.00	4.60	5.60	0.00	0.00	6.40	—	17.84	14.96	10.40	—
	20–50	1.34	1.91	0.60	—	8.25	4.11	0.40	—	0.00	1.00	0.80	7.8	0.00	0.60	—	4.98	8.22	4.20	—
3.0	0–5	17.19	14.15	18.70	5.10	52.01	27.77	16.40	6.30	26.80	27.10	17.20	14.73	7.33	9.20	17.30	55.13	73.70	37.60	33.30
	5–20	15.02	4.22	14.80	—	36.48	10.09	2.20	—	25.80	5.60	4.60	0.00	0.00	5.40	—	52.73	13.92	9.80	—
	20–50	3.68	3.74	16.80	55.40	11.36	4.55	<0.20	22.30	2.20	2.40	6.00	3.90	1.68	7.40	34.10	9.02	11.12	27.20	59.00
3.5	0–5	11.52	14.79	14.40	—	25.76	21.84	10.60	—	25.40	18.00	16.20	10.91	9.02	7.80	—	46.06	114.26	47.40	—
	5–20	2.91	6.19	4.60	—	11.26	13.04	0.80	—	3.40	2.80	4.20	7.15	6.52	6.80	—	28.47	19.23	40.80	—
	20–50	2.49	3.73	1.80	23.40	5.56	2.98	0.20	8.30	0.00	0.00	1.00	0.00	0.00	2.20	90.30	7.09	9.01	9.00	60.80
4.0	0–5	15.52	4.13	13.00	—	33.14	16.35	19.40	—	23.10	11.10	16.20	8.39	0.00	13.80	—	29.57	58.31	29.10	—
	5–20	4.34	4.09	7.60	—	22.86	8.18	0.40	—	11.60	5.50	4.00	10.05	4.57	2.20	—	22.27	25.13	4.20	—
	20–50	3.47	4.04	7.00	29.80	23.42	7.97	0.60	7.30	4.60	3.80	2.20	15.61	0.00	7.50	31.70	26.24	12.71	3.80	38.00
5.0	0–5	62.78	3.41	5.40	—	39.70	12.33	7.60	—	25.30	21.70	15.00	38.71	0.00	4.00	—	60.79	34.72	14.60	—
	5–20	1.38	2.71	5.40	—	7.81	8.68	4.00	—	1.60	26.20	16.20	5.01	0.00	5.20	—	10.26	33.09	12.20	—
	20–50	12.39	6.83	6.60	10.10	8.57	6.83	6.20	26.00	0.00	20.10	4.60	0.00	8.94	4.80	126.30	8.83	21.77	5.60	45.80
5.5	0–5	9.37	11.01	13.40	—	31.47	14.99	7.60	—	25.40	18.00	13.80	0.00	8.08	8.80	—	64.96	32.79	32.00	—
	5–20	2.72	2.69	1.80	—	10.88	5.52	0.40	—	3.40	2.80	2.00	0.00	1.83	2.00	—	34.92	9.08	8.40	—
	20–50	6.76	0.79	2.40	1.10	5.91	3.05	<0.20	3.20	0.00	0.00	1.00	0.00	0.00	2.40	10.10	6.57	4.74	11.80	50.00
6.0	0–5	2.74	6.70	12.80	—	15.61	9.32	5.20	—	12.40	15.20	12.60	0.00	0.00	6.80	—	35.64	62.63	47.40	—
	5–20	1.93	3.42	2.00	—	7.98	5.46	2.20	—	0.00	4.60	6.00	0.00	0.00	2.00	—	11.82	9.76	13.60	—
	20–50	0.00	1.55	7.00	4.70	5.28	7.33	<0.20	15.30	0.00	1.70	1.40	0.00	0.00	2.00	17.20	15.52	11.51	<2	32.00
7.0	0–5	8.79	10.08	6.20	—	31.45	34.66	8.40	—	23.10	17.60	11.60	9.25	6.77	6.60	—	73.08	113.26	166.60	—
	5–20	2.91	6.50	8.40	—	8.73	7.41	1.60	—	4.10	6.10	4.60	4.48	0.00	3.40	—	11.39	19.78	13.60	—
	20–50	3.34	11.69	5.60	23.80	9.57	4.13	<0.20	5.30	0.00	3.70	2.20	0.00	8.13	1.80	27.50	18.13	28.33	4.60	97.50
	0–5	4.52	16.78	2.60		22.86	22.15	2.80		15.80	25.30	4.60		10.31	8.80		39.20	198.07	20.60	

Station[a]	Range	1	2	3	4	5	6	7	8	9	10	11	12	13	14	15	16	17	18	19
7.5	5–20	2.82	11.23	2.40	—	12.77	4.74	0.40	—	6.30	2.50	2.80	0.00	4.43	6.40	—	14.27	18.87	4.20	—
	20–50	2.57	1.67	1.20	—	10.68	2.78	<0.20	—	0.00	1.80	0.80	0.00	0.00	0.80	—	6.29	10.86	9.00	—
	0–5	3.02	6.58	9.20	7.90	14.04	40.10	6.40	5.30	12.70	16.80	7.40	0.00	6.13	16.20	27.80	58.53	218.88	131.40	138.80
8.0	5–20	1.21	3.61	1.20	—	5.74	5.18	3.60	—	0.00	4.30	1.20	0.00	3.33	1.80	—	9.26	14.16	22.20	—
	20–50	0.00	1.14	1.00	—	4.49	2.89	<0.20	—	0.00	0.00	0.60	0.00	0.00	1.20	—	5.22	7.52	4.20	—
	0–5	5.99	16.48	6.20	4.00	28.89	19.96	8.40	4.30	10.40	148.30	6.80	17.62	16.82	7.20	17.70	48.63	50.78	15.00	109.30
9.0	5–20	0.00	8.75	5.80	—	12.62	11.14	4.20	—	8.50	8.20	4.60	0.00	8.02	5.60	—	18.60	29.16	7.40	—
	20–50	2.49	5.31	4.40	—	10.12	4.80	2.40	—	5.80	3.60	2.60	0.00	0.00	3.60	—	9.39	7.79	6.40	—
	0–5	0.00	6.55	3.20	3.70	15.74	12.02	7.40	3.80	0.00	9.90	8.00	0.00	4.39	6.20	15.60	14.72	19.97	27.60	132.30
9.5	5–20	0.00	1.88	4.00	—	8.12	3.18	1.80	—	0.00	1.50	5.40	0.00	0.00	6.20	—	9.33	6.42	6.40	—
	20–50	0.00	0.60	2.20	—	5.49	2.01	<0.20	—	0.00	0.00	1.80	0.00	0.00	1.60	—	5.26	3.71	3.20	—
	0–5	3.20	16.90	2.60	15.00	25.35	21.07	3.00	4.00	108.60	15.40	8.00	0.00	17.39	4.60	28.70	21.30	45.32	17.90	34.50
10.0	5–20	0.00	7.98	2.00	—	10.84	11.55	1.40	—	0.00	5.10	3.00	0.00	4.83	4.00	—	14.07	7.42	5.60	—
	20–50	2.48	11.18	3.80	—	8.84	2.71	1.60	—	0.00	3.00	2.60	0.00	4.63	3.80	—	11.85	5.14	6.00	—
	0–5	0.00	9.79	14.20	34.20	12.31	7.18	8.00	11.00	0.00	13.70	4.60	0.00	8.16	17.80	27.80	12.07	19.15	18.00	30.50
11.0	5–20	1.72	7.02	2.00	—	7.00	7.56	0.40	—	0.00	5.00	1.40	0.00	6.58	3.00	—	6.14	9.21	2.20	—
	20–50	0.00	33.95	<0.20	—	4.07	6.40	0.80	—	0.00	8.90	0.40	0.00	11.81	0.60	—	4.37	5.94	5.00	—
	0–5	9.45	7.02	8.00	11.10	25.63	14.86	9.60	4.50	16.20	8.30	4.60	13.45	5.04	19.60	24.90	31.30	13.87	23.60	42.40
11.5	5–20	2.23	2.89	7.80	—	5.26	3.05	1.40	—	0.00	1.80	2.60	0.00	2.17	5.20	—	5.08	5.50	4.20	—
	20–50	0.00	2.00	4.20	—	4.47	2.50	0.40	—	0.00	2.10	2.40	0.00	0.00	4.40	—	5.93	3.78	10.40	—
	0–5	4.36	9.98	15.00	18.70	29.28	11.97	16.40	5.80	17.40	11.50	6.80	0.00	8.56	19.60	26.10	26.62	17.46	39.40	47.00
12.0	5–20	2.07	7.56	3.80	—	7.88	3.87	0.80	—	2.90	2.60	3.00	0.00	4.98	5.00	—	6.36	4.49	6.00	—
	20–50	2.06	2.02	0.60	—	6.05	2.16	0.20	—	0.00	2.30	0.60	11.20	0.00	1.80	—	7.74	25.05	4.20	—
	0–5	2.83	8.54	3.00	10.40	16.11	19.08	4.00	4.30	12.80	9.40	4.60	0.00	7.96	7.00	33.00	25.04	15.68	49.40	34.50
13.0	5–20	0.00	15.68	6.60	—	13.09	29.58	3.80	—	8.10	8.40	4.60	0.00	12.54	9.40	—	19.04	23.31	10.40	—
	20–50	7.84	12.19	6.80	—	14.58	13.22	7.20	—	7.40	8.40	4.60	0.00	6.66	9.20	—	14.37	6.85	17.00	—
	0–5	2.30	0.00	1.80	4.10	14.87	8.89	2.40	5.00	6.00	4.60	4.60	10.67	3.25	4.00	19.10	12.92	7.73	8.80	68.00
13.5	5–20	2.84	2.36	5.20	—	11.67	4.85	1.60	—	0.00	2.70	3.40	0.00	0.00	4.80	—	8.20	3.13	3.60	—
	20–50	1.44	1.25	0.20	—	6.15	1.99	<0.20	—	0.00	1.60	0.80	0.00	0.00	0.40	—	5.84	2.90	3.60	—
	0–5	5.68	2.00	3.80	5.30	22.74	8.34	2.80	2.00	7.40	7.10	4.60	0.00	0.00	7.00	21.40	26.02	63.82	28.40	53.80
14.0	5–20	1.32	3.98	0.40	—	4.49	5.02	<0.20	—	0.00	4.10	0.20	8.09	3.00	1.20	—	4.84	11.44	2.20	—
	20–50	3.83	1.67	<0.20	—	3.61	1.80	<0.20	—	0.00	2.10	0.20	2.95	6.11	0.60	—	4.93	2.21	3.20	—
	0–5	2.81	7.07	2.20	6.30	14.92	16.07	3.40	6.30	30.70	7.60	5.80	0.00	5.64	4.60	19.30	34.17	10.28	11.80	62.80
15.0	5–20	2.37	5.18	5.40	—	8.03	10.90	4.20	—	10.80	5.80	4.60	0.00	4.41	6.80	—	13.50	6.34	9.40	—
	20–50	3.52	6.32	6.60	—	10.16	6.66	6.80	—	4.10	4.30	4.60	8.40	4.20	4.40	—	21.87	4.28	10.40	—
	0–5	2.86	3.17	1.80	8.70	10.14	6.68	1.20	1.00	4.90	4.50	4.20	7.15	0.00	3.80	20.30	13.64	10.62	8.80	50.00
16.0	5–20	1.90	0.47	0.80	—	5.25	1.79	<0.20	—	0.00	0.00	0.40	0.00	0.00	0.60	—	10.27	3.21	4.20	—
	20–50	1.95	2.18	0.80	—	5.17	2.37	0.20	—	0.00	3.00	0.60	0.00	0.00	0.60	—	9.07	2.09	3.90	—

[a] Stations increase in distance from points of discharge to the LOC.

(*Lemna* sp.), dotted duckweed (*Spirodela punctata*), and saber-shape bogmat (*Wolffiella oblonga*). In addition, the floating ferns water spangles (*Salvinia minima*) and mosquito fern (*Azolla caroliniana*) are widespread throughout the wetlands. The most dominant emergent species in the wetland were common cattail (*Typha latifolia*), river seedbox (*Ludwigia leptocarpa*), and maidencane (*Panicum hemitomon*).

The woody plant community nomenclature is based on the most important tree-size class (dbh >1 in.) individuals measured in the quantitative vegetation quadrats. In terms of tree-size class individuals, the wetland is best described as a wax myrtle–sweetbay–pond cypress association (*Myrica cerifera–Magnolia virginiana–Taxodium ascendens*). The general nomenclature of the plant community types in North Wide Cypress Swamp has not changed from 1989 to 1993, but the dominance of certain species has, in some cases, changed over this time period (Wallace, 1996).

The future trends in vegetation community development in the wetland will be governed by flooding duration. As long as leachate mixed with storm water is discharged to the wetland and the present hydroperiod is maintained at the discharge structures, the wetland will remain a deep-water marsh. The control wetland will continue to reflect conditions determined by periodic drought or heavy rainfall years.

7.3.1.4 Macroinvertebrates and Fish

Table 7.7 provides a comparison of the composite seasonal diversity index and the pollution tolerance class distribution for the macroinvertebrate communities at each of six wetland stations. Diversity indexes less than 1.0 are found in heavily polluted areas; between 1.0 and 3.0 in areas of moderate pollution; and greater than 3.0 in clean water areas (Wilhm and Dorris, 1968). Although there are limited data establishing diversity levels and their interpretation in wetlands, of the 36 diversity indexes presented in Table 7.7, 5 were less than 2.0 and 19 were greater than 3.0. These results reflect good water quality in Wide Cypress Swamp. In addition, changes in the overall macroinvertebrate diversity have not occurred.

All macroinvertebrate species were classified according to their level of pollution tolerance or intolerance: Class I — intolerant of organic pollution; Class II — tolerant of a moderate amount of organic pollution; Class III — tolerant of gross amounts of organic pollution; Class IV — air-breathing forms; Class V — unknown (Beck, 1954). The results indicate that there were pollution-sensitive species (Class I and II) collected at each station for most sample events. In addition, the total number of species at each station was well distributed among the pollution tolerance classes for all sampling events. No samples displayed an exclusive Class III distribution, which would be indicative of a stressed or degraded system. The ability of each station to support pollution-sensitive species is indicative of a healthy environment.

The effect of the discharge of dilute leachate on macroinvertebrates seems positive, as reflected in the diverse populations present and the number of pollution-sensitive species. Dipteran larvae (including Chironomidae) were the most abundant macroinvertebrates.

Table 7.8 indicates the fish species sampled at Wide Cypress Swamp. There were nine species of fish in five different families. Six species were collected with regular frequency. The wetland is dominated by omnivorous forage fish species, predominantly mosquito fish (*Gambusia affinis*) and least killifish (*Heterandria formosa*). The species diversity of fish in the wetland was within the range for natural wetlands (Gruendling and Gaines, 1996). This implies that the fish community structure has not been degraded beyond the bounds of similar natural systems (Gruendling and Gaines, 1996). No trends in fish density were apparent with increasing distance from the points of discharge to the wetland.

7.3.1.5 Hydrology

The water budget for North Wide Cypress Swamp indicates that over a 24-month period precipitation and pumpage from borrow pond 2 to the wetland accounted for the 44 and 54% of

Table 7.7 Composite Seasonal Diversity Indexes and Pollution Tolerance Distribution of Macroinvertebrates Collected from Wide Cypress Swamp

Station	Sampling Date	Composite Diversity Index	Number of Species, Class				
			I	II	III	IV	V
Control	May 1992	2.92	2	3	6	2	10
Control	Sept. 1992	2.5	0	1	5	4	1
Control	Jan. 1993	3.63	0	3	7	1	11
Control	May 1993	3.2	2	2	5	3	6
Control	Sept. 1993	2	0	0	2	0	0
Control	Jan. 1994	1.99	1	1	1	1	2
LF-200-N	May 1992	1.08	0	3	3	0	2
LF-200-N	Sept. 1992	0.65	0	1	3	3	4
LF-200-N	Jan. 1993	2.74	2	6	10	3	14
LF-200-N	May 1993	1.62	0	2	6	2	5
LF-200-N	Sept. 1993	2.01	0	1	1	1	2
LF-200-N	Jan. 1994	0.79	0	3	7	2	8
LF-1400	May 1992	2.6	0	4	3	7	4
LF-1400	Sept. 1992	3.93	1	3	2	6	11
LF-1400	Jan. 1993	3.27	0	4	4	5	7
LF-1400	May 1993	3.72	0	5	3	4	6
LF-1400	Sept. 1993	1.46	0	1	0	1	1
LF-1400	Jan. 1994	3.42	0	2	3	3	8
LF-2400	May 1992	3.2	0	3	4	5	8
LF-2400	Sept. 1992	3.28	0	2	2	3	4
LF-2400	Jan. 1993	3.46	1	5	8	5	11
LF-2400	May 1993	3.91	0	2	9	7	10
LF-2400	Sept. 1993	2.53	0	1	2	6	6
LF-2400	Jan. 1994	3.4	0	2	4	4	10
LF-3900	May 1992	3.01	1	2	5	1	7
LF-3900	Sept. 1992	3.6	1	3	3	4	6
LF-3900	Jan. 1993	3.62	1	5	11	5	17
LF-3900	May 1993	3.04	0	2	7	3	6
LF-3900	Sept. 1993	2.41	0	0	0	1	3
LF-3900	Jan. 1994	2.14	0	0	0	1	3
LF-5100	May 1992	2.16	0	3	7	3	6
LF-5100	Sept. 1992	3.74	0	3	2	4	6
LF-5100	Jan. 1993	3.23	3	5	8	7	16
LF-5100	May 1993	3.01	0	2	7	3	6
LF-5100	Sept. 1993	2	0	0	0	1	0
LF-5100	Jan. 1994	3.82	1	3	8	2	14

inflow to the wetland, and evapotranspiration, surface discharge, subsurface outflow, and vertical seepage accounted for 34, 49, 16, and 1% of the outflow from the wetland, respectively.

Water quality–monitoring data and the water budget were used to develop a mass balance for North Wide Cypress Swamp. For the purpose of the mass balance calculations, the catchment was considered as two hydrologic compartments, north and south of the mid-berm (Figure 7.1). The mass balance for North Wide Cypress Swamp is presented in Table 7.9. Over a 24-month period (1992 to 1993) North Wide Cypress Swamp has assimilated 83% of the total nitrogen load, 73% of the total phosphorus load, and 61% of the BOD_5 load. The majority of the assimilation occurred in compartment 1 (north of MBRM). Assimilation of these constituents indicates that the wetland is providing treatment of the dilute landfill leachate. The 24-month flow-weighted average concentrations of total nitrogen, total phosphorus, and BOD_5 discharging from the wetland treatment system were 1.42, 0.08, and 2.18 mg/L, respectively.

The hydraulic residence time for North Wide Cypress Swamp is approximately 25 days, indicating that this wetland has a significant amount of storage, which promotes infiltration.

Table 7.8　Fish Species at Wide Cypress Swamp

Species	Common Name	Wetland Stations						Feeding Guild	Fish Type
		200 or 200-N	1400	2400	3900	5100	Control		
Centrarchidae									
Enneacanthus gloriosus	Blue-spotted sunfish	3	—	—	—	3	2	Insectivore	Forage
Micropterus salmoides	Largemouth bass	1	—	—	—	—	—	Omnivore	Sport
Lepomis macrochirus	Bluegill	—	—	—	—	—	3	Omnivore	Sport
Elassoma evergladei	Everglades pygmy sunfish	—	1,2,3	2,3	2,3	2,3	2,3	Insectivore	Forage
Clariidae									
Clarias batrachus	Walking catfish	2	—	—	—	—	—	Omnivore	Rough (exotic)
Cyprinodontidae									
Fundulus chrysotus	Golden topminnow	1	2	1,2	1,2	3	2,3	Insectivore	Forage
Lepisosteidae									
Lepisosteus platyrhincus	Florida gar	3	—	—	—	3	—	Piscivore	Rough
Poeciliidae									
Gambusia affinis	Mosquitofish	1,2,3	1,2,3	1,2,3	1,2,3	1,2,3	2,3	Omnivore	Forage
Heterandria formosa	Least killifish	1,2,3	1,2,3	1,2,3	1,2,3	1,2,3	3	Omnivore	Forage

Note: 1 = year 1; 2 = year 2; 3 = year 3.

LEACHATE WETLAND TREATMENT SYSTEM IN ORANGE COUNTY, FLORIDA 121

Table 7.9 Nitrogen, Phosphorus, and BOD Mass Balances for the Second- and Third-Year Study Periods in North Wide Cypress Swamp

	Nitrogen				Phosphorus				BOD$_5$			
	Total Loading	Total Export	Total Assimilation	Percent (%) Assimilation	Total Loading	Total Export	Total Assimilation	Percent (%) Assimilation	Total Loading	Total Export	Total Assimilation	Percent (%) Assimilation
					Units are in kg/ha							
First compartment (152 acres)	678.2	150.6	527.6	78	24.3	11.9	12.4	51	467.2	216.2	251.1	54
Second compartment (238 acres)	110.5	75.8	34.7	31	6.8	3.8	3.0	44	138.1	116.5	21.5	16
Total site (390 acres)	273.1	46.3	226.8	83	8.4	2.3	6.1	73	182.1	71.1	111.0	62
					Units are in lb/acre							
First compartment (152 acres)	605.0	134.3	470.6	78	21.7	10.6	11.0	51	416.8	192.8	224.0	54
Second compartment (238 acres)	98.6	67.6	31.0	31	6.8	3.8	3.0	44	123.1	104.0	19.2	16
Total site (390 acres)	243.6	41.3	202.3	83	8.4	2.3	6.1	73	162.5	63.4	99.0	61

Therefore, this system may be managed to provide treatment for higher flows associated with large storm events.

7.3.1.6   Wetland Microcosms

There was extensive alteration of organic compounds and little demonstrated retention of these compounds within the sediment, water, or biota in the wetland microcosms. This is an important finding as the potential effects of VOCs are minimized by poor retention within wetland systems. The primary loss route for these constituents was probably through volatilization and, secondarily, through microbial degradation.

Lead and cadmium were retained to a significant extent in the wetland microcosms and were the primary focus of the investigation. Both the cattail and duckweed microcosms removed approximately half of the lead from the amended leachate. Lead concentration averaged 0.396 mg/L in the inflow leachate, and 0.196 and 0.219 mg/L in the outflows from the cattail and duckweed systems, respectively (Figure 7.6). Duplicate microcosms of cattail and duckweed removed similar amounts of lead. Inflow cadmium concentrations averaged 0.105 mg/L and the outflow from both the cattail and duckweed microcosms averaged 0.052 mg/L (Figure 7.7). The mean metal removal (retention) efficiency was comparable between the wetland plant types, at about 45 to 51% for lead and 50% for cadmium. Lead and cadmium retention could be increased by increasing hydraulic retention times in the wetland microcosms. Mass cadmium removal by the cattail and duckweed microcosms averaged approximately 7.7 mg Cd/m^2/day during the study, and lead removal averaged 29.1 and 25.8 mg Pb/m^2/day in the cattail and duckweed microcosms, respectively. Results of this study were reported by DeBusk et al. (1996).

A clear demarcation of metal enrichment was apparent in different components of the wetland microcosms after 13 months. As indicated on Figures 7.8 and 7.9, the highest lead and cadmium burdens (on a concentration basis) occurred in the cattail roots and in the duckweed. Sediments exhibited intermediate burdens, and cattail shoots, rhizomes, and fish exhibited the lowest tissue burdens (DeBusk et al., 1996). These results indicate that sediment processes rather than water column events govern accumulation of lead and cadmium in the sediments. In the sediments, nearly all the lead and cadmium were present as metal sulfides. It is important to note that metal complexation with sulfides, measured as the ratio of metals to AVSs, limits bioavailability and toxicity of these metals.

Comparison of metal uptake by cattail and duckweed indicates that duckweed, on a whole-plant basis, accumulates lead and cadmium more effectively than does cattail. The high productivity and comparative ease of harvesting suggests that duckweed may be a more appropriate macrophyte for use in biological remediation strategies. Metals, which do not degrade, can accumulate in sediment and biomass. Results from this wetland microcosm study indicate the potential fate of metal and organic compounds and suggest that constructed and natural wetland systems can be managed to maximize treatment of landfill leachate constituents.

7.3.2   Water Quality-Based Effluent Limits Study

A comprehensive presentation of the results is provided in "Orange County Landfill Level II WQBEL" (Camp Dresser & McKee, 1997). An overview of the results is provided below. Based on the results of the water quality monitoring, the DO concentration in the LOC was lower than the Class III surface water quality standard of 5.0 mg/L (Tables 7.10a and 7.12a). North Wide Cypress Swamp is the headwaters for the LOC, and wetlands typically have a low DO. Also, the landfill and other off-site non-point sources contribute BOD_5 to the LOC and lower the DO.

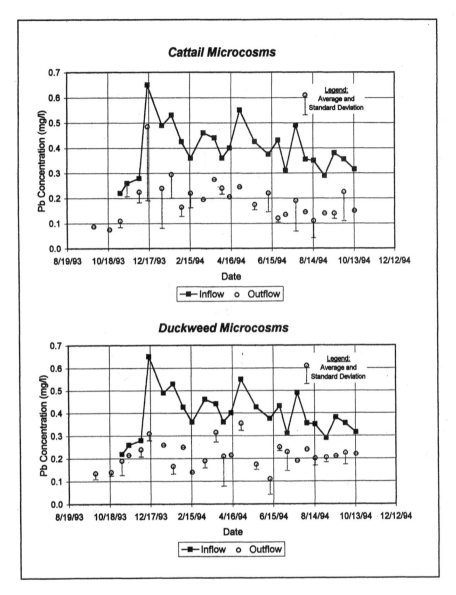

Figure 7.6   Total lead in microcosm inflow and outflow.

7.3.2.1 Results of Intensive Sampling Events

There were no point-source discharges from the landfill during the low-flow survey. The water quality for the low-flow sampling event are presented in Tables 7.10a and b and Table 7.11. Water quality results showed that DO and pH were low throughout the LOC and in the point-source inflows. This is not unusual for surface waters receiving water from wetlands. The pH values in the LOC, between stations OD-1 and LEX, were less than the Class III standard and the values reported for the background station, even with no discharge from the wetland treatment system. However, DO and pH increased with distance downstream.

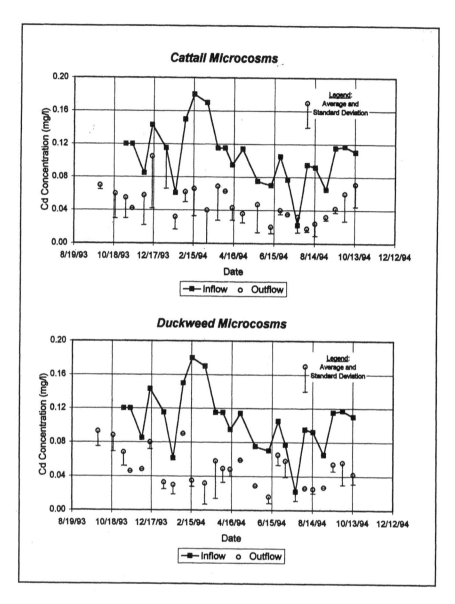

Figure 7.7 Total cadmium in microcosm inflow and outflow.

The BOD$_5$ concentrations were less than 3 mg/L in the discharge from the wetland treatment system and the storm water pond for the lined portion of the landfill. BOD$_5$ concentrations ranged from 0.9 to 6.6 mg/L in the LOC and generally decreased from upstream to downstream. The concentrations of total phosphorus, *ortho*-phosphate, ammonia nitrogen, nitrite nitrogen, nitrate nitrogen, and total Kjeldahl nitrogen (TKN) were low throughout the LOC and within the wetland treatment system and pond 4.

During the low-flow survey, the concentrations of cadmium, chromium, and iron were below their respective Class III surface water quality standards. The concentrations of copper and lead were below detection limits. Detection limits for zinc exceeded Class III standards in the wetland treatment system and six of the stations in the LOC. Although no definitive conclusions can be made for copper, lead, and zinc relative to the Class III surface water quality standards, the results of routine surface water monitoring at the landfill indicate that the landfill discharges meet the Class III standards for these metals.

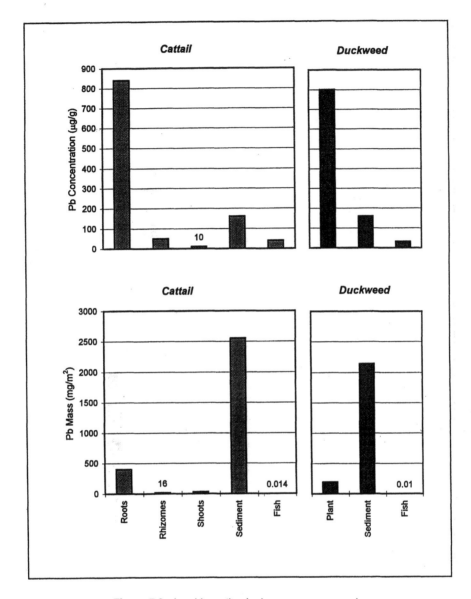

Figure 7.8 Lead in wetland microcosm components.

The measured water quality for the high-flow sampling event are presented on Tables 7.12 and 7.13. Field measurements showed that DO was below the Class III surface water standard of 5 mg/L throughout the system. pH followed the same trend as recorded during the low-flow intensive survey; i.e., pH increased with distance downstream. The average BOD_5 concentrations in the surface water discharges from the wetland treatment system and pond 4 were 3.5 and 1.5 mg/L, respectively. BOD_5 concentrations ranged from 1.4 to 4.8 mg/L in the LOC and decreased from upstream to downstream. The concentrations of total phosphorus, ortho-phosphate, ammonia nitrogen, nitrate nitrogen, and TKN were low throughout the LOC and the landfill discharges.

During the high-flow survey, the concentrations of cadmium, chromium, and iron were below their respective Class III surface water quality standards. The concentrations of copper and lead were below their limits of detection. Detection limits for zinc exceeded Class III standards in the wetland treatment system and six of the stations in the LOC. Again, although no definitive conclusions can be made for copper, lead, and zinc relative to the Class III surface water quality

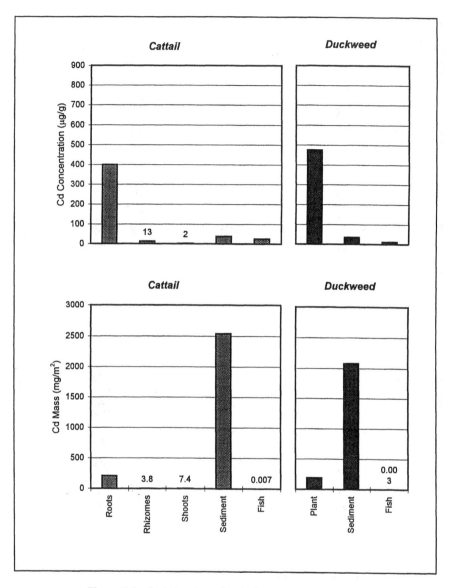

Figure 7.9 Cadmium in wetland microcosm components.

standards, the results of the routine surface water monitoring at the landfill indicate that the landfill discharges meet the Class III standards for these metals.

Based on the results of the intensive surveys, it is necessary to add DO at station LEC1 to improve water quality downstream in the LOC.

7.3.2.2 Water Quality Modeling

The results of the modeling simulations showed that the DO in landfill discharge and incremental (non-point-source) inflows alone does not significantly affect the DO in the LOC between its headwater to just downstream of the Greenway reach. The DO in this portion of the LOC is naturally low because there are substantial wetland inflows to the canal and reaeration potential is very low. The results of the modeling and the intensive surveys indicate that an aeration system should be installed in the canal (upstream of station LEC1) to raise the DO above the Class III surface water quality standard of 5.0 mg/L.

When sufficient DO is provided at station LEC1, the DO gradually decreases to below the standard in the direction of flow in the LOC — ultimately reaching a low of 3.5 mg/L, 2.8 miles (4.5 km) downstream of station LEC1 — due to BOD_{ult} in the landfill discharge, off-site point-source inflows, and incremental inflows. Therefore, reaeration in the LOC, particularly in the portion of the canal between the landfill to just downstream of the Greenway reach, must be increased with improvements to the LOC in combination with construction of an aeration system so that there is reasonable assurance that the landfill discharge will not cause or contribute to a violation of the Class III surface water quality standard for DO. Given these improvements, the landfill discharge WQBELs determined from the modeling are 10 mg/L BOD_{ult} (4.4 mg/L BOD_5) under low-flow conditions and 20 mg/L BOD_{ult} (8.8 mg/L BOD_5) under high-flow conditions.

The DO and BOD_{ult} profiles for the low-flow and high-flow design simulations are shown on Figures 7.10 and 7.11, respectively. Low-flow baseline conditions are defined as no flow from the landfill and all non-point-source discharges meeting standards. High-flow baseline conditions are defined as high flow in the canal, existing surface water loading from the landfill, and all off-site non-point-source discharges meeting standards. The low- and high-flow BOD limits were established by systematically increasing the BOD until the DO in the LOC was not depressed below its respective baseline DO concentrations. The differences in DO between each simulation and the baseline condition for the low-flow and high-flow simulations are presented on Figures 7.12 and 7.13, respectively.

These WQBELs compare favorably with BOD data collected at station LEC1 between 1991 and 1996. Improvements to the LOC and routine vegetative maintenance will help reduce BOD loading to LOC.

7.3.2.3 *Improvements to the Surface Water Discharge from the Landfill*

The results of the water quality analyses and modeling indicate that the addition of DO to the surface water discharge from the landfill will help maintain the Class III water quality standard of 5.0 mg/L for DO at the approved point of compliance (LEC1) under both design low-flow and high-flow conditions. From a design perspective, the high-flow condition was most limiting because of a higher BOD loading to the LOC. A proposed aeration system would consist of four (three plus a standby) floating aerators placed upstream of station LEC1 in the LOC.

The water quality modeling also shows that there will be a decrease in DO in the LOC downstream of the landfill between the 90° bend and the East Orlando Canal. This decrease in DO sag is due to poor reaeration in the LOC between the landfill and the 90° bend, with a corresponding sudden increase in reaeration capacity downstream of the 90° bend. The poor reaeration capacity upstream of the 90° bend is attributable to an overgrowth of vegetation in the LOC and a low head that results in lower velocities in this portion of the canal. The higher reaeration capacity downstream of the 90° bend is attributable to an overgrowth of vegetation and the deposition of eroded sediments from the canal banks in the LOC that result in a higher velocity in this portion of the LOC. It is therefore desirable to create uniform flow and reaeration capacity in the LOC. To accomplish this, the vegetation should be removed and a portion of the on-site LOC and the off-site portion of the canal between the landfill property boundary and the East Orlando Canal must be regraded. Additionally, the side banks along a portion of the LOC that is downstream of the 90° bend should be regraded and stabilized to minimize future erosion. Approximately 16,000 lineal ft (4.9 km) of the LOC should be improved for this purpose.

Table 7.10a Summary of Field Parameters, Flow, and Water Depth for the Low-Flow Intensive Survey (May 20, 1993)

Station	Distance Downstream from Pond 4 (ft)	pH	Temperature C	Temperature F	DO (mg/L)	Specific Conductance (mmho/cm)	Secchi Disk Depth (ft)	Flow (cfs)	Total Water Depth (ft)	Sample Depth (ft)	Time of Sample Collection
Detection Limit		0.05	0.1		0.1	10	0.1	0.01			
CLASS III Std.		NB + 1 unit			>5	NB + 50%	NB + 10%				
Point-Source Discharges											
Pond 4	0	7.5	25.8	78.4	6.8	128	2.2	0.00	2.3	1.0	8:40 am
SBRM-East	3,300	5.2	24.6	76.3	6.0	210	>1.5	0.00	1.5	0.8	9:47 am
SBRM-West	3,485	5.2	25.1	77.2	6.0	210	1.8	0.00	2.0	1.0	10:00 am
Landfill Outfall Canal											
OD-1	820	5.1	26.1	79.0	0.8	90	1.1	0.00	1.9	0.9	9:08 am
LEC-1	5,650	4.5	24.8	76.6	1.4	120	1.1	0.49	2.2	1.1	10:57 am
PC9Y	6,855	4.6	24.7	76.5	0.6	130	1.4	0.54	3.3	1.7	10:22 am
PB	10,570	4.4	23.9	75.0	1.1	145	1.2	0.71	1.3	1.1	11:28 am
LEX	15,210	4.6	25.6	78.1	5.1	120	1.1	0.91	1.3	0.6	2:10 pm
DUP	15,210	4.6	25.6	78.1	5.1	120	NA	NA	NA	NA	2:15 pm
LEE(S)	23,090	5.4	26.8	80.2	7.0	120	>0.7	1.82	0.7	0.3	3:30 pm
LED	Inflow @ 24,300	6.9	29.9	85.8	7.8	205	>1.5	2.77	1.5	0.7	3:15 pm
LEDE	25,210	6.7	29.0	84.2	7.1	180	>1.0	5.12	1.0	0.5	3:51 pm
LEE(N)	27,885	6.5	29.3	84.7	7.3	180	>0.5	5.00	0.5	0.3	4:20 pm
LEF	Inflow @ 28,520	6.9	30.0	86.0	8.3	185	>1.8	8.08	1.8	1.0	4:35 pm
BDR	37,000	NA	NA	NA	NA	NA	NA	8.00	NA	NA	12:30 pm

Note: NA = not analyzed; NB = natural background; BDR = USGS Gauging Station at Berry Dease Road.

Table 7.10b Summary of the Nutrient and BOD Analysis for the Low-Flow Intensive Survey (May 20, 1993)

Station	Distance Downstream from Pond 4 (ft)	Ammonia Nitrogen[a] (mg/L)	Nitrate Nitrogen (mg/L)	Nitrite Nitrogen (mg/L)	TKN (mg/L)	TP (mg P/L)	ortho-Phosphate (mg/L)	BOD$_5$ (mg/L)	BOD$_{10}$ (mg/L)	BOD$_{15}$ (mg/L)	BOD$_{20}$ (mg/L)	Chlorophyll a (mg/m³)	Time of Sample Collection
Detection Limit CLASS III Std.		<0.03 <0.02[b]	<0.02	<0.02	<0.04	<0.05	<0.005	<0.05	<0.05	<0.05	<0.05	<0.5	
Point-Source Discharges													
Pond 4	0	<0.03	<0.02	<0.02	0.93	<0.05	<0.005	2.80	NA	NA	5.90	3.6	8:40 am
SBRM-East	3,300	<0.03	<0.02	<0.02	1.40	<0.05	<0.005	2.80	NA	NA	6.60	4.0	9:47 am
SBRM-West	3,485	<0.03	<0.02	<0.02	1.35	<0.05	<0.005	2.70	NA	NA	7.10	4.2	10:00 am
Landfill Outfall Canal													
OD-1	820	<0.03	<0.02	<0.02	1.70	0.09	<0.005	6.60	7.90	>8.0	>8.0	71.0	9:08 am
LEC-1	5,650	<0.03	<0.02	<0.02	1.62	0.06	<0.005	3.50	4.40	5.20	7.00	5.8	10:57 am
PC9Y	6,855	<0.03	<0.02	<0.02	1.51	0.05	<0.005	3.40	NA	NA	7.30	12.0	10:22 am
PB	10,570	<0.03	<0.02	<0.02	2.00	0.05	<0.005	5.40	NA	NA	6.70	2.4	11:28 am
LEX	15,270	<0.03	<0.02	<0.02	1.54	0.05	<0.005	1.90	2.60	3.30	4.10	7.3	2:10 pm
DUP	15,210	<0.03	<0.02	<0.02	1.35	0.06	<0.005	1.90	2.90	3.90	4.60	1.1	2:15 pm
LEE(S)	23,090	<0.03	<0.02	<0.02	1.31	<0.05	<0.005	2.40	NA	NA	4.10	1.1	3:30 pm
LED	Inflow @ 24,300	<0.03	<0.02	<0.02	0.43	0.14	0.101	0.80	2.50	1.80	2.40	0.8	3:15 pm
LEDE	25,210	<0.03	<0.02	<0.02	0.70	0.08	0.066	1.00	NA	NA	2.60	0.8	3:51 pm
LEE(N)	27,885	<0.03	<0.02	<0.02	0.76	0.08	0.064	0.90	NA	NA	2.80	<0.5	4:20 pm
LEF	Inflow @ 28,520	<0.03	<0.02	<0.02	0.66	0.09	0.061	1.40	2.30	2.50	4.00	1.6	4:35 pm

[a] Reported as total ammonia nitrogen.

[b] As un-ionized ammonia.

Note: NA = not analyzed.

Table 7.11 Summary of the Heavy Metal, Chloride, and Alkalinity Analyses for the Low-Flow Intensive Survey (May 10, 1993)

Station	Distance Downstream from Pond 4 (ft)	Cd	Cr	Cu	Fe	Pb	Zn	Chlorides	Total Alkalinity	Time of Sample Collection
Detection Limit		<0.0001	<0.005	<0.010	<0.030	<0.005	<0.030	<5	<2.0	
CLASS III Std.		See table	0.011	See table	<1.0	See table	See table	NB + 10%	>20	
Point-Source Discharges										
Pond 4	0	<0.0001	<0.005	<0.010	0.080	<0.005	<0.030	9	39.0	8:40 am
SBRM-East	3,300	<0.0001	<0.005	<0.010	0.523	<0.005	<0.030	49	5.5	9:47 am
SBRM-West	3,485	<0.0001	<0.005	<0.010	0.449	<0.005	<0.030	48	7.4	10:00 am
Landfill Outfall Canal										
OD-1	820	<0.0001	<0.005	<0.010	0.628	<0.005	<0.030	18	4.4	9:08 am
LEC-1	5,650	<0.0001	<0.005	<0.010	0.670	<0.005	<0.030	27	<2.0	10:57 am
PC9Y	6,855	<0.0001	<0.005	<0.010	0.680	<0.005	<0.030	25	<2.0	10:22 am
PB	10,570	<0.0001	<0.005	<0.010	0.540	<0.005	<0.030	32	<2.0	11:28 am
LEX	15,210	<0.0001	<0.005	<0.010	0.581	<0.005	<0.030	25	<2.0	2:10 pm
DUP	15,210	<0.0001	<0.005	<0.010	0.596	<0.005	<0.030	25	<2.0	2:15 pm
LEE(S)	23,090	<0.0001	<0.005	<0.010	0.826	<0.005	<0.030	23	24.0	3:30 pm
LED	Inflow @ 24,300	<0.0001	<0.005	<0.010	0.209	<0.005	<0.030	19	45.0	3:15 pm
LEDE	25,210	<0.0001	<0.005	<0.010	0.395	<0.005	<0.030	20	40.0	3:51 pm
LEE(N)	27,885	<0.0001	<0.005	<0.010	0.466	<0.005	<0.030	20	34.0	4:20 pm
LEF	Inflow @ 28,520	<0.0001	<0.005	<0.010	0.441	<0.005	<0.030	19	34.0	4:35 pm

Note: NA = not applicable; NB = natural background. Units in mg/L unless otherwise noted.

Class III Surface Water Standards (Chapter 62-302.530 FAC)

Station	Cd (mg/L)	Cu (mg/L)	Pb (mg/L)	Zn (mg/L)
Point Source Discharge				
Pond 4	0.00047	0.00449	0.00075	0.04058
SBRM-East	0.00028	0.0026	0.00033	0.02362
SBRM-West	0.00028	0.0026	0.00033	0.02362
Landfill Outfall Canal				
OD-1	0.00019	0.00170	0.00018	0.01545
LEC-1	0.00026	0.00242	0.00030	0.02196
PC9Y	0.00026	0.00238	0.00029	0.02160
PB	0.00027	0.00244	0.00030	0.02220
LEX	0.00023	0.00208	0.00024	0.01894
LEE(S)	0.00020	0.00175	0.00018	0.01595
LED	0.00050	0.00480	0.00083	0.04334
LEDE	0.00036	0.00334	0.00048	0.03029
LEE(N)	0.00037	0.00344	0.00051	0.03118
LEF	0.00044	0.00425	0.00069	0.03843

Table 7.12a Summary of Field Parameters, Flow, and Water Depth for the High-Flow Intensive Survey (September 13, 1993)

Station	Distance Downstream from Pond 4 (ft)	pH	Temperature C	Temperature F	DO (mg/L)	Specific Conductance (mmhos/cm)	Secchi Disk Depth (ft)	Flow (cfs)	Total Water Depth (ft)	Sample Depth (ft)	Time of Sample Collection
Detection Limit		0.05	0.1		0.1	10	0.1	0.01			
CLASS III Std.		MB + 1 unit			>5	NB + 50%	NB + 10%				
Point-Source Discharges											
Pond 4	0	6.6	27.4	81.3	4.0	110	>2.8	0.01	2.8	1.4	7:30 am
SBRM-East	3,300	5.5	26.6	79.9	2.2	180	1.2	7.00	3.7	1.8	8:10 am
SBRM-West	3,485	5.5	26.4	79.5	1.9	185	1.2	7.00	4.3	2.2	8:45 am
Landfill Outfall Canal											
OD-1	820	4.7	26.3	79.3	2.9	80	2.0	0.02	3.3	1.7	7:45 am
LEC1	5,650	5.1	26.6	79.9	1.3	150	1.2	8.10	4.5	2.3	8:40 am
PC9Y	6,855	5.1	26.5	79.7	0.3	140	3.0	8.40	5.0	2.5	9:01 am
PB	10,570	5.2	25.6	78.1	1.1	110	1.3	9.20	2.5	1.3	9:14 am
LEX	15,210	4.9	25.8	78.4	0.6	90	2.2	10.20	2.7	1.3	9:36 am
LEE(S)	23,090	4.8	25.5	77.9	7.7	60	1.1	41.30	2.0	1.0	10:47 am
LED	Inflow @ 24,300	6.3	25.6	78.1	8.0	120	0.8	29.70	2.7	1.3	10:17 am
LEDE	25,210	6.1	25.8	78.4	10.6	80	0.2	71.80	3.2	1.6	11:15 am
LEE(N)	27,885	6.1	26.2	79.2	2.8	80	1.2	73.00	3.0	1.5	11:35 am
LEF	Inflow @ 27,885	6.4	28.6	83.5	3.2	100	1.3	89.80	4.0	2.0	11:50 am
BDR	37,000	NA	NA	NA	NA	NA	NA	103.20	NA	NA	12:15 pm

Note: NA = not analyzed; NB = natural background; BDR = USGS Gauging Station at Berry Dease Road.

Table 7.12b   Summary of the Nutrient and BOD Analyses for the High-Flow Intensive Survey (September 13, 1993)

Station	Distance Downstream from Pond 4 (ft)	Ammonia Nitrogen[a] (mg/L)	Nitrate Nitrogen (mg/L)	Nitrite Nitrogen (mg/L)	TKN (mg/L)	TP (mg P/L)	ortho-Phosphate (mg/L)	BOD5 (mg/L)	BOD10 (mg/L)	BOD15 (mg/L)	BOD20 (mg/L)	Chlorophyll a (mg/m³)	Time of Sample Collection
Detection Limit CLASS III Std.		<0.03 <0.02[b]	<0.02	<0.02	<0.04	<0.05	<0.005	<0.05	<0.05	<0.05	<0.05	<0.5	
Point-Source Discharges													
Pond 4	0	0.06	0.02	<0.02	0.97	<0.05	<0.005	1.5	NA	NA	4.3	4.9	7:30 am
SBRM-East	3,300	<0.03	<0.02	<0.02	1.57	<0.05	<0.005	2.9	NA	NA	7.8	4.0	8:10 am
SBRM-West	3,485	<0.03	<0.02	<0.02	1.69	<0.05	<0.005	4.0	NA	NA	10.0	4.0	8:45 am
Landfill Outfall Canal													
OD-1	820	<0.03	<0.02	<0.02	0.81	<0.05	<0.005	4.8	3.6	3.8	5.2	7.8	7:45 am
LEC1	5,650	<0.03	<0.02	<0.02	1.90	<0.05	<0.005	4.0	4.7	6.6	8.6	6.4	8:40 am
PC9Y	6,855	<0.03	<0.02	<0.02	1.84	<0.05	<0.005	3.3	NA	NA	7.9	3.8	9:01 am
PB	10,570	0.03	<0.02	<0.02	2.78	0.17	0.005	2.3	5.3	6.2	9.5	2.4	9:14 am
LEX	15,210	<0.03	<0.02	<0.02	1.32	<0.05	<0.005	1.4	NA	NA	3.2	1.1	9:36 am
LEE(S)	23,090	<0.03	<0.02	<0.02	1.12	<0.05	0.009	1.0	2.7	3.2	10.9	0.8	10:47 am
LEE(S)-dup	23,090	<0.03	<0.02	<0.02	1.11	<0.05	0.007	1.0	NA	NA	5.6	<0.5	
LED	Inflow @ 25,210	0.10	0.11	<0.02	1.27	0.26	0.063	1.7	NA	NA	12.7	7.9	10:17 am
LEDE	25,500	0.03	0.05	<0.02	1.13	0.09	0.041	1.7	NA	NA	21.6	1.0	10:15 am
LEE(N)	27,885	0.03	0.06	<0.02	1.08	0.10	0.041	1.4	3.7	6.0	19.6	2.5	11:35 am
LEF	Inflow @ 28,520	0.04	0.07	<0.02	1.01	0.10	0.057	1.4	NA	NA	7.0	1.8	11:50 am
Field Blank	—	<0.03	<0.02	<0.02	0.05	<0.05	<0.005	1.5	NA	NA	4.3	<0.5	

[a] Reported as total ammonia nitrogen.
[b] As unionized ammonia.
Note: NA = not analyzed.

Table 7.13 Summary of the Heavy Metal, Chloride, and Alkalinity Analyses for the High-Flow Intensive Survey (September 13, 1993)

Station	Distance Downstream from Pond 4 (ft)	Cd (mg/L)	Cr (mg/L)	Cu (mg/L)	Fe (mg/L)	Pb (mg/L)	Zn (mg/L)	Total Hardness (mg/L as CaCO³)	Chlorides (mg/L)	Total Alkalinity (mg/L)	Time of Sample Collection
Detection Limit		<0.0001	<0.005	<0.010	<0.030	<0.005	<0.030	<1.25	<5	<2.0	
CLASS III Std.		See table	<0.011	See table	<1.0	See table	See table	NA	NB + 10%	>20	
Point-Source Discharges											
Pond 4	0	<0.0001	<0.005	<0.010	0.069	<0.005	<0.030	32.2	11	26.0	7:30 am
SBRM-East	3,300	<0.0001	<0.005	<0.010	0.825	<0.005	<0.030	17.0	42	7.0	8:10 am
SBRM-West	3,485	<0.0001	<0.005	<0.010	0.816	<0.005	<0.030	17.0	42	7.4	8:45 am
Landfill Outfall Canal											
OD-1	820	<0.0001	<0.005	<0.010	0.392	<0.005	<0.030	10.3	17	3.0	7:45 am
LEC1	5,650	<0.0001	<0.005	<0.010	0.752	<0.005	<0.030	15.6	34	5.0	8:40 am
PC9Y	6,855	<0.0001	<0.005	<0.010	0.773	<0.005	<0.030	15.3	32	5.0	9:01 am
PB	10,570	<0.0001	<0.005	<0.010	0.661	<0.005	<0.030	15.8	21	5.0	9:14 am
LEX	15,210	<0.0001	<0.005	<0.010	0.751	<0.005	<0.030	13.1	20	3.0	9:36 am
LEE(S)	23,090	<0.0001	<0.005	<0.010	0.691	<0.005	<0.030	10.7	11	2.0	10:47 am
LEE(S)-dup	23,090	<0.0001	<0.005	<0.010	0.695	<0.005	<0.030	10.7	12	2.0	
LED	Inflow @ 24,300	<0.0001	<0.005	<0.010	1.04	<0.005	<0.030	34.8	13	26.0	10:17 am
LEDE	25,210	<0.0001	<0.005	<0.010	0.752	<0.005	<0.030	22.8	13	14.0	10:15 am
LEE(N)	27,885	<0.0001	<0.005	<0.010	0.752	<0.005	<0.030	23.6	12	14.0	11:35 am
LEF	Inflow @ 28,520	<0.0001	<0.005	<0.010	0.711	<0.005	<0.030	30.2	13	24.0	11:50 am
Field Blank	—	<0.0001	<0.005	<0.010	<0.030	<0.005	<0.030	<1.25	<5	<2.0	

Note: NA = not applicable; NB = natural background.

Class III Surface Water Standards (Chapter 62-302.530 FAC)

Station	Cd (mg/L)	Cu (mg/L)	Pb (mg/L)	Zn (mg/L)
Point-Source Discharges				
Pond 4	0.00047	0.00449	0.00075	0.04058
SBRM-East	0.00028	0.00260	0.00033	0.02362
SBRM-West	0.00028	0.00260	0.00033	0.02362
Landfill Outfall Canal				
OD-1	0.00019	0.00170	0.00018	0.01545
LEC-1	0.00026	0.00242	0.00030	0.02196
PC9Y	0.00026	0.00238	0.00029	0.02160
PB	0.00027	0.00244	0.00030	0.02220
LEX	0.00023	0.00208	0.00024	0.01894
LEE(S)	0.00020	0.00175	0.00018	0.01595
LEE(S)-dup	0.00020	0.00175	0.00018	0.01595
LED	0.00050	0.00480	0.00083	0.04334
LEDE	0.00036	0.00334	0.00048	0.03029
LEE(N)	0.00037	0.00344	0.00051	0.03118
LEF	0.00044	0.00425	0.00069	0.03843

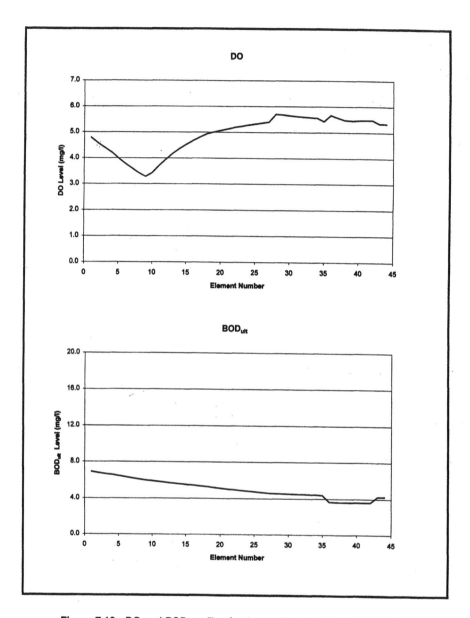

Figure 7.10 DO and BOD profiles for the low-flow design (Simulation 8).

7.4 CONCLUSIONS

The following conclusions were reached based on the results of this study:

1. There were significant differences between mean concentrations of selected water quality parameters for pollutant sources and successive stages of the landfill leachate treatment process.
2. Treatment occurs through the process of collecting and diluting the leachate in the borrow ponds and applying the dilute leachate to the wetland.
3. Over a 24-month period, there was 83, 73, and 61% assimilation of the total nitrogen, total phosphorus, and BOD_5 loaded to the wetland treatment system, respectively.

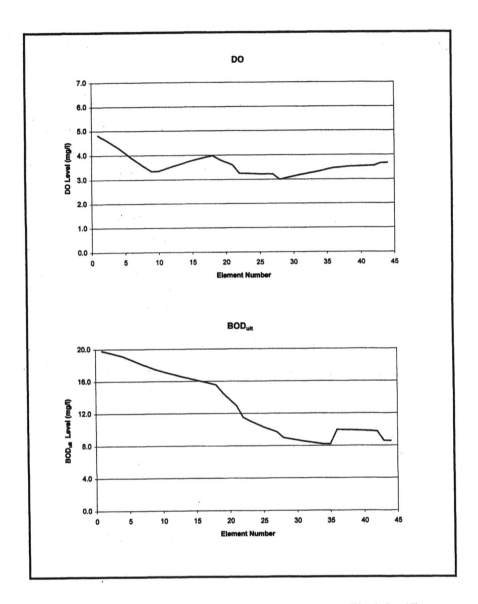

Figure 7.11 DO and BOD profiles for the high-flow design (Simulation 15).

4. Accumulation of heavy metals occurred in soil and in vegetation to a lesser extent in the wetland treatment system and in wetland microcosms, and the immobilization of the heavy metals with acid volatile sulfides may reduce the risk for bioaccumulation.

5. Some accumulation of BNAEs may be occurring in the wetland treatment system, but the concentration of VOCs in the surface water in the wetland were low and VOCs were not retained in wetland microcosms, indicating that their potential effects are minimized in wetland systems.

6. Although OPPs have been detected in surface water, soil, fish, and vegetation samples, they were at levels generally below existing standards and criteria.

7. Except for DO and pH, which are naturally low in wetlands, the receiving waters for the wetland treatment system meet the Class III water quality standards, and the water quality of these receiving waters is good.

8. Macroinvertebrate populations in the wetland were indicative of good water quality, and fish populations were within the range for natural wetlands.

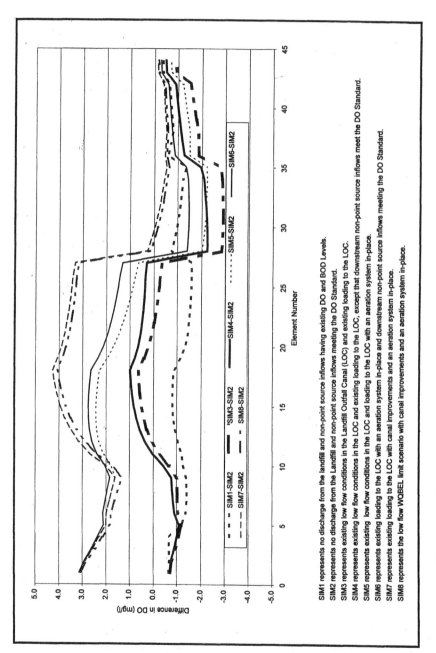

SIM1 represents no discharge from the landfill and non-point source inflows having existing DO and BOD Levels.
SIM2 represents no discharge from the Landfill and non-point source inflows meeting the DO Standard.
SIM3 represents existing low flow conditions in the Landfill Outfall Canal (LOC) and existing loading to the LOC.
SIM4 represents existing low flow conditions in the LOC and existing loading to the LOC, except that downstream non-point source inflows meet the DO Standard.
SIM5 represents existing low flow conditions in the LOC and loading to the LOC with an aeration system in-place.
SIM6 represents existing loading to the LOC with an aeration system in-place and downstream non-point source inflows meeting the DO Standard.
SIM7 represents existing loading to the LOC with canal improvements and an aeration system in-place.
SIM8 represents the low flow WQBEL limit scenario with canal improvements and an aeration system in-place.

Figure 7.12 Change in DO in the LOC under low-flow with existing and proposed conditions.

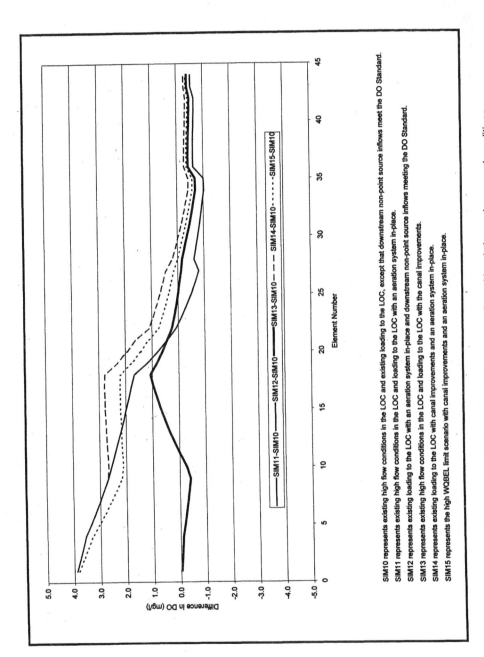

SIM10 represents existing high flow conditions in the LOC and existing loading to the LOC, except that downstream non-point source inflows meet the DO Standard.
SIM11 represents existing high flow conditions in the LOC and loading to the LOC with an aeration system in-place.
SIM12 represents existing loading to the LOC with an aeration system in-place and downstream non-point source inflows meeting the DO Standard.
SIM13 represents existing high flow conditions in the LOC and loading to the LOC with the canal improvements.
SIM14 represents existing loading to the LOC with canal improvements and an aeration system in-place.
SIM15 represents the high WQBEL limit scenario with canal improvements and an aeration system in-place.

Figure 7.13 Change in DO in the LOC under high-flow with existing and proposed conditions.

9. Viable biological communities have adapted to new hydrologic conditions created with the discharge from the borrow ponds.
10. Long hydraulic residence time in the wetland treatment system and the results from the wetland microcosm study indicate that this system has the potential to be managed to provide treatment for higher flows associated with large storm events and to maximize treatment of landfill leachate.
11. WQBELs for BOD_5 are required by the FDEP for the discharges from the storm water pond from the lined portion of the landfill and the wetland treatment system discharge at the Orange County Landfill. Compliance for these WQBELs will be established in the LOC at station LEC1.
12. The WQBELs for BOD_5 were established with an in-line aeration system and improvements to the LOC in place. Given these improvements, the landfill discharge WQBELs are 4.4 mg/L BOD_5 (10 mg/L BOD_{ult}) under low-flow conditions and 8.8 mg/L BOD_5 (20 mg/L BOD_{ult}) under high-flow conditions.
13. The landfill wetland treatment system in combination with an in-line aeration system and improvements to the LOC provide sufficient assimilative capacity to maintain existing downstream receiving water quality in the Little Econlockhatchee River.

REFERENCES

Beck, W. M., Jr., 1954. Studies in stream pollution biology: I. A simplified ecological classification of organisms, *Journal of Florida Academy of Sciences* 17:211–227.

Brown, L. C. and Barnwell, T. O., Jr., 1987. The Enhanced Stream Water Quality Models QUAL2E and QUAL2E-UNCAS: Documentation and User Manual. U.S. Environmental Protection Agency, Office of Research and Development, Athens, GA.

Camp Dresser & McKee, 1996. Orange County Landfill Wide Cypress Swamp Wetland Research Program Final Summary Report. Prepared for Orange County Public Utilities.

Camp Dresser & McKee, 1997. Orange County Landfill Level II WQBEL. Prepared for Orange County Utilities.

DeBusk, T. A., Laughlin, R. B., and Schwartz, L. N., 1996. Retention and compartmentalization of lead and cadmium in wetland microcosms, *Water Research* 30:2707–2716.

Di Toro, D. M., Mahony, J. D., Hansen, D. J., Scott, K. J., Carlson, A. R., and Ankley, G. T., 1992. Acid volatile sulfide predicts the acute toxicity of cadmium and nickel in sediments, *Environmental Science and Technology* 26:96–101.

Gruendling, G. K. and Gaines, F. F., 1996. Fish of Wide Cypress Swamp, Section 5, in Orange County Landfill Wide Cypress Swamp Wetland Research Program Final Summary Report. Prepared for Orange County Public Utilities.

Staubitz, W. W., Surface, J. M., Steenhuis, T. S., Peverly, J. H., and Lavine, M. J., 1989. Use of constructed wetlands to treat landfill leachate, in Hammer, D. A., Ed., *Constructed Wetlands for Wastewater Treatment: Municipal, Industrial and Agricultural*, Lewis Publishers, Chelsea, MI, 735–742.

Wallace, P. M., 1996. Vegetation in wide cypress swamp, Section 4 in Orange County Landfill Wide Cypress Swamp Wetland Research Program Final Summary Report. Prepared for Orange County Public Utilities.

Wilhm, J. L. and T. C. Dorris, 1968. Biological Parameters for Water Quality Criteria, *Bioscience* 18:477–481.

CHAPTER 8

Attenuation of Landfill
Leachate by a Natural Marsh System

François La Forge, Leta Fernandes, and Mostafa A. Warith

CONTENTS

ABSTRACT: Since the early 1980s, leachate originating from the Alice and Fraser municipal landfill has been discharging into a natural marsh located some 300 m downgradient from the waste disposal site. However, monitoring of the water quality within the marsh indicates that the contaminant level has not yet surpassed background concentration downstream from the immediate discharge area. A combination of two existing mathematical models was used in an attempt to predict the mobility of several contaminant species within the marsh environment. Parameters needed for the predictive model were gathered based on the physical configuration of the landfill and the marsh, as well as on laboratory-derived data on the attenuation capacity of the marsh soil matrix.

8.1 INTRODUCTION

The unique combination of physical and chemical properties of marsh soils enables them to adsorb significantly contaminants that may be present in wastewater. In the past few decades, planned use of natural wetlands for treatment of secondary wastewaters effluent (secondary-treated effluents) has been studied and implemented in several countries (Coupal and Lalancette, 1976; Brodrick et al., 1988; Hammer, 1989). The treatment of landfill leachate by natural marsh soil was further investigated by Cameron (1978), as well as by McLellan and Rock (1988), with varying degrees of success.

Investigation of the migration of pollutants through a natural marsh located downgradient from the Alice and Fraser landfill site in eastern Ontario is expected to assist in the development of a predictive mathematical model that could simulate the long-term attenuation capacity of the marsh. This mathematical model is to incorporate data collected through laboratory experiments on the adsorption capacity of the marsh soils along with the water chemistry of the marsh, which was continuously monitored over the past decade. This paper will address only the mobility of some heavy metals (lead and zinc) and major cations and anions (sodium, calcium, and chloride). Such mathematical models could eventually be used to simulate the short- and long-term attenuation of various landfill contaminants during their migration through a landfill/marsh system. These models are required to evaluate better the environmental viability of establishing landfills, or other waste-producing operations, in the vicinity of marshes. A more-detailed study can be found in La Forge (1994).

8.2 SITE SETTING

The Alice and Fraser landfill site, which has been in operation since 1977, is located approximately 12 km southwest of the city of Pembroke in the province of Ontario, Canada. The landfill covers an area of 1.25 km² including three waste containment cells (industrial, dry, and municipal waste), as well as a buffer zone. Since there is no liner underlying the landfill, the generated leachate is free to migrate in the underlying sand and gravel deposits. Once the leachate reaches the water table, it is transported with the bulk flow of the water toward the Beaver Meadow marsh, which is located 300 m downgradient from the municipal landfill (Figure 8.1). Site monitoring indicates that leachate-impacted water has been discharging into the Beaver Meadow marsh since the early 1980s. However, monitoring of the surface water quality within the marsh indicates that the contaminant level has not yet surpassed background concentration downstream from the discharge area. This confirms that the on-site marsh north of the landfilled areas has, so far, been capable of attenuating the leachate generated by the landfill.

8.3 LABORATORY SETUP

The assimilation potential of soil is often site dependent, and laboratory analyses are generally required to evaluate the attenuation capability of a particular marsh soil. To characterize the attenuation potential of the Beaver Meadow marsh adequately, it was essential to evaluate the physicochemical properties of the native marsh soil. A comprehensive laboratory analysis program was setup to investigate the properties, which included:

1. Evaluation of the basic characteristics of the Beaver Meadow marsh soil using standard laboratory analysis procedures;

2. Soil column experiments to establish the breakthrough characteristics and the retardation factors for the heavy metals and major cations and anions as they leach through cells containing marsh soil;
3. Determination of the migration profiles of the contaminant species by analyzing the pore water and the soil matrix within the columns.

The ultimate goal of the laboratory analysis was to estimate some of the primary parameters needed to implement the contaminant migration model.

8.3.1 Soil Sampling

The soil used in the laboratory experiments was collected from a section of the Beaver Meadow marsh located 700 m upstream from the main leachate stream. The soil sample location was chosen based on its accessibility and on the relatively natural state of the soil, unaffected by the landfill leachate. The soil had a distinctive dark brown color, a spongy texture, and an organic odor. The basic properties of the Beaver Meadow marsh soil are typical of organic soil deposit, having an organic matter content of approximately 30%, a high moisture content (80%), and porosity (85%) with a relatively low dry bulk density (450 kg/m^3) (McLellan and Rock, 1988).

8.3.2 Stock Solution for Column Experiment

In view of the strong variability of the leachate concentration observed at the Alice and Fraser landfill site, it was decided that synthetic stock solutions be prepared for the column experiments. By employing a laboratory-controlled spiked solution, the contaminant level was precisely adjusted and produced required volumes.

The heavy metal solution was prepared with laboratory-grade lead nitrate (Pb(NO$_3$)$_2$) and zinc nitrate (Zn(NO$_3$)2.6H$_2$O). Both compounds were dissolved in distilled water to produce an approximate concentration of 100 mg/L each of lead and zinc. The cations solution consisted of a mixture

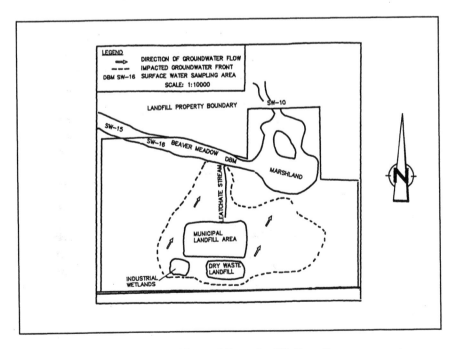

Figure 8.1 Alice and Fraser landfill site setting.

of laboratory-grade calcium chloride (CaCl$_2$) and sodium chloride (NaCl) dissolved in distilled water. A resulting concentration of approximately 100 mg/L each of calcium and sodium, as well as a chloride concentration of 250 mg/L, was obtained. All stock solutions were prepared to produce between 10 and 20 l of synthetic wastewater. Additional stock solutions were prepared in the same fashion when needed during the course of the experimental phase.

8.3.3 Column Study

Although batch laboratory studies can provide useful information on the adsorption potential of contaminants by the marsh soil, continuous-flow column experiment is a more reliable method to simulate natural environments (Eckenfelder, 1989). Column-type continuous-flow systems try to recreate as close as possible the soil/water ratio encountered in nature (Domenico and Schwartz, 1990). In addition, moving pore fluids in the column actively redistributes the products of reaction, unlike batch experiment.

In this investigation, laboratory leaching columns were designed to simulate saturated flow of leachate through the Beaver Meadow marsh soil. A total of ten columns were constructed. Each column consisted of a hollow Plexiglas cylinder 3 mm thick, with an inside diameter of 100 mm and a total length of approximately 100 mm. Plexiglas top cap and bottom plate were screwed to the top and bottom of each cylinder and fitted with rubber O-rings to prevent water leakage. The columns were initially saturated with distilled water, and the hydraulic conductivity of each column was monitored for a period of several days until the flow rate was stabilized. The experiment was then started by switching the column input from distilled water to a heavy metal stock solution for columns 1 to 4 and to the major cations and anions solution for columns 6 to 9. Columns 5 and 10 were used as controls, allowing only distilled water to permeate.

Column effluent was collected on a daily basis and analyzed using an inductively coupled argon plasma (ICP) spectrograph, Model 975. A schematic representation of a laboratory leaching column experiment is presented in Figure 8.2.

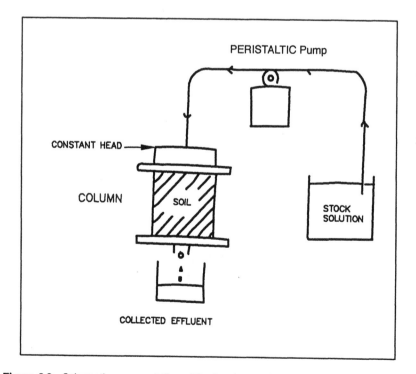

Figure 8.2 Schematic representation of the leachate column experiment configuration.

8.4 LABORATORY RESULTS

Results from the leaching test indicate that the major cations and anions solution required little time to migrate through the columns. Figure 8.3 shows a plot of the breakthrough curve obtained from leaching column 6. This plot is typical of the breakthrough curves observed during the cations and anions leaching test. As expected, the breakthrough of chloride, which is a nonreactive contaminant, was observed after approximately 1 pore volume, indicating that the attenuation of the chloride ions was negligible across the length of the soil column. The breakthrough of calcium was observed at approximately 1.4 pore volumes, while for sodium the breakthrough appeared at approximately 2.1 pore volumes.

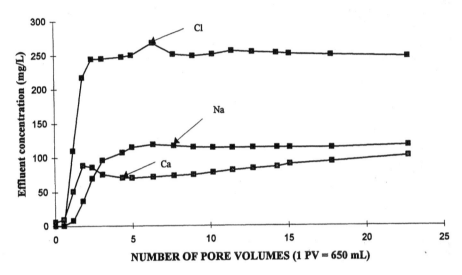

Figure 8.3 Breakthrough curves associated with column 6.

Results from the leaching tests confirmed that the heavy metals are strongly retained by the marsh soil matrix. Throughout the course of the leaching experiment, which lasted approximately 3 months, lead or zinc was not detected in the effluents. All of the heavy metals that permeated through the cell were retained by the soil matrix. It is, however, interesting to observe that the concentration of calcium and magnesium tends to increase with time even though no such compounds were present in the influent solution. This phenomenon can be explained by the process of preferential adsorption. A typical breakthrough curve observed for column 4 is presented in Figure 8.4.

The pore fluid of columns 2 and 4 was extracted in order to evaluate the variation in concentrations of lead and zinc with depth within the soil column. Results obtained from the chemical analysis of the extracted pore fluid for zinc are presented in Figure 8.5. The results showed that a low concentration of zinc was detected down to a depth of 30 mm in the top layers of the soil columns. By contrast, lead was effectively retained on the marsh soil, with only trace amounts (<0.1 mg/L) detected in the pore fluid (Figure 8.6).

In order to establish the fate of the heavy metals within the marsh soil leaching column, a digestion analysis of the soil was undertaken. The data obtained from the digestion analysis confirmed that both lead and zinc ions were preferentially adsorbed by the top layer of the soil column. Results obtained from the digestion analysis are presented in Figures 8.7 and 8.8 for zinc and lead, respectively. Mass balance analyses were carried out to establish the validity of the method used. By comparing the total influx of contaminant entering the soil leaching columns with the mass of contaminants detected in the effluent, pore fluids, and soil matrix, it was possible to estimate the total recovery of between 75 and 95% for both lead and zinc.

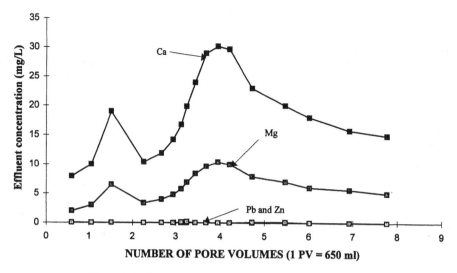

Figure 8.4   Breakthrough curves associated with column 4.

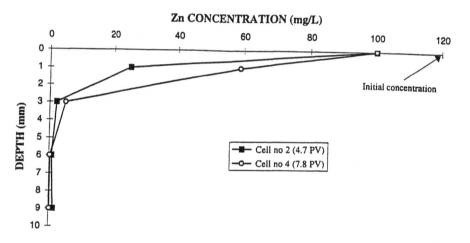

Figure 8.5   Zinc migration profiles within the pore fluid.

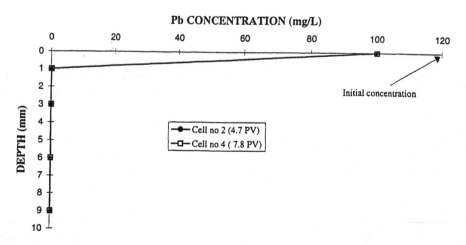

Figure 8.6   Lead migration profiles within the pore fluid.

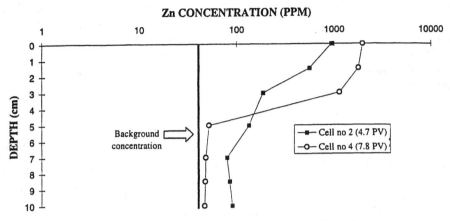

Figure 8.7 Zinc migration profiles within the soil matrix.

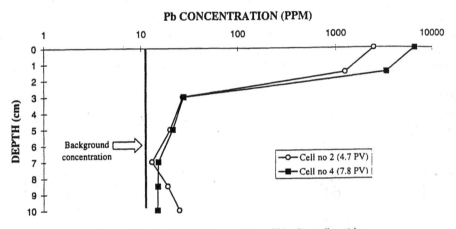

Figure 8.8 Lead migration profiles within the soil matrix.

8.4.1 Migration Potential

By knowing the contaminant distribution through space and time, it was possible to back-calculate the retardation coefficient (R_F) characteristics of each contaminant species using a numerical model. By adapting the advection–dispersion equation (ADE) with sorption to a spreadsheet layout, it was possible to back-calculate numerically, by trial and error, the contaminant distribution profiles obtained in the laboratory column experiments. By doing so, retardation factors of 1, 1.25, and 2 were calculated for chloride, calcium, and sodium, respectively, while an R_F of approximately 100 and 2000 wase established for zinc and lead, respectively.

8.5 MODELING

Contaminant transport simulation is a three-step process in which the governing equations of flow and mass transport are employed sequentially. First, the flow equation is solved for the head distribution. Next, the head distribution is used with the hydraulic conductivity and effective porosity to calculate the seepage velocity distribution using Darcy's law. Finally, the contaminant transport equation is solved using the previously calculated velocity distribution.

To achieve these tasks, two distinct models were utilized. In a first step, the local flow system within the Beaver Meadow marsh was calibrated using the Waterloo Hydrogeologic Software

FLOWPATH flow and pathline model (Franz and Gulguer, 1992). The hydraulic head distribution gathered from FLOWPATH for the marsh system was then used as input for the velocity and contaminant transport model (DID2XY). The DID2XY model was initially written by Samani (1987). This program, which resolves unsteady two-dimensional solute and fluid transport in clay soil, was used to calculate the velocity distribution and concentration variation as a function of time and space.

8.5.1 FLOWPATH Model

In order to simulate the subsurface water flow pattern accurately within the Beaver Meadow marsh system, the FLOWPATH model was designed in such a way as to reproduce the geometric and boundary conditions of the system correctly. To achieve this, a grid design was constructed to reproduce the aerial configuration of the marsh as well as the variable topographical relief.

The grid network constructed to simulate the Beaver Meadow marsh consisted of over 500 cells, the majority of them having dimensions of 13×13 m. The model covers an area of 67700 m^2. Constant head boundaries were assumed at both the western and northern limits of the marsh, while the perimeter of the marsh was simulated by an impermeable boundary, since no groundwater flux is assumed to enter the system. For modeling purposes, it was assumed that the water table was located at the surface, which is typically the case at most wetland sites. The resulting grid network is presented in Figure 8.9. The FLOWPATH model calculated the hydraulic head distribution across the Beaver Meadow marsh, as well as the preferred transport pathway of contaminants. The FLOWPATH output was then used as input for the DID2XY model for both the velocity and contaminant transport simulation.

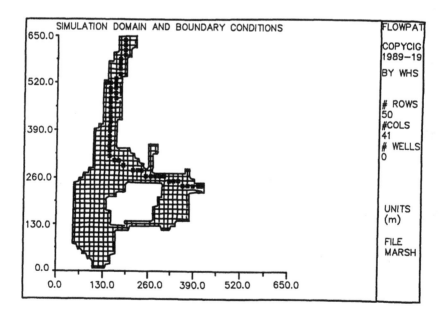

Figure 8.9 The FLOWPATH grid network representing the Beaver Meadow marsh system.

8.5.2 DID2XY Model

The DID2XY model was used to simulate the migration potential of lead, zinc, calcium, sodium, and chloride within the subsurface water phase of the Beaver Meadow marsh system. The analysis was performed to predict migration after 1, 10, 50, and 100 years.

Modeling results indicate that the migration potentials of the targeted contaminants were somewhat limited. Even after a period of 100 years, the migration front of sodium, calcium, and chloride reached respective distances of 120, 200, and 260 m. The concentration contours for calcium after 100 years are presented in Figure 8.10. This slow migration potential is directly related to the low hydraulic conductivity of the marsh soil media and the low hydraulic gradient observed within this segment of the marsh. Modeling results gathered for zinc indicate that the mobility of this contaminant would be limited to fractions of a meter per year, while lead would be in effect immobile within the water-saturated phase of the organic soil. In comparing the relative mobility of the contaminants investigated, it is shown that the attenuation capacity of a marsh soil is highly variable, depending on the targeted contaminants.

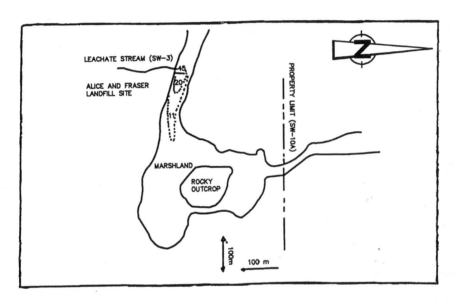

Figure 8.10   Concentration contours (mg/L) for calcium after 100 years.

8.6 DISCUSSION

As the practice of using natural marsh ecosystems for the treatment of contaminated water increases in popularity in remote locations, the need to understand fully the pollution attenuation processes and the migration potential of the pollutants is becoming indispensable. However, the complex interactions among soil, plants, and microbial activity, as well as an often ill-defined hydrology associated with this type of ecosystem, seriously challenge engineers and scientists to forecast the attenuation potential of natural marshes adequately. In recent years, several mathematical models have been advanced to simulate contaminant migration within marshes; however, they tend to be quite site specific, and they rarely incorporate all possible flow paths endemic to marshes (groundwater, overland, and channel flow).

The behavior, transport, and ultimate fate of contaminants in organic soil are greatly affected by their participation in sorption reactions. The adsorption potential of heavy metals and major cations and anions was evaluated in the laboratory through column experiments. Extreme variations in retardation factors were observed, from values approaching 1 for chloride to a retardation factor of over 2000 for lead.

A combination of two existing computer models was used to predict the migration potential for zinc and calcium. The model was adapted to the hydrologic characteristics of the marsh, emphasizing subsurface flow, as well as the adsorption capacity of the organic soil. Initial modeling results indicate that marsh soil has the capacity to retard substantially the migration of several contaminant species typically found in landfill leachate. The predictive capacity of the model will eventually be evaluated based on the contaminant migration plume observed at the actual marsh site. Further calibration of the model will undoubtedly be necessary to better reflect the complex reactions associated with the Beaver Meadow marsh system.

ACKNOWLEDGMENTS

Thanks are due to Golder Associates Ltd. for providing the background information of the Alice and Fraser landfill site. The authors also wish to thank the National Science and Engineering Research Council, F.C.A.R., and the University of Ottawa for the award of postgraduate scholarships.

REFERENCES

Brodrick, S. J., Cullen, P., and Maller, W., 1988. Denitrification in a natural wetland receiving secondary treated effluent, *Water Resources* 22:431–439.

Cameron, R. D., 1978. Treatment of a complex landfill leachate with peat, *Canadian Journal of Civil Engineering* 5:83–97.

Coupal, B. and Lalancette, J. M., 1976. The treatment of waste waters with peat, *Water Research* 10:1071–1076.

Domenico, P. A. and Schwartz, F. W., 1990. *Physical and Chemical Hydrogeology*, John Wiley & Sons, New York.

Eckenfelder, W. W., 1989. *Industrial Water Pollution Control*, 2nd ed., McGraw-Hill, New York.

Franz, T. and Guiguer, N., 1992. FLOWPATH Version 4 — User's Manual, Waterloo Hydrogeologic Software, Waterloo, Ontario.

Hammer, D. A, 1989. *Constructed Wetlands for Wastewater Treatment — Municipal, Industrial and Agricultural*, Lewis Publishers, Chelsea, MI.

La Forge, F., 1994. Attenuation of Landfill Leachate by a Natural Marshland System, M.A.Sc. thesis. Department of Civil Engineering, University of Ottawa, 278 pp.

McLellan, J. K. and Rock, C. A., 1988. Pretreating landfill leachate with peat to remove metals, *Water, Air and Soil Pollution* 37:203–215.

Samani, H. M., 1987, Mathematical Modeling of Contaminant Transport through Clay Soil Using Irreversible Thermodynamics, Ph.D. thesis, Department of Civil Engineering, McGill University, Montreal, Quebec, 250 pp.

CHAPTER **9**

Leachate Treatment in Extended Aeration Lagoons and Constructed Wetlands in Norway

Trond Mæhlum, William S. Warner, Per Stålnacke, and Petter D. Jenssen

CONTENTS

ABSTRACT: Constructed wetlands (CWs) used to treat raw landfill leachates may not function effectively because of the high concentrations of organic matter, metals, and other pollutants that are toxic. Deposition of solids and blockage of pore space will reduce the permeability of the media in subsurface-flow CWs. To mitigate these effects, two integrated leachate treatment systems were designed: an extended aeration lagoon with aspirator propeller aerators/mixers, and sedimentation and treatment in subsurface-horizontal-flow (SHF) CWs. The wetlands were planted with *Typha latifolia* and *Phragmites australis*. The media consisted of washed gravel and lightweight aggregates in the size range of 10 to 20 mm. The mean theoretical hydraulic retention time in the filters was 3 and 5 days (mean hydraulic loading rate 12 and 5 cm/day, respectively, for the two systems). The influences of temperature and input leachate flow rates on treatment efficiency of physical, chemical, and biological processes in the aerated lagoons and CWs were studied. Due to random and systematic variations of influent and effluent water quality, different removal efficiency calculation methods were performed. Significant differences in lagoon treatment efficiency due to temperature and flow rate were detected for Fe, chemical oxygen demand (COD), and nitrogen. In one system, phosphoric acid was added occasionally in the aerated lagoon to achieve increased biological degradation and nitrification. At water temperatures >10°C, 50% to 99% nitrification was achieved in the lagoon. The dynamics in NO_3–N concentrations in the lagoon were positively correlated to the temporal variability in water temperature. At temperatures <5°C, nitrification almost ceased even with the addition of phosphoric acid. Results from this study indicate efficient leachate polishing in SHF CWs at hydraulic loading rates <5 cm/day. Operational problems and low treatment efficiency because of low temperatures and frost in the CWs were not observed. Clogging of the coarse media has been observed. Nitrification and major removal of organic matter and suspended solids are recommended before treatment of landfill leachate in SHF CWs.

9.1 INTRODUCTION

On-site treatment of landfill leachate is not widespread in Norway. In 1997, fewer than 30 of the 350 municipal sanitary waste (MSW) landfills used on-site biological treatment systems. About 35 MSW landfills discharge raw leachate to sewers without on-site pretreatment. Regulations now require that most landfills must be lined to control leachates and propose that leachate treatment be mandatory in the future. On-site "high-tech" leachate treatment systems are often avoided because of the high cost to construct and operate them. Therefore, low-cost leachate treatments suitable for cold climatic conditions are preferred. Several studies in Europe and the U.S. conclude that extended aeration lagoons are highly efficient in removing COD and NH_4–N (Maris and Harrington, 1984; Robinson and Maris, 1985; Robinson and Grantham, 1988; Robinson, 1990; Metcalf & Eddy, Inc., 1991). Extended aeration lagoons can readily cope with a wide range of leachate flows and strengths, including shock loads due to the normally large lagoon volume. Constructed wetlands (CWs) have been used successfully for secondary and tertiary treatment of effluent from aerated lagoons in the U.K. (Robinson, 1993) and the U.S. (Martin and Moshiri, 1995). Properties that make CWs suitable for wastewater treatment include extensive adsorptive surfaces (sediments, plants, and roots), aerobic–anaerobic interfaces, and diverse active microbial populations that will translocate, metabolize, or use the various contaminants (Kadlec and Knight, 1996). This paper examines two cold climate leachate treatment systems that integrate extended aeration lagoons (partly mixed and without recycling), sedimentation, and CWs (Mæhlum et al., 1995; Mæhlum and Haarstad 1997). Particular attention is given to the influence of temperature and input leachate flow rates on treatment efficiency of physical, chemical, and biological processes associated with treating organic compounds, nitrogen, and iron.

9.2 MATERIALS AND METHODS

9.2.1 Study Sites

The landfill sites at Esval and Bølstad located in the southeast of Norway (60°N, 40 km from Oslo) have a cool temperate continental climate, with about 800 mm annual precipitation and a January mean air temperature around –6°C (Table 9.1). Although leachate temperature and production rates are similar, the Esval site is almost 50% larger than the Bølstad site, 7.0 vs. 5.0 ha. The landfills and treatment systems are located in ravines formed in deep marine clay deposits. The landfills and the treatment systems are not lined because of the low permeability of the clay soil. Until recently, only a small fraction of the waste has been sorted for various recyclable materials, like paper, ferrous metals, glass, and plastics.

Table 9.1   Description of Esval and Bølstad Municipal Sanitary Waste Landfills and Natural Leachate Treatment Systems in Norway

General Information	Esval Landfill	Bølstad Landfill
Start operation	1972	1962
Landfill area (1997)	7.0 ha	4.5 ha
Landfill catchment area	7.5 ha	5.0 ha
Annual waste flow (1995)	50 000 t	7000 t
Total waste received	500 000 t (1 000 000 m³)	190 000 t
Landfill depth	25–30 m	20 m
Waste type	MSW	MSW
Annual mean precipitation	825 mm	785 mm
Mean air temperature	7°C	6°C
Mean air temperature January	–7°C	–5°C
Landfill sealing	Natural clay	Natural clay
Drainage system	Gravel ditches	Pipe
Leachate production, mean (range)	120 (30–500) m³/day	80 (30–300) m³/day
Leachate temperature January/June	1°/20°C (lagoon)	1°/20°C (lagoon)

Leachate Treatment Systems

Established	1993	1994
1. Stage — pretreatment	Anaerobic pond	None
Area (volume)	450 m²/650 m³	—
2. Stage — oxidation	Extended aeration lagoon	Extended aeration lagoon
Area/volume	2700 m²/4100 m³	1000 m²/2200 m³
Aeration equipment	3 × 3.5 kW floating propeller/mixers	2 × 4.0 kW floating propeller/mixers
3. Stage — sedimentation (S3)	Quiet zone in aerated lagoon	Sedimentation pond (140 m²)
4. Stage — filter (S4 and S5)	Two parallel SHF CW[a]	Four parallel SHF CW, mesocosmc system[c]
Area	2 × 400 m²	4 × (1 × 10) m²
5. Stage	FWS CW[b]	None
Area/volume	2000 m²/1000 m²	—
Total costs	U.S. $200 000	U.S. $150 000

[a] SHF CW — horizontal subsurface flow constructed wetland.

[b] FWS CW — free-water surface constructed wetland (not included in the monitoring program).

[c] Mesocosm pilot CWs, HLR 5 cm/day, monitored since June 1996.

9.2.2   System Description

Both treatment plants are similar with some variation in design features (Table 9.1). A common design feature is that leachates are treated in extended aeration lagoons, with aspirator propeller aerators/mixers, followed by sedimentation and subsurface horizontal flow (SHF) CWs. The wetland media consisted of washed gravel (10 to 20 mm diameter) and lightweight aggregates (Filtralite™, expanded clay aggregates manufactured by Norsk Leca, Inc.). A 10-cm layer of mulch was placed over this filter media. The wetlands were planted with *Typha latifolia*, *Phragmites australis*, and *Phalaris arundin-acea* which were available from a nearby natural wetland. It was intended that lagoon aeration would oxidize COD, NH_4–N, and Fe(II). The mean hydraulic retention time (HRT) for both lagoons was about 30 days, which provided a high buffering capacity to cope with the variable nature of leachate. Estimated HRT varied from 3 days, during wet periods with snowmelt and high precipitation, to 60 days, during dry periods with high evapotranspiration. The required dissolved oxygen levels in the ponds were maintained by adjusting the numbers of aerators operating, thus conserving energy.

There are four major differences between the two treatment systems. (1) Esval is a five-stage system, beginning with an anaerobic pond for pretreatment and ending with a free-water surface CW, whereas Bølstad has neither of the above features in its three-stage design. At Bølstad there is a small sedimentation pond after the aeration. At Esval, sedimentation takes place in a quiet zone of the lagoon. (2) The volume of Esval aeration lagoon is more than twice the size of Bølstad lagoon and sedimentation pond. (3) After aeration, the Esval lagoon captures effluent in two 400-m^2 SHF CWs; however, Bølstad uses four small (10 m^2) SHF CWs. The latter are part of a mesocosm pilot project, with a fixed hydraulic loading rate (HLR) of 5 cm/day for each wetland cell during the 1996–97 monitoring period. (4) The N:P ratio at Bølstad and Esval was about 300:1. To improve the nutrient balance and optimize microbial degradation processes and nitrification, 5 to 15 kg phosphoric acid (as 80% H_3PO_4) was added occasionally to the Bølstad lagoon during 1995–97 (see Figure 9.3).

9.2.3 Monitoring, Sampling Program, and Chemical Analyses

Random and systematic variations of water quality can be expected in landfill leachates as a result of changing conditions in the landfill and the weather. Systematic variations may be either trends or cyclic variations, or a combination of the two. The time of sampling is important for cyclic variations and to detect minimum and maximum concentrations. The study monitoring strategy covered (1) the statistical parameters that characterize concentration central tendency and its variability, e.g., mean, median, and standard deviation of the individual treatment units; (2) potential concentration trends; (3) abnormal seasonal situations like flood peaks and warm and dry periods; and (4) loading rates for planning and design purposes. Effluents from different treatment stages — raw leachate (S1), aeration and sedimentation lagoon (S3), and effluent CW composite sample (S4/5) — have been monitored with a sampling frequency of 2 to 10 weeks. The water was taken as both grab (spot) samples and flow-proportional, composite water samples (S1).

Variables measured included water flow, water temperature, pH, electrical conductivity, BOD_7 (manometric), CODcr (spectrophotometric), TOC (combustion), SS (filtration), NH_4–N (autoanalyzer), NO_3–N (autoanalyzer), Tot-N (autoanalyzer), PO_4–P (FIA/ion chromatograph), Tot-P (FIA/ICP-AES), chloride (ion chromatograph), alkalinity, heavy metals (ICP-AES/EAAS): Tot-Fe, Cd, Cr, Cu, Mn, Zn, Pb, and Ni. The samples were filtered (0.45 μm) and acidified for selected parameters in the laboratory 1 to 2 days after sampling. The BOD_7 samples were frozen after sampling. The chemical analysis were performed 1 to 4 weeks after sampling at an accredited laboratory according to Norwegian standards for water analysis with some minor modifications (see Mæhlum, 1998, for additional information). Temperature and oxygen in the lagoon effluent were measured continuously. Leachate flow was measured with a flow recorder in a V- or H-notch wire, and in periods with defective flow recorder at S1 only weekly point measurements were conducted at Esval.

9.2.4 Statistical Analysis

Influences of water temperature, input water flow rates, and seasonality on the removal efficiency of different water quality parameters in the two lagoons and CWs were analyzed. Treatment efficiency was tested for significant differences during low temperatures (0 to 4°C), normal temperatures (4 to 14°C), and high temperatures (≥14°C). Similar analyses were performed for three input wastewater flow rate situations (low flow, moderate flow, and high flow) and for four season variation: December to February, March to May, June to August, and September to November. Treatment efficiency was calculated as the pairwise difference (in percent of influent concentrations) between the momentary grab samples between two stages (mean and median). Exploratory analysis indicated that nonparametric tests were most suitable for these data, because of occasionally low number of observations, non-Gaussian distributions, and nonequal variances.

9.3 RESULTS AND DISCUSSION

Table 9.2 shows mean and median composition, standard deviation, minimum, and maximum values for influent leachate characteristics at Esval and Bølstad landfills. Table 9.3 presents a summary of treatment efficiency in each treatment stage and total removal efficiency at Esval and Bølstad. Alternative calculation methods are presented: concentration mean values with and without dilution, based on changes in chloride concentrations, and pairwise difference between momentary grab samples (mean and median). Table 9.4 presents analyses of variance for treatment efficiency in the aerated lagoons and CWs at different seasons, intervals of water temperature, and input water flow rates. Figures 9.1 and 9.2 present changes in influent and effluent concentrations of COD, BOD_7, TOC, SS, electrical conductivity, chloride, Tot-N, NH_4–N, NO_3–N, Tot-Fe, pH, and input leachate flow rate at Esval and Bølstad, respectively. Figure 9.3 presents lagoon temperature and the effect of P addition to the lagoon with concentrations of NO_3–N and NH_4–N in raw leachate and effluent from the sedimentation stage and SHF CWs of the Bølstad treatment plant. Figure 9.4 presents treatment efficiency in the lagoon for selected parameters from Table 9.4 at three input leachate flow rate intervals and three water temperature intervals.

Both Esval and Bølstad leachates can be characterized as low strength (Table 9.2) because of dilution of surface water runoff from the catchment area into the leachate collection system. Leachate strength at Esval has increased considerably since 1993, as indicated by an increase in COD concentration from 500 to 6000 mg/L (1996, Figure 9.1). Nitrogen and iron also increased in the same period. The higher leachate strength is attributed to the expansion of the landfill area, with a larger proportion of acetogenic leachate from the new part of the landfill. The Tot-P content was very low and varied from 0.2 to 6.1 mg/L (Table 9.2). Leachate strength from both sites changed markedly in relation to rainfall events, snowmelt, and dry periods during warm summers and cold winters with snow and ice.

Cold periods (air temperature <0°C, water temperature <4°C) had a notable effect on effluent quality, such as the reduced oxidation of organic matter, NH_4–N, and Fe in both systems. The effect of water temperature on treatment efficiency in S1 and S3 due to water temperature is statistically significant for COD and Tot-N at Bølstad, for BOD at Esval and for Fe at both landfill lagoons (Table 9.4). Some of the most striking differences are illustrated in Figure 9.4A. For example, treatment efficiency for Fe was significantly higher at high water temperatures (>14°C) than at moderate (4 to 14°C) and low temperatures (0 to 4°C). The same holds true for BOD at Esval and for Tot-N at Bølstad. There was no significant difference in treatment efficiency between low and moderate water temperatures.

Similar results were obtained in the analyses for seasonality and treatment efficiency (Table 9.4). Generally, summer and autumn showed better treatment efficiency than winter and spring, particularly for BOD and Fe at Esval (significance $p < 0.05$). The Wilcoxon/Kruskal–Wallis test detected statistically significant differences in treatment efficiency at various input wastewater flow rate intervals for nitrogen and Fe (Table 9.4). The high treatment efficiency of Fe at low-input wastewater flow rates and the low treatment efficiency at high-input wastewater flow rates (Figure 9.4B) suggests the effect of dilution. Raw leachates were diluted with surface water as a result of flooding during spring and fall. During these periods, the removal efficiency in general was often low (<30%), probably because of short biological solids detention time (sludge age), where bacteria are removed faster than they can reproduce.

Lagoon treatment efficiency for Tot-N without correction was in the range of 30 to 40% (Table 9.4). In the Bølstad lagoon, 50 to 99% nitrification was achieved when water temperatures were >10°C (Figure 9.3). At water temperatures <5°C, nitrification was low even with the addition of P. The water temperature was shown to have a significant effect on the NO_3–N concentrations. More precisely, a Kruskal–Wallis test on differences in NO_3–N concentrations at three water temperature intervals (0 to 4, 4 to 14, >14°C) at Bølstad revealed a p-value less than 0.0001 (Table 9.4A).

Table 9.2 Mean and Median Composition, Standard Deviation (SD), Minimum and Maximum Concentrations of Raw Leachates at Esval (1993–97) and Bølstad (1995–97) Municipal Sanitary Landfills in Norway

	Esval						Bølstad					
	Mean	Median	SD	Min	Max	n	Mean	Median	SD	Min	Max	n
pH	—	7.1	—	6.3	7.8	47	—	7.5	—	6.9	8.0	52
El. cond. (mS/m)	641	662	181	182	1013	44	290	294	78	105	416	44
COD, mg/L	2267	1710	1686	540	6770	48	311	314	82	155	500	37
BOD_7, mg/L	735	307	690	125	2515	23	78	63	53	7	201	29
TOC, mg/L	678	329	692	41	2130	31	125	124	60	52	331	30
SS, mg/L	147	120	130	22	790	44	275	180	261	11	900	19
Tot-P, mg/L	1.1	0.9	0.9	0.2	6.1	47	0.4	0.3	0.3	0.02	1.1	29
PO_4–P, mg/L	0.14	0.10	0.12	0.03	0.5	21	0.19	0.05	0.2	0.01	0.8	27
Tot-N, mg/L	219	223	86	34	412	48	126	131	47	34	257	45
NO_3–N, mg/L	0.2	0.2	0.1	0.02	0.4	22	1.5	0.5	3.0	0.1	16	37
NH_4–N, mg/L	179	179	80	30	397	48	96	100	40	27	171	45
Cl, mg/L	964	1000	296	250	1460	49	220	212	80	73	361	39
HCO_3^-, mg/L	2868	2727	934	610	4349	28	1547	1452	861	542	4331	15
Na, mg/L	624	619	191	164	885	19	185	169	67	73	303	24
K, mg/L	241	212	97	52	452	19	149	135	50	51	205	13
Ca, mg/L	432	338	225	139	917	19	188	163	36	150	258	13
Mg, mg/L	90	84	33	22	181	19	53	44	17	20	73	13
Tot-Fe, mg/L	45	36	31	4	131	44	32	24	35	1.7	150	42
Mn, mg/L	4.4	4.2	2.0	1.8	10	19	1.9	1.6	0.7	0.6	2.4	21
Cu, µg/L	20	15	17	0.1	70	20	10	4	8	0.5	30	20
Zn, µg/L	312	140	352	20	1190	19	53	60	34	20	135	20
Pb, µg/L	8.1	5.0	7.4	1.0	26	11	5.0	1.0	7.0	1.0	35	27
Cd, µg/L	0.9	1.0	0.9	0.01	3.2	10	0.3	0.4	1.0	0.1	0.6	27
Ni, µg/L	39	30	18	3.0	83	20	23	23	12	9	50	18
Cr, µg/L	39	29	22	20	100	18	15	8	5	7	20	9
Ba, µg/L	307	280	82	201	466	17	165	132	34	127	210	9
Hg, µg/L	0.2	0.2	0.1	0.1	0.4	6	0.3	0.1	0.3	0.1	1.3	14

Note: El. cond. = electrical conductivity.

Table 9.3 Treatment Efficiency (%) at Esval (1993–1997) and Bølstad (1995–1997) Integrated Landfill Leachate Systems with Reference to Figures 9.1 and 9.2

Parameter[a]	Concentration Mean				Pairwise Mean and Median							
	S1–S3,[a] nc[b]	S3–S4/5, nc	S1–S4/5 nc	S1–S4/5 c[b]	S1–S3 nc Mean	n	SD	Median[c]	S3–S4/5 nc Mean	Median[c]	n	SD
A. Esval Treatment Plant												
COD	60	4	61	38	56	45	48	69***	24	26***	39	16
BOD$_7$	74	41	85	76	65	21	39	77***	42	48***	18	22
TOC	59	10	63	41	50	29	30	54***	19	20***	24	16
SS	35	45	66	45	7.2	42	114	51**	45	59***	36	48
Tot-N	35	7	39	2	27	45	52	44***	13	11***	39	13
NH$_4$-N	38	13	46	14	33	45	35	45***	14	12***	39	16
Fe	78	8	80	68	73	41	37	86***	-25	40+	37	131
Zn	68	39	81	69	—	—	—	—	—	—	—	—
Chloride[c]	29	11	38	0	26	45	36	42***	11	11***	38	15
HCO$_3^-$	50	13	56	30	—	—	—	—	—	—	—	—
B. Bølstad Treatment Plant												
COD	41	34	61	37	37	36	24	40***		+	5	
BOD$_7$	61	11	65	44	32	24	89	66***		—	1	
TOC	48	25	61	37	40	28	26	45***		*	6	
SS	71	45	84	74	56	19	32	58***		+	4	
Tot-N	37	45	65	45	31	44	28	37***		*	6	
NH$_4$-N	60	88	95	92	50	44	44	55***		+	8	
Fe	79	80	96	93	59	42	43	73***		+	6	
HCO$_3^-$	58	50	79	66	—	—	—	—		—	—	
Cl-	27	16	38	0	21	38	26	26***		+	4	

[a] S1 raw leachate; S3, effluent extended aerated and sedimentation lagoon; S4/5, effluent of subsurface horizontal flow constructed wetlands (composite sample of gravel filter and lightweight aggregates).

[b] Figures 9.1 and 9.2 present the chloride concentrations, which indicate the level of dilution (dilution factor 1.6).

[c] Assumption of same water in as out and no hydraulic retention time.

* Statistical significance levels according to Wilcoxon Matched-Paired Signed-Ranks test: *** ($p < 0.001$), ** ($p < 0.01$), * ($p < 0.05$), + ($p > 0.05$).

Note: Treatment efficiency in stages S1–S3, S3–S4/5, and S1–S4/5 were calculated based on concentration mean values with (c) and without (nc) correction due to dilution, and alternative as pairwise mean and median difference between the momentary grab samples at two stages. Number of observations for the pairwise observations and standard deviation (SD) are also presented.

Table 9.4 Analysis of Variance for Treatment Efficiency in (A) Lagoon (S1–S3) and (B) Constructed Wetlands (S3–S4/5) at Esval and Bølstad MSW Landfill Leachate at Different Seasons,[a] Intervals of Water Temperature,[b] and Input Leachate Flow Rates[c]

	Water Temperature (3 intervals)		Input Wastewater Flow Rates (3 intervals)		Quarter (4 alt seasons)	
	Esval	Bølstad	Esval	Bølstad	Esval	Bølstad
A. Lagoon						
COD	0.06	0.01	0.10	0.22	0.003	0.01
BOD_7	0.009	0.09	0.23	0.33	0.002	0.33
Tot-N	0.54	0.006	0.03	0.003	0.45	0.004
Fe	<0.001	0.001	0.02	0.006	0.005	0.005
B. Constructed Wetland						
COD	0.84		0.42		0.28	
BOD_7	0.39		0.93		0.51	
Tot-N	0.95		0.05		0.56	
Fe	0.03		0.51		0.10	

[a] Seasons (quarter): Dec–Feb, Mar–May, Jun–Aug, Sep–Nov.

[b] Water temperature intervals: 0–4, 4–14 and >14°C.

[c] Input leachate flow rates; Esval: <0.9, 0.9–2.0, >2.0 L/s (wetland HLR <10, 10–22, >22 cm/day). Bølstad: 0–0.4, 0.4–1.0, >1.0 L/s.

Notes: Values in table represents *p*-values for Wilcoxon/Kruskal Wallis Rank Sums test. Statistical significances marked in bold font ($p < 0.05$). Statistical analysis of the influence of temperature on treatment efficiency at the CWs in Bølstad were not possible to calculate statistical due to the short monitoring period and low number of observations.

Low Tot-N removal at Esval and restricted nitrification in the lagoon seemed to be due to insufficient oxidation and lack of phosphorus (the P:COD ratio was less than 1:5000). Nitrification consumes alkalinity, but lack of HCO_3^- (Tables 9.2 and 9.3) or pH (Figures 9.1 and 9.2) did not restrict nitrification in the lagoons. In theory N removal (nitrification and denitrification) is expected in extended aerated lagoons where anaerobic, anoxic, and aerobic zones prevail. In addition to the influence of low temperatures on the treatment processes, aerator malfunction at the Esval system from December to March may explain low performance during cold periods. The floating 3.3 kW aerators were not suited to low temperatures with snow and ice, and only one aerator was totally operational (Figure 9.1). Energy consumption for the aerators was on average 4.5 and 8.5 kWh, respectively, at Esval and Bølstad in the monitoring periods.

Depending on the calculation methods, treatment efficiency at Esval CWs was in the range of 4 to 48% for organic matter parameters and 7 to 14% for nitrogen parameters (Table 9.3). At the wetland stage in Esval, there was much less statistical evidence of influence of water temperature, water flow rates, and seasonality on the treatment efficiency (Table 9.4). Only Fe showed a statistically significant temperature influence ($p < 0.05$), which was related to lower removal rates for Fe during high water temperatures compared with low or moderate ones. This could be explained by lower effluent concentration from the lagoon during the warmer periods. Support for this could also be found in the indications of lower treatment efficiency for Fe in the CWs at low flows (normally in summer) compared with normal or high flow situations (particularly cold snowmelt water).

Preliminary results from the Bølstad mesocosm SHF CWs (Table 9.3, Figure 9.3) with HLR 5 cm/day indicate 20 to 40% Tot-N removal, which can be explained by nitrate reduction (denitrification), and 25 to 35% removal of TOC and COD. There was no significant difference in treatment efficiency between the lightweight aggregate (S4) and the gravel (S5) at Esval and Bølstad CWs relative to the removal of Tot-N, COD, BOD_7, TOC, or Tot-Fe (data not shown). Problems with media clogging appeared at Esval after 3 years because of insufficient oxidation and sedimentation

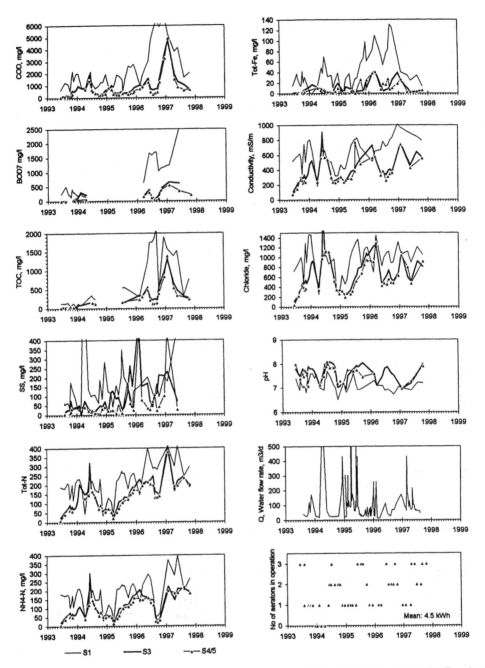

Figure 9.1 Temporal variability in concentrations of COD, BOD$_7$, TOC, SS, Tot-N, NH$_4$–N, Tot-Fe, electrical conductivity, chloride, pH, flow rate, and number of aerators in operation at Esval treatment plant during the investigation period July 1993 to December 1997: raw leachate (S1-solid), effluent aerated lagoon/sedimentation pond (S3-bold), effluent SHF-CWs, gravel and lightweight aggregates (S4/5).

prior to the CWs and overloading of the filters during flood peaks (HLR > 30 cm/day). Hydraulic problems due to frost penetration in the CWs were not observed. SHF CWs can remove several constituents in the leachate (organic matter, nitrogen, heavy metals) provided pretreatment involves sufficient oxidation and sedimentation. The SHF CWs tested at both Esval and Bølstad seemed most efficient for polishing during dry and warm periods with HLR < 5 cm/day.

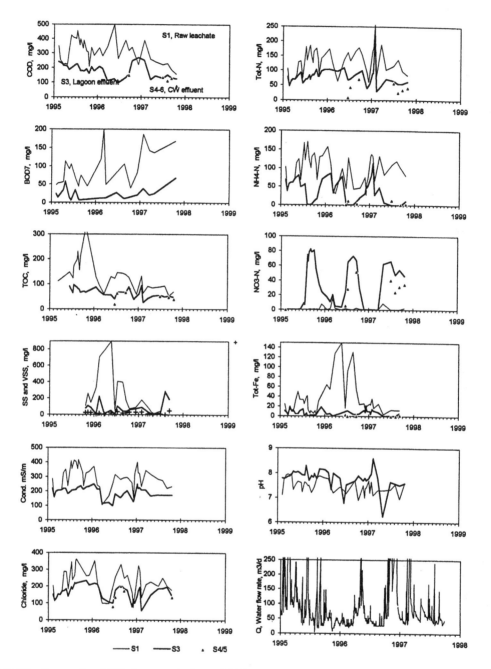

Figure 9.2 Temporal variability in concentrations of COD, BOD₇, TOC, SS, electrical conductivity, chloride, Tot-N, NH₄–N, NO₃–N, Tot-Fe, pH, and flow rate at Bølstad treatment plant during the investigation period January 1995 to December 1997: raw leachate (S1-solid), effluent aerated lagoon/sedimentation pond (S3-bold), effluent SHF mesocosm CWs, gravel and lightweight aggregates (S4/5). S4/5 were monitored since June 1996.

Depending on the calculation method (Table 9.3), total treatment efficiency (lagoon and CWs) for both sites were in the range of 38 to 78% COD, 40 to 80% BOD₇, 40 to 65% TOC, 2 to 65% Tot-N, and 68 to 96% Fe. Removal of other heavy metals, like Zn, was in the same range as for Fe (Table 9.3). The systems satisfied the treatment requirement — 75% COD removal, 90% Fe removal, 40% Tot-N removal, and 50% nitrification — only during periods with high temperature

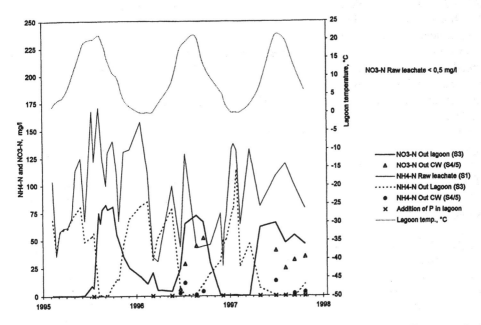

Figure 9.3 Lagoon temperature and concentrations of NO_3–N and NH_4–N in influent and effluent aeration/sedimentation lagoon and CW mesocosm system of Bølstad treatment plant as a response to changing temperatures and addition of supplemental phosphorus.

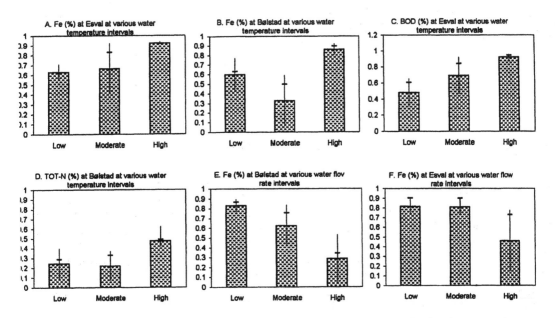

Figure 9.4 Treatment efficiency in lagoon for selected parameters at Bolstad and Esval at three water-temperature intervals and three water flow rate intervals (see Table 9.4 for additional information about the intervals). The figures show mean (bar), standard deviation, and median.

(>14°C), low input water flow rates (HRT > 30 days in the lagoons and >7 days in the SHF CWs), and with addition of sufficient oxygen and phosphoric acid. Since the fraction of harmless humic and fulvic acids contribute significantly to the lagoon effluent COD (data not shown), the relevance of this parameter is questionable in leachate treatment. Further BOD measurements can be hindered by metals, organic micropollutants, and high ammonia and chloride concentrations. Therefore, TOC may be better suited as a parameter for organic degradation.

Reduction of chloride and electrical conductivity in both systems indicate dilution of leachate by surface water and groundwater inputs from the catchment area (Figures 9.1 and 9.2, Table 9.3). Calculating the treatment efficiency without control of the hydrology and water balance should be interpreted with caution, due to (1) the large catchment area, (2) the large surface area of the treatment systems, and (3) the variation in hydraulic loading.

Based on the experiments described, it is expected that purification performance in the studied systems and other leachate lagoon/CWs systems in cold climates could be improved by the following measures:

1. Diverting surface water entering the landfills and lagoons would increase the sludge age in the lagoons and the HRT in all treatment stages.
2. Intermittent operation of the aerators and intermittent loading (rest-and-dry periods) of the SHF CWs would improve aerobic and anaerobic degradation processes.
3. Use of aeration equipment tolerant of snow and ice conditions and varying leachate strengths would be beneficial.
4. Monthly addition of phosphoric acid in the extended aeration lagoons is required.
5. Insulation of the aerobic pretreatment units are needed. In general, insulated sequenced batch reactor (SBR) systems may be better suited to aerobic leachate treatment than open lagoon systems in cold climatic regions. Future leachate treatment systems in cold climates should make use of heat in undiluted raw leachate and conserve it in the treatment system. Additional heat from the landfill may be transferred to reactors and filters.
6. Improvement of sedimentation prior to CW units should be done by providing, for example, flocculation/chemical precipitation.
7. The length of the SHF CW treating leachate should be shorter than the width to reduce clogging at the inlet zone, and the HLR is recommended to be <5 cm/day.
8. Use of free-water surface CWs, separately or in combination with subsurface flow wetlands, to prevent clogging problems is recommended.

9.4 CONCLUSIONS

Extended aeration lagoons appear to be a cost-effective first stage for landfill leachate treatment in cold climatic regions, although annual high treatment efficiency for organic matter, nitrogen, and iron was not obtained in this study. Significant differences in lagoon treatment efficiency were detected for Fe, COD, and nitrogen, because of differences in water temperatures and flow rates. The variations in NO_3–N concentrations in the lagoon correlated positively with the temporal variability in water temperature. High ammonia removal (>75%) can be achieved at temperatures >10°C by adding phosphoric acid to the lagoon, continuously or one to two times per month. The HRT should be maintained >20 days to keep nitrifying bacteria in the lagoon during periods of low temperatures. Extended aeration lagoons, in conjunction with SHF CWs, can improve effluent quality if the wetland HLR is <5 cm/day. In this study, clogging in the wetland filter units appeared after 3 years because of insufficient pretreatment prior to wetlands and overloading of the filters during flood peaks. High removal of biodegradable matter, NH_4–N, and suspended solids is recommended before polishing pretreated leachate in natural systems using filter media. Use of free-water surface CWs, separately or in combination with subsurface flow wetlands, appears best suited as second- or third-stage leachate treatment to minimize clogging problems.

ACKNOWLEDGMENTS

Monitoring of the Esval and Bølstad landfill leachate treatment plants was funded jointly by the Natural Systems Technology for Wastewater Treatment (NAT) research program (1993 to 1997): Department of the Environment, Centre for Soil and Environmental Research (Jordforsk), the Norwegian Research Council, Ås and Nes municipalities, the County Governor of Oslo and Akershus and Norsk Leca. Water quality analyses were carried out by Jordforsk (Landbrukets Analysesenter) and the Department of Soil and Water Sciences (Agricultural University of Norway). Thanks are due to Tong Zhu for laboratory support and Øystein Johansen, Geir Tveiti, Trond Paulsen, Kristin Handeland, Katrine Lynne, Astrid Eikeland, and Anne Charlotte Moen, who provided technical assistance.

REFERENCES

Handeland, K., 1997. Sigevann fra kommunale fyllplasser. En studie av hydrologi og renseevne ved Bølstad renseanlegg i Ås, Cand. agric. thesis. Agricultural University of Norway, Ås (in Norwegian).

Kadlec, R. H. and Knight, R. L., 1996. *Treatment Wetlands*, CRC Press, Boca Raton, FL.

Maris, P. J., and Harrington, D. W., 1984. Leachate treatment with particular reference to aerated lagoons, *Water Pollution Control* 83:521–538.

Mæhlum, T., Haarstad, K., and Kraft, P., 1995. On-site treatment of landfill leachate in natural systems, in Christensen, T. H., Cossu, R., and Stegmann, R., Eds. *Proceedings of Sardinia 1995, Fifth International Landfill Symposium*, Cagliari, Italy, 1:463–468.

Mæhlum, T. and Haarstad, K., 1997. Leachate treatment in pond and constructed wetlands in cold climate. *Proceedings at Sardinia 1997*, Fifth International Landfill Symposium, Cagliary, Italy, CISA, 11:337–344.

Mæhlum, T., 1998. Cold-climate constructed wetlands' Aercbic pretreatment and subsurface flow wetlands for domestic sewage and landfill leachate purification. Doctor Scientarum Thesis. 1998–9. Agricultural University of Norway.

Martin, C. D. and Moshiri, G. A., 1995. Nutrient reduction in an in-series constructed wetland system treating landfill leachate, *Water Science and Technology* 29:267–272.

Metcalf & Eddy, Inc., 1991. *Wastewater Engineering, Treatment, Disposal, and Reuse*, 3rd ed., revised by G. Tchobanoglous and F. L. Burton, McGraw-Hill, New York.

Robinson, H. D., 1990. On-site treatment of leachates from landfilled wastes, *Journal of the Institution of Water and Environmental Management* (London) 4:78–89.

Robinson, H. D., 1993. The treatment of landfill leachates using reed bed systems, in Christensen, T. H., Cossu, R., and Stegmann, R., Eds., *Proceedings of Sardinia 1993, Fourth International Landfill Symposium*, Cagliari, Italy, 1:907–922.

Robinson, H. D. and Grantham, G., 1988. The treatment of landfill leachates in on-site aerated lagoon plants: experience in Britain and Ireland, *Water Resources* 22:733–747.

Robinson, H. D. and Maris, P. J., 1985. The treatment of leachates from domestic waste in landfill sites, *Journal Water Pollution Control Federation* 57:30–38.

CHAPTER 10

Treatment of Leachate from a Landfill Receiving Industrial, Commercial, Institutional, and Construction/Demolition Wastes in an Engineered Wetland

Majid Sartaj, Leta Fernandes, and Normand Castonguay

CONTENTS

ABSTRACT: The Huneault Landfill, located in Gloucester, Ontario, has been in operation since the early 1960s. Since 1971, this landfill has received industrial, commercial, and institutional (IC&I) wastes, as well as construction/demolition materials. The landfill produces an estimated 57 000 m^3/year of leachate, which is distinctly different from most landfill leachates in Ontario and which is attributed to the type of waste landfilled at this site. An engineered wetland system consisting of a vertical flow peat filter followed by three surface-flow (SF) wetlands was selected as the leachate treatment method before discharging this effluent into the receiving environment. The engineered wetland treatment system came into operation in August 1995. Leachate is collected in two retention ponds and then pumped and distributed over the peat filter by way of spray irrigation. The filtrate is then collected at the bottom by a network of subsurface drain systems that flow into the three SF wetlands. The peat filter has a surface area of 5580 m^2 and depth of 1.4 m. The three engineered SF wetland cells operate in series with a total surface area of 4374 m². Field monitoring of the engineered wetland system demonstrates that this system is very effective in treating landfill leachate. Removal efficiencies for some of the selected parameters were 95, 58, and 90% for NH_3, total organic carbon (TOC), and boron, respectively.

10.1 INTRODUCTION

Landfill facilities are potential threats to groundwater and surface water quality, the primary concern being the production of leachate (Howard, 1997). The current land disposal practice of engineered landfilling consists of two main components, an impermeable or low-permeable barrier at the base of the waste pile to prevent or reduce leachate migration through the base and an underdrain system to collect the leachate generated within the landfill. Part of the leachate may be recirculated within the waste pile to enhance stabilization of the organic contents of the waste (Townsend et al., 1996). The excess leachate could be highly toxic, and, if discharged directly into water bodies, it could impact aquatic life and degrade water quality (Cameron and Koch, 1980). Many landfill operators are now considering nonconventional systems such as constructed wetlands that can address their leachate management problems. Wetlands have been identified as a promising technology for treatment of a variety of waste waters, the main attractions being much lower capital, operating, and maintenance costs. The operating costs of engineered wetlands are very low because they are based on technologies that do not use expensive energy and chemical inputs and require low maintenance. They also have the potential for long-term performance without the need for modifications or alterations (Pries, 1994).

Both natural and constructed wetland systems have been used to treat a variety of wastewaters including agricultural and surface mine runoff, irrigation return flows, secondary-treated sewage effluents, leachates, urban storm water, and other sources of water pollution (Pries, 1994). U.S. EPA (1987) reported the use of constructed wetlands treating acid mine runoff in more than 100 sites. The use of constructed, rather than natural, wetlands is generally preferred since all natural wetlands are considered part of natural water resources and have to comply with the water quality requirements of regulatory agencies. Other advantages of constructed wetlands include a greater degree of control of substrate, vegetation types, flow characteristics, flexibility in sizing, and the potential to treat more wastewater. Constructed wetlands are categorized into two main groups: surface flow (SF) and subsurface flow (SSF). A third type of wetland design, which is at present evolving to allow manipulation of the flow, is vertical flow subsurface systems (VFSS). In VFSS wetlands, water flows vertically through the porous media and is collected at the base.

The performance of wetlands is measured by removal efficiencies of contaminants. Wetland systems can significantly reduce biochemical oxygen demand (BOD_5), suspended solids (SS), nitrogen, trace organics, heavy metals, and pathogens. Removal efficiencies in the range of 70 to 90% for BOD_5 and SS, 60 to 86% for nitrogen, and between 97 and 99% for Cu, Zn, and Cd were observed (Gersberg et al., 1985; Pries, 1994).

This paper reports on the design and implementation of a constructed wetland for treatment of leachate produced at the Huneault Landfill, in Gloucester, Ontario. The physical characteristics of the landfill site, physicochemical characteristics of the leachate produced, and the design of the system, as well as its performance, are presented and discussed.

10.2 SITE CHARACTERISTICS OF THE ENGINEERED WETLAND

The Huneault Waste Management facility, occupying an area of 40 ha located in the City of Gloucester, Ontario, has been in operation and serving the Regional Municipality of Ottawa-Carleton since the early 1960s. Since 1971 this landfill has been operating under a certificate of approval from the Ministry of Environment and Energy (MOEE) to receive construction/demolition waste, industrial/commercial/institutional (IC&I) waste, as well as miscellaneous inert materials. In accordance with the certificate of approval, there is no domestic (putrescible) or industrial liquid waste being discharged into this landfill. In 1996, a total of 177,169 t of materials entered the landfill site of which 50.5% was landfilled and the rest was either converted or used as final cover. The average annual precipitation for the Ottawa region is 917 mm. The average annual water surplus (i.e., precipitation minus evaporation) is 320 mm. The water surplus provides an estimate of the amount of water available for infiltration into the site and for surface runoff during a 12-month period. The average temperatures vary from a low of –10.6°C in January to 20.7°C in July with an annual average of 6°C.

10.3 CHARACTERISTICS OF THE HUNEAULT LANDFILL LEACHATE

Leachate composition varies significantly among landfills, depending on waste composition, waste age, and landfilling technology. Leachate generated at the Huneault Landfill has been monitored on a continuous basis since 1990. The leachate characteristics of the Huneault Landfill and other Ontario landfills are presented in Table 10.1. Concentrations of some selected contaminants were on average 16.5 mg/L of ammonia, 40 mg/L of BOD_5, 10.2 mg/L of boron, 2.8 mg/L of iron, 0.03 mg/L of lead, 0.04 mg/L of zinc, and 51 mg/L of total suspended solids (TSS). This landfill is unique in the sense that the collected leachate is distinctly different from most other landfill leachates in Ontario, with most constituents having much lower concentrations. The average concentrations of ammonia and TSS are almost 10 times lower and the average BOD_5 is about 100 times lower than the other Ontario landfills.

Table 10.1 Characteristics of Leachate at Huneault Landfill and Other Ontario Landfills

Parameter	Huneault Landfill			Ontario Landfills		
	Min.	Max.	Avg.	Min.	Max.	Avg.
Ammonia (mg/L)	0.1	31	16.5	7.6	1820	175
BOD_5 (mg/L)	2	173	39.9	1	66000	4975
Copper (mg/L)	0.002	4.5	0.43	0.007	7	0.045
Boron (mg/L)	0.4	16.9	10.2	0.8	52	10.4
Iron (mg/L)	0.003	8.09	2.8	0.01	1300	58
Conductivity (ms/cm)	669	4540	3533	475	26100	6088
TDS (mg/L)	1920	2572	2234	196	9030	4327
Sodium (mg/L)	54	514	369	13.8	16000	936
Manganese (mg/L)	0.13	8.56	1.67	0.030	793	3.54
Organic N (mg/L)	1.27	37.6	12.54	—	—	—
COD (mg/L)	69	600	289	1.0	47300	7855
TSS (mg/L)	2.0	130.0	51.2	3.0	8130	445
Chlorides (mg/L)	76.5	533	366	0.4	12000	270
Lead (mg/L)	0.0	0.28	0.03	0.001	2.1	0.07
Zinc (mg/L)	0.01	0.13	0.03		16	1.42

Data from Howard and Livingston (1997) and Castonguay (1997).

10.4 DESIGN OF THE PEAT FILTER AND WETLAND SYSTEM

The Huneault Landfill is situated upgradient from a natural wetland, the Mer Bleue Bog. A constructed wetland system was considered to be an ideal buffer for treated effluent prior to its introduction into a natural environment. Therefore, the goal of the constructed wetland at the Huneault Landfill was to obtain an effluent of a quality acceptable to be discharged off site and of

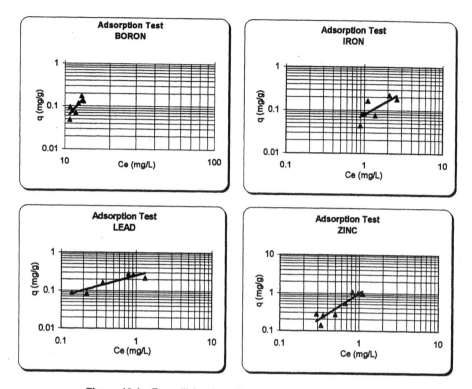

Figure 10.1 Freundlich adsorption isotherms for boron, iron, lead, and zinc.

a volume sufficient that the remaining untreated leachate could be easily handled with the other water management systems such as irrigation. It was decided that as much as one third of the annual volume of leachate generated, estimated at 1.8 L/s, be treated for off-site disposal, i.e., 0.6 L/s or 18,900 m³/year.

The constructed wetland is expected to be operational only during the frost-free period of the year. For the Ottawa area the average date of the last frost in spring is May 11 and the first frost in fall is October 1, giving an average frost-free period of 142 days (Castonguay, 1997). Therefore, the design influent leachate rate is calculated as follows:

$$Q_i = 18,900 \text{ m}^3/142 \text{ days} = 133.1 \text{ m}^3/\text{day}$$

With limited surface area available for an engineered wetland, it was established early in the design stage that the leachate should first flow through a porous media to promote treatment by way of filtering, adsorption, and biodegradation. Organic soils such as peat provide great potential for contaminant removal. Peat is a natural, inexpensive, and widely available substance. In addition, it has a large specific area and is highly porous, which makes it a good adsorbent material (McLellan and Rock, 1988). Peat has proved to be effective in removing Pb, Cd, Zn, and Cr from raw wastewater and Fe, Mn, Zn, and K from landfill leachate (Cameron, 1978; Zhipie et al., 1984). A set of batch adsorption tests with leachate spiked for Fe, Pb, and Zn was carried out to assess the adsorption of peat from a local source. Freundlich isotherms were developed for boron, iron, lead, and zinc, as shown in Figure 10.1. The solid phase concentration (q) is in the order of 0.1 (mg/g) for boron, iron, and lead and greater than 0.1 for zinc. This means that each gram of peat is capable of removing at least 0.1 mg of these metals. For the most conservative design, i.e., for boron, which has the highest concentration with an average concentration of 10 mg/L and an estimated volume of the annual leachate production of 18,900 m³, the peat requirement will be as follows:

Peat requirement = (18,900 m³/year × 1000 L/m³ × 10 mg/L)/ 0.1 mg/g of peat

= 1.89 × 10⁹ g of peat per year (dry basis)

= 1.89 × 10³ t of peat per year (dry basis)

The design of the engineered wetland system consists of four cells, the first one comprising the peat filter serving as VFSS system and the other three cells serving as SF wetlands. As shown in Figure 10.2, untreated leachate is collected in the two existing ponds (ponds 1 and 2) and then pumped and distributed over the peat filter by way of spray irrigation. The filtrate is then collected at the bottom of the peat filter by a network of subsurface drains that flow into the engineered wetlands as surface water. The natural topography of the constructed wetlands site slopes downward toward the south, with a relief of approximately 2 m between the northwest and southeast corners.

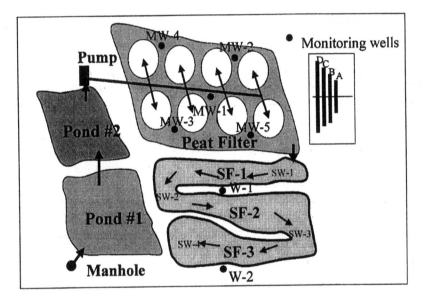

Figure 10.2 Schematic diagram of the engineered wetland.

The peat filter has an average length and width of 103 and 57 m, respectively, providing a surface area of 5580 m². The peat filter is underlain by native clay soil. A pervious media consisting of native sand and a series of perforated drain pipes were placed over the base of the peat filter in order to collect the leachate at the base. The total depth of the peat filter is approximately 2.0 m, the peat layer is 1.4 m, and the sand layer is 0.6 m. The drainage pipes used were 100-mm-diameter perforated agricultural drainage pipe wrapped in a geotextile filter cloth and spaced 4.6 m apart. Peat obtained from a local source in eastern Ontario was placed over the sand layer. A total of 7,725 t of peat was used for the construction of the peat filter. The hydraulic conductivity of the peat measured at 15 observation wells installed in the peat filter ranges from 1.06 × 10⁻⁵ to 1.97 × 10⁻³ cm/s and averages 3.0 × 10⁻⁴ m/sec. The upper layer of the peat generally has a higher hydraulic conductivity averaging 7.65 × 10⁻⁴ cm/s. A very low hydraulic conductivity can result in the leachate backing up above the peat filter, while in the case of a high hydraulic conductivity the leachate passes through the peat filter very quickly without enough contact time with the peat.

The three engineered SF wetland cells operate in series and range in size from 1209 to 1595 m², for a total of 4374 m². The total length of SF wetlands is 268 m and the average channel width is 12.7 m. The presence of clay soil at shallow depths and the availability of clay for the construction of perimeter berms made it possible to design the constructed wetlands to operate in series and to enhance biological processes. Such a design should also assist in reducing problems with channelization and excessive infiltration into the permanent pools. Water level control structures ensure that a minimum water level is maintained in each cell through the operating period of the SF wetlands. The water channels and shallow pools have a combined design volume of 705 m³, allowing for a retention time of 5.3 days at a design flow rate of 133 m³/day (1.5 L/s). The physical characteristics of the constructed Huneault wetlands are summarized in Table 10.2.

Table 10.2 Summary of Physical Characteristics of the Constructed Huneault Wetland

Cell No.	SF 2	SF 3	SF 4	Total SF
Surface area (m²)	1570.4	1595.0	1209.0	4374.40
Length, L, (m)	77.1	100.6	90.0	267.7
Width, W (m)	10.2	15.9	13.4	12.7
Length/width	7.6	6.3	6.7	21.1
Volume (m³)	116.1	315.8	273.3	705.2
Retention time (days)	0.87	2.37	2.05	5.29

The monitoring system shown in Figure 10.2 includes five monitoring wells within the peat filter, MW-1, ... , MW-5; four locations within the SF wetlands, SW-1, ... , SW-4; one location at each pond, pond 1 and pond 2; and two locations to check the background concentration of groundwater within the site, W-1 and W-2. For each monitoring well within the peat filter, there are four sampling levels as shown in Figure 10.2. One level is located in the sand layer at the bottom of the peat filter, designated as D, at a distance of 1.7 m from the surface, and the other three levels are located within the peat filter at 0.5, 0.9, and 1.2 m from the surface, designated as A, B, and C, respectively.

The Huneault engineered wetlands were designed and constructed in 1994–95. The wetlands became operational in August 1995. The peat filter was saturated with leachate from the bottom to promote escape of entrapped air. This was accomplished by pumping leachate into a manhole connected to the header pipe of the base drainage system. An estimated 1327 m³ of leachate was required to saturate the peat filter completely.

10.5 RESULTS AND DISCUSSION

The results from the monitoring of the peat filter and wetland system during 1996 are presented in this section. The monitoring program consisted of the following:

1. Laboratory analysis of the samples for inductively coupled plasma (ICP) metal scan (heavy metals, calcium, sodium, boron), total dissolved carbon (TC), total dissolved organic carbon (TOC), BOD_5, chloride, ammonia nitrogen, and total phosphorus (TP);
2. Measuring of pH, conductivity, total solids (TS), and temperature in the field;
3. Measuring of hydraulic conductivity of the peat filter in the field.

10.5.1 Metal Analysis

The collected samples were filtered (2 μm glass microfiber GF/C filter) and then analyzed by the ICP method for B, Ca, Cd, Cr, Cu, Fe, Na, Ni, Pb, and Zn. It should be noted that these measurements are concentrations of dissolved solids. Concentrations of Cd, Cr, Cu, Ni, and Pb were very low and were below the detection limit most of the time.

Boron is one of the most troublesome trace elements in soil management. While low concentrations of boron are essential for plant growth, it becomes phytotoxic at concentrations only slightly higher than the optimal range. Maas (1984) classified the tolerance of a wide variety of crops based on the threshold concentration, which is the maximum concentration of boron in the soil solution that does not restrict the yield. According to this classification, crops are categorized as sensitive, such as apricot and cherry, with a threshold of <1.0 mg/L; moderately sensitive, such as carrot and potato, with a threshold of 1.0 to 2.0 mg/L; moderately tolerant, such as cabbage and corn, with a threshold of 2.0 to 4.0 mg/L; tolerant, such as tomato and beet, with a threshold of 4.0 to 6.0 mg/L; and very tolerant, such as cotton, with a threshold of 6.0 to 15.0 mg/L. The threshold concentrations for the grain yields of wheat, barely, and sorghum were measured to be 0.3, 3.4, and 7.4 mg/L of boron, respectively (Bingham et al., 1985).

The initial concentration of boron in the holding ponds, based on the 1996 monitoring data, varied from 11 to 18 mg/L, with an average of $\mu = 15$ mg/L and a standard deviation of $\sigma = 2.2$ based on 16 measurements. An increase in boron concentration has been noted at this site that may be attributed to the recirculation of leachate into the waste pile, another leachate management system in use at the Huneault Landfill. Boron concentration was reduced to values between 3 and 13.6 mg/L at the top level of peat surface (level A) and <0.05 to 4.9 mg/L at the bottom of peat filter (level D) with an average of $\mu = 1.34$ mg/L and $\sigma = 1.28$ based on 28 measurements. This shows that boron removal efficiency was 91% $[100 \times (15 - 1.34)/15]$ for the peat filter. However, boron concentration increased to values from 2.4 up to 6.6 mg/L, 4.9 mg/L on average, once the leachate from the peat filter entered the surface water wetland. The average concentrations of boron in the ponds, at the top of the peat filter, at the bottom of the peat filter and in the SF wetlands are shown in Figure 10.3. The peat filter consistently showed high removal efficiencies.

A reason for the observed increase in boron concentration between the bottom of the peat filter and the SF wetland could be the leakage of untreated leachate through channeling or short-circuiting within the engineered peat filter. One possible location for short-circuiting is a well installed for monitoring the water level at the edge of the peat filter. This location was confirmed to be a source

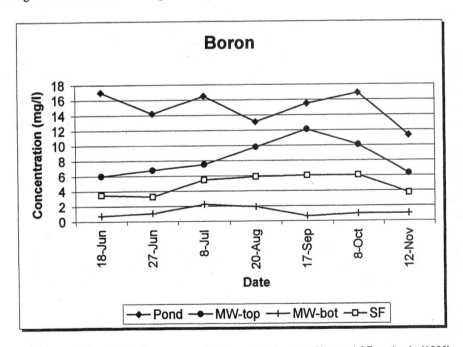

Figure 10.3   Boron concentrations within the peat filter and SF wetlands (1996).

of leakage since a sample taken at this location indicated a concentration of 13.7 mg/L for boron at the bottom of peat filter. This result suggests that some leachate entered the collection system within the sand layer without receiving any treatment from the peat filter, thus causing the boron concentration to rise in the SF wetlands.

The background concentration of boron in the groundwater within the constructed wetland was 0.6 to 1.0 mg/L as measured at W-1 and W-2. This shows that groundwater has not been contaminated by the untreated leachate. With the exception of the last sampling event, there is an increasing trend in boron concentration at the top level of the engineered peat filter. This could be an indication that the adsorption capacity of the peat for boron is getting exhausted in the top layer. The decrease in boron concentrations at the top layer for the last sampling event could be due to a decrease in initial concentration in the untreated leachate from the pond, see Figure 10.3.

Distribution of iron and zinc within the peat filter and SF wetlands does not show any specific trend. Iron concentrations varied from values as low as <0.06 up to 2.3 mg/L, with most measurements being below 1.0 mg/L. Zinc concentrations were within the range of <0.06 to 0.5 mg/L (with just one occurrence of 2.1 mg/L). The initial concentration of calcium in the ponds was between 250 to 300 mg/L, which was reduced to values in the range of 50 to 150 mg/L within the peat filter and 30 to 130 mg/L within the SF wetlands. Sodium concentrations were reduced from initial values of 300 to 450 mg/L to values in the range of 200 to 300 mg/L in the SF wetlands.

10.5.2   BOD$_5$, TOC, Cl, NH$_3$, TP, TDS, Conductivity, pH, and Dissolved Oxygen

The leachate produced at this landfill is weak in terms of organic matter concentrations. The initial dissolved BOD$_5$ in the ponds was less than 5 to 18 mg/L, with an average of $\mu = 9.1$ mg/L and a standard deviation of $\sigma = 5.0$ based on 14 measurements. The BOD$_5$ was reduced to values between <5 and 6 mg/L through the peat filter and <5 mg/L within the SF wetlands. This means a minimum reduction of 45% [$100 \times (9.1 - 5.0)/9.1$]. Figure 10.4 shows the variation of concentrations of TOC within the system. The initial concentration of total dissolved organic carbon in the pond varied from 80 to 150 mg/L, with an average of 109 mg/L based on 16 measurements. The TOC was reduced to values in the range of 45 to 100 mg/L at the top of peat filter and 20 to 80 mg/L ($\mu = 30$, based on 34 measurements) at the bottom of the peat filter. The concentration of TOC within the SF wetlands were in the range of 30 to 65 mg/L, and 45 mg/L on average. This implies an overall removal efficiency of 58% [$100 \times (109 - 45)/109$] for the system.

Chloride could be used as an indicator to trace the movement of contaminants. Initial concentration of chloride was 300 to 450 mg/L within the pond. Chloride concentrations within the SF wetlands were in the range of 200 to 400 mg/L, which does not show a significant reduction. The range of initial ammonia concentration was quite wide with concentrations as low as 4.5 mg/L and as high as 35 mg/L. The peat filter was quite effective in removing ammonia as concentrations in the range of 0.1 to 3.5 mg/L were measured at the bottom of the peat filter. The removal efficiency of peat filter is 95% [$100 \times (19 - 1)/19$]. There was no significant amount of phosphorus present in the leachate, and most of the measurements were close or below the detection limit of 0.05 mg/L.

Total dissolved solids (TDS) were reduced from an initial concentration of 2000 to 2200 mg/L to values in the range of 300 to 1200 mg/L at the bottom of the peat filter. SF wetlands had on average 850 mg/L of TDS; this shows a 60% removal efficiency. As expected, conductivity had the same trend as TDS. pH was in the range of 6.5 to 7.6 within the pond and peat filter and then increased to values between 7 and 9 as the leachate entered the SF wetlands. Dissolved oxygen (DO) was initially between 0.1 and 2.5 mg/L, which then slightly increased to values in the range of 2 to 3.5 mg/L within the peat filter. SF wetlands had an oxygen level of 8 to 12 mg/L.

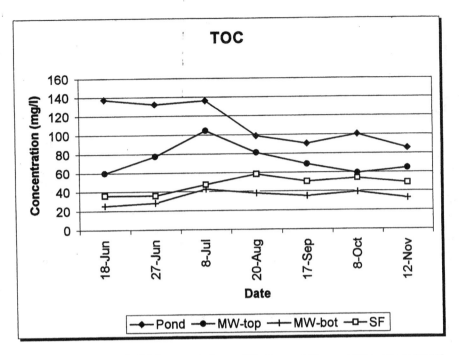

Figure 10.4 TOC concentrations within the peat filter and SF wetlands (1996).

10.5.3 Hydraulic Conductivity

Hydraulic conductivity of the peat filter was measured in the field at the beginning (26 June 1996) and at the end (25 November 1996) of the monitoring program. There was no specific change in the hydraulic conductivity of the peat filter. It ranged between 10^{-4} to 10^{-5} cm/s. However, the hydraulic conductivity of the sand filter was one to two orders of magnitude higher than the peat material.

10.6 CONCLUSIONS

The leachate produced at the Huneault Landfill is distinctly different from other municipal landfills in Ontario and, in general, has low concentrations of contaminants. A wetland system consisting of one VFSS wetland followed by three SF wetlands is used for treatment of the leachate. Results from field monitoring and laboratory adsorption tests showed that this system is capable of removing a variety of pollutants including heavy metals and inorganic and organic substances from landfill leachate.

Overall, a significant decrease in the concentrations of contaminants was observed and the treatment system made of peat filter and SF wetlands proved to be quite effective. It is worth noting that most of the removals for all parameters of concern occur within the peat filter. Because 90% of boron, 58% of TOC and TDS, 95% of ammonia, and at least 45% of BOD_5 removal occurs within the peat filter itself, the provision of a peat filter can significantly reduce the large land area generally required for constructed wetlands.

As expected, the concentrations of contaminants were reduced throughout the depth of the peat filter as the leachate moves toward the bottom of the filter. Adsorption, biodegradation, and filtration are most probably the main mechanisms for contaminant removal. Except the last sampling event, there was an increasing trend in the concentration of boron at the top level of the peat filter based on seven sampling events, which could be an indication that the adsorption capacity of peat for boron at the top level is reaching the saturation level.

REFERENCES

Bingham, F. T., Strong, J. E., Rhoades, J. D., and Keren, R., 1985. An application of Maas–Hoffman response model for boron toxicity; *Soil Science Society of America Journal* 49:672–674.

Cameron, R. D., 1978. Treatment of a complex landfill leachate with peat, *Canadian Journal of Civil Engineering* 5:83–97.

Cameron, R. D. and Koch, F. A., 1980. Toxicity of landfill leachates, *Journal of Water Pollution Control Federation* 52:760–769.

Castonguay, N., 1997. 1996 Operation and Monitoring Report for Huneault Landfill by Castonguay Technologies, Ottawa.

Gersberg, R. M., Lyon, S. R., Elkin, B., and Goldman, C. R., 1985. The removal of heavy metals by artificial wetlands, in *Proceedings of the Water Reuse Symposium III*, San Diego, CA. AWWA Research Foundation, Denver, CO, 1985.

Howard, K. W. F., 1997. Impact of urban development on groundwater, in Eyles, N., Ed., *Environmental Geology of Urban Areas*, Geological Association of Canada, Scarborough, Ontario, Canada. pp. 93–103.

Howard, K. W. F. and Livingston, S., 1997. Contaminant source audits and groundwater quality, in Eyles, N., Ed., *Environmental Geology of Urban Areas*, Geological Association of Canada, Scarborough, Ontario, Canada, pp. 104–116.

Maas, E. V., 1984. Salt tolerance of plants, in Christie, B., Ed., *Handbook of Plant Science*, CRC Press, Cleveland, OH.

McLellan, J. K. and Rock, C. A., 1988. Pretreating landfill leachate with peat to remove metals, *Water, Air and Soil Pollution* 37:203–215.

Pries, J. H., 1994. Wastewater and Stormwater Applications of Wetlands in Canada, North American Wetland Conservation Council, Issue Paper No. 1994-1, Ottawa, Ontario.

Townsend, T. G., Miller, W. L., Lee, H., and Earle, J. F. K., 1996. Acceleration of landfill stabilization using leachate recycle, *Journal of Environmental Engineering* 122:263–268.

U.S. Environmental Protection Agency, 1987. Report on the Use of Wetlands for Treatment and Municipal Wastewater, U.S. Environmental Protection Agency, EPA/430/09-88-005.

Zhipie, Z., Zenghni, Y., and Piya, C., 1984. A preliminary study of removal of Pb, Cd, Zn, Ni and Cr from wastewater with several Chinese peat, in *Proceedings of 7th International Peat Congress*.

CHAPTER 11

Evaluation of a Constructed Wetland for Treatment of Leachate at a Municipal Landfill in Northwest Florida

William F. DeBusk

CONTENTS

ABSTRACT: The Perdido Landfill, located in Escambia County, Florida, is the principal solid waste disposal facility for the greater Pensacola area. Individual cells in the landfill are completely lined, and a leachate collection system is incorporated into the facility. Since 1991, leachate has been pretreated using a constructed wetland system prior to discharge to percolation ponds. The treatment wetland system consists of a primary treatment lagoon and ten wetland cells, connected in series, with dimensions of 35 by 300 ft (11 by 91 m) each. Each of the ten cells is a surface flow wetland containing a variety of emergent macrophyte species. Monitoring and analysis of surface water in the Perdido Landfill leachate treatment wetland has been carried out by the Wetlands Research Laboratory of the University of West Florida since the inception of the system. Chemical and biological data generated during the monitoring program indicate that the wetland treatment system has provided a high level of treatment of the landfill leachate. For example, the 5-day biochemical oxygen demand (BOD_5) was reduced by an average of 96%, while total suspended solids (TSS), iron (Fe), and total Kjeldahl nitrogen (TKN) concentrations were decreased by >98%, on average. Mean concentration of total phosphorus, a nutrient widely associated with eutrophication of natural water bodies, was reduced to 0.6 mg L^{-1} in the wetland outflow, compared with 2.1 mg L^{-1} in the landfill leachate. The constructed wetland at the Perdido Landfill has proved to be a sustainable means for removal of nutrients and metals from leachate during an operational period of over 5 years. The performance of this system strongly supports the use of wetlands as efficient, low-cost alternatives to conventional treatment of landfill leachate.

11.1 INTRODUCTION

The Perdido Landfill, located on a 424-acre site in west-central Escambia County, FL (Figure 11.1), is operated by the county Division of Solid Waste, within the Department of Environmental Resources Management (ECDERM). The landfill receives an average of approximately 600 ton of municipal solid waste (MSW) per day. The average composition of MSW entering the landfill, based on screening performed in 1991, is shown in Table 11.1. The landfill cells are completely lined and fitted with a drainage system, which provides centralized leachate collection.

Figure 11.1   Location of the leachate treatment wetland at Perdido Landfill, near Pensacola, FL.

An on-site constructed wetland system for leachate treatment was completed and placed into operation in 1991. The Wetlands Research Laboratory of the Institute for Coastal and Estuarine Research (ICER) at the University of West Florida has conducted monitoring and analysis of surface water in the Perdido Landfill wetland treatment system since 1991. The objective of this paper is to present an evaluation of wetland performance to date, based on the existing water quality data.

11.2 SYSTEM DESIGN

The hybrid wetland system consists of a plastic-lined primary treatment pond of ~2 acres (0.8 ha), a series of ten surface flow wetland cells, 300 × 35 ft (91 × 11 m) each, containing emergent macrophytes, and a final detention pond for treated leachate (Figure 11.2). The total area of the ten wetland cells is 2.4 acres (0.98 ha). The wetland cells are underlain by a clay layer naturally occurring at the site, which was formerly used as a clay mine. During construction, a layer of organic soil was placed over the clay, prior to planting. Additional details concerning the design and layout of the constructed wetland system can be found in Martin et al. (1993).

The path of leachate flow through the sequence of wetland ponds and cells is illustrated in Figure 11.2. Raw leachate from the landfill drainage system is collected in the primary treatment pond L1. Direct rainfall and surface runoff from the immediate vicinity are also collected in L1, serving to dilute the leachate. The L1 pond is equipped with an aeration system and, prior to 1996,

Table 11.1   Composition of Municipal Solid
Waste at Perdido Landfill

Component	Percent of Total
Paper and paperboard	33.2
Yard wastes	18.3
Textiles	12.4
Food wastes	10.9
Glass	8.5
Plastics	6.7
Metals	4.0
Miscellaneous	6.0

Note: Based on 1991 data.

was stocked with a dense cover of water hyacinths (*Eichhornia crassipes*). Diluted, pretreated leachate is transported to the vegetated wetland cells by daily pumping from L1 to a small detention pond (S1, a.k.a. the "surge pond") at the head of the wetland. The wetland cells lie uphill from L1, thus the need for pumping the leachate to the wetland. Flow of water through the wetland cells is gravity assisted; a 2% grade exists between the wetland inflow and outflow. This results in a 2-ft (60-cm) elevation difference between successive cells. Water flows through the wetland cells in a serpentine fashion (Figure 11.2), facilitated by a slope toward the outflow end. The final pond (SF in Figure 11.2) is primarily open water, with densely vegetated banks, and serves as a holding pond for the treated leachate stream. Water from the SF pond (wetland outflow) is pumped to a nearby percolation pond for groundwater recharge.

The wetland cells (W1 to W10) have received an average flow of about 125,000 gpd (473 m³ d⁻¹) from the L1 pond during the operational period. Given a maximum water depth of 18 in. (45.7 cm) in the wetland cells, the mean HRT for the constructed wetland (W1 to W10) is estimated to be about 9.3 days, or slightly less than 1 day for each cell. Note that the water budgets for the wetland cells are only rough approximations, since data for pumping rates from the SF (wetland outflow) pond were not collected on a regular basis. Furthermore, hydraulic loading rates for leachate and

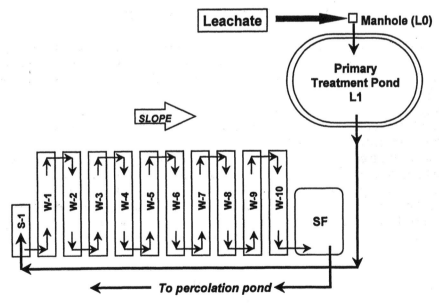

Figure 11.2   Schematic diagram of the treatment wetland system at the Perdido Landfill. L0 = leachate stream, L1 = primary treatment pond, S1 = surge pond, W1 – W10 = wetland cells, SF = wetland outflow detention pond.

storm water to L1 are not known. Therefore, even though the pumping rate from L1 to S1 is known, a hydrologic budget cannot be calculated for the L1 pond.

Wetland cells were planted with a variety of emergent macrophytes, with plant selection based on contaminant removal capability as well as tolerance to a high-ionic-strength waste stream and potentially high levels of heavy metals. The dominant macrophyte species in the wetland cells are shown in the schematic in Figure 11.3. Additional plant species have appeared in the wetland due to recruitment from local seed sources and the existing seed bank in the organic soil. As a result, each wetland cell contains a relatively high diversity of vegetation in addition to the dominant species shown in Figure 11.3.

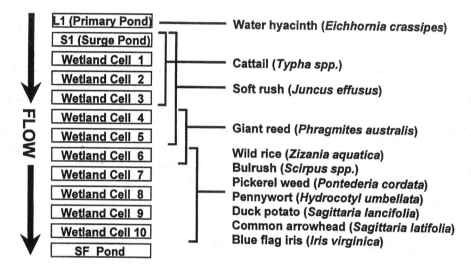

Figure 11.3 Overview of predominant species of wetland macrophytes in the treatment system.

11.3 SURFACE WATER MONITORING

Surface water in the wetland treatment system has been monitored since July 1991. The physical, chemical, and biological parameters listed in Table 11.2 were monitored on a regular basis, with the exception of lead and cadmium which were analyzed on an infrequent basis in wetland outflow samples only.

Sampling frequency for wetland surface water was somewhat variable, dependent to a great extent on availability of funding. Sampling of wetland outflow at the SF pond ranged in frequency from weekly to monthly, although total P and TKN were not analyzed for all sampling events. Sampling of raw leachate (site L0, Figure 11.2), primary treatment pond (L1), surge pond (S1), and wetland cells W1, W3, W5, W7, and W9 was performed four times per year. Water samples from the wetland cells were taken at the outflow end of each cell, while SF and L1 were sampled near the edge of the ponds. The leachate sampling point (L0) was located within the outflow pipe, which discharged into the L1 primary treatment pond, and was accessed from above through a manhole.

11.4 RESULTS

The Perdido Landfill leachate is characterized by high levels of dissolved and suspended solids (Table 11.3). Concentrations of analytes were highly variable, presumably dependent on antecedent

Table 11.2   Methods for the Analysis of Surface Water Samples Collected at the Perdido Landfill Leachate Treatment Wetland System

Analyte	Method Description	U.S. EPA Ref.
Total suspended solids (TSS)	Gravimetric	Method 160.2
Biogeochemical oxygen demand (BOD)	5-day, 20°C	Method 405.1
Total organic carbon (TOC)	Ultraviolet promoted, persulfate oxidation	Method 415.2
Nitrate + nitrite	Automated cadmium reduction	Method 353.1
Total ammonia	Automated phenate	Method 353.1
Total Kjeldahl nitrogen (TKN)	Semiautomated block digestion	Method 351.2
Total phosphorus (TP)	Automated ascorbic acid	Method 365.4
Total iron (Fe)	AAS, direct aspiration	Method 236.1
Total manganese (Mn)	AAS, direct aspiration	Method 243.1
Lead (Pb)	AAS, furnace	Method 239.2
Cadmium (Cd)	AAS, furnace	Method 213.1
Chloride (Cl⁻)	Automated ferricyanide	Method 325.2

rainfall. Among the analytes monitored in the leachate, nitrogen, organic C, and iron compounds were particularly abundant. The ubiquity of C and N compounds in the leachate was related to the high proportion of organic matter, such as paper products and yard and food wastes in the MSW. Much of the organic C in the leachate was readily available for microbial degradation, based on the average BOD_5 analysis. Nitrogen was present primarily in reduced inorganic form (NH_4^+), representing about 86% of the total N in the leachate. Essentially all of the remaining leachate N was incorporated in dissolved or particulate organic matter. Nitrate and nitrite would not be expected to occur in significant amounts in the highly reduced environment within the landfill cells. The high concentration of total Fe (dissolved + particulate) reflects the presence of steel and other iron-containing items buried in the landfill. Reduced (ferrous) iron is potentially more mobile in a landfill than oxidized (ferric) iron, due to the generally higher solubility of the former. In contrast to the high concentrations of C and N, total P content of the leachate was relatively low; for example, the average mass N:P ratio in the leachate was 219.

Concentration of toxic heavy metals in the leachate (e.g., cadmium) was relatively low, since batteries and other objects containing these metals are prohibited from the landfill. Routine surface water monitoring for lead and cadmium was performed only at the wetland outflow (SF); therefore, these elements were not considered in evaluation of the constructed wetland. Outflow concentrations of both metals were typically less than 6 µg L^{-1}. Toxic organic compounds such as pesticides and solvents were not monitored in the leachate or wetland system, but were analyzed in samples from

Table 11.3   Average Chemical Composition of Perdido Landfill Leachate during the Operational Period of the Constructed Wetland (1991–1997)

Analyte	Units	Concentration			
		Mean	Std. Dev.	Max.	Min.
Total ammonia N	mg L^{-1}	398	144	630	12
Nitrate + nitrite N	mg L^{-1}	0.15	0.10	0.50	0.10
Total Kjeldahl N	mg L^{-1}	460	166	730	27
Total phosphorus	mg L^{-1}	2.1	2.2	11.0	0.3
Iron	mg L^{-1}	294	280	1100	5
Manganese	mg L^{-1}	1.81	2.11	7.90	0.04
Cadmium	mg L^{-1}	0.017	0.011	0.025	0.009
BOD_5	mg L^{-1}	209	135	480	10
Total organic C	mg L^{-1}	423	288	1400	180
pH	units	7.02	0.16	7.27	6.88
Conductivity	µS cm^{-1}	6644	2381	9150	3200
Total suspended solids	mg L^{-1}	3414	3617	13000	3
Turbidity	NTU	636	498	1000	77

Table 11.4 Surface Water Concentration Data Summary for Selected Physical, Chemical, and Biological Parameters in the Wetland Treatment System

Site	Statistic	BOD$_5$	TSS	TOC	Total P	NH$_3$–N	NO$_3$–N	TKN	Fe	Mn	Cl$^-$
	Mean	209	3414	423	2.08	398.0	0.1	460.0	294.2	1.74	958
L0	Std. dev.	135	3617	288	2.24	144.3	0.1	166.3	279.7	2.08	429
	n	9	25	19	26	31	30	30	25	23	20
	Mean	37	141	71	3.53	4.5	8.9	12.4	8.2	0.18	210
L1	Std. dev.	38	1906	58	4.42	9.6	71.0	11.0	19.3	0.20	104
	n	5	34	36	42	48	19	45	27	26	34
	Mean	16	16	50	1.94	2.4	9.5	8.6	2.8	0.16	175
S1	Std. dev.	11	15	18	1.92	5.4	13.3	8.3	3.5	0.36	56
	n	2	33	35	41	48	18	45	27	25	32
	Mean	11	14	47	1.98	2.5	7.2	7.8	1.2	0.09	184
W1	Std. dev.	9	30	11	1.74	5.0	11.2	6.6	1.2	0.05	42
	n	6	33	35	41	47	18	44	25	23	32
	Mean	10	8	43	1.94	1.5	5.0	5.8	1.1	0.10	N.D.
W3	Std. dev.	2	10	10	1.91	3.8	7.9	4.5	1.2	0.06	N.D.
	n	2	26	27	33	39	17	39	19	19	0
	Mean	17	7	40	1.62	1.3	2.1	4.9	0.7	0.11	170
W5	Std. dev.	31	8	9	1.68	3.7	4.0	4.4	0.6	0.12	43
	n	6	33	35	40	47	19	45	26	24	32
	Mean	8	8	38	1.40	1.2	1.8	4.4	1.4	0.14	N.D.
W7	Std. dev.	0	11	9	1.29	3.5	3.4	3.9	1.7	0.19	N.D.
	n	2	28	29	35	41	17	39	23	21	0
	Mean	8	8	33	1.12	0.8	0.8	4.1	1.6	0.10	146
W9	Std. dev.	0	8	11	1.19	2.6	1.7	3.8	2.0	0.07	47
	n	2	34	36	42	48	19	43	25	23	33
	Mean	8	9	43	0.64	0.3	4.0	4.3	0.8	0.08	140
SF	Std. dev.	10	11	13	0.75	0.4	6.2	3.3	1.0	0.16	43
	n	130	160	158	62	89	127.0	70	130	131	34

Note: Sample sites are arranged sequentially from top to bottom according to water flow (refer to Figure 11.2 for locations). Leachate analyses are repeated from Table 11.3 to facilitate direct comparison with downstream sites.

compliance monitoring wells around the landfill. These were also not significant constituents of the landfill leachate.

Surface water analytical data for the wetland treatment system are summarized in Table 11.4. High variability was associated with L0, the leachate inflow stream, and L1, the primary treatment pond, for all parameters. A significant portion of this variability was probably a function of dilution by rainfall within the landfill itself and within L1, based on Cl$^-$ concentration data for L0 and L1. Since it was not possible to calculate a water budget for L1 with available data, Cl$^-$ was used as a semiquantitative indicator of dilution. If it can be assumed that input of Cl$^-$ to L1 in runoff was negligible, and plant uptake and other losses of Cl$^-$ from the water column in L1 were also negligible, then the average dilution factor for leachate in L1 was roughly 4.5. Thus, it is highly probable that dilution in L1 accounted for a significant portion of the concentration reduction for certain chemical parameters. Nevertheless, it was readily apparent that L1 provided substantial removal of total N, Fe, Mn, and suspended solids (Table 11.4). Concentration data also suggested that removal of organic C — BOD$_5$ and total organic carbon (TOC) — occurred in L1, while net removal of P was not achieved.

Substantial reduction of TSS and Fe concentrations occurred in the wetland cells, as well as in L1. Total P concentration was also reduced in the wetland, in contrast to L1. Total N decrease (TKN + NO$_2^-$ + NO$_3^-$), however, was modest in the wetland, compared with the high degree of reduction reported for L1 (prior to 1996; see discussion below). Overall, net removal (or reduction in concentration) was achieved in the wetland for all measured parameters. Furthermore, outflow

(SF) concentrations of these parameters were typically much lower than the respective criteria for discharge to groundwater (via the percolation pond). The overall reductions in concentration achieved by the leachate treatment system (from L0 to SF, Table 11.4) ranged from 69% for total P to >98% for TSS, Fe, and total N (TKN + NO_2^- + NO_3^-). Estimation of mass removal in the wetland was performed by constructing a water budget, as follows. In addition to surface inflow and outflow, the only significant inputs and losses of water to be considered were direct rainfall and evapotranspiration (ET). For the purposes of rough calculation, it was assumed that rainfall and ET cancel out when considered on an annual basis. In fact, Cl⁻ concentration data for the wetland cells (Table 11.4) indicate that a minimal degree of net dilution may have occurred between S1 and SF. Estimated mass balance for measured parameters in the wetland cells (S1 through SF) showed that percent removal ranged from 39.7 to 93.5 for TOC and TSS, respectively (Table 11.5). Relative removal of Fe and total P was greater than 80%. The estimated budget in Table 11.5 assumes no net change in flow through the wetland; therefore, the figures representing percent removal are identical to percent change in concentrations.

Table 11.5 Estimated Mass Balance for Selected Parameters in the Constructed Wetland Cells, Exclusive of the L1 Pond (total area of approximately 1 ha)

Analyte	Loading	Removal	Percent Removal
	kg d⁻¹		
BOD₅	17.82	13.79	77.4
TSS	67.54	63.12	93.5
TOC	33.86	13.44	39.7
Total P	1.69	1.38	81.8
Total N	10.18	6.24	61.3
Fe	3.93	3.52	89.7
Mn	0.09	0.04	52.0

Note: Based on the assumption of zero water balance within the wetland.

Temporal trends in concentration of selected chemical parameters at the wetland outflow (SF) are shown in Figure 11.4. High-frequency variability was generally due to short-term changes in dilution among sampling events. However, several other factors affecting outflow concentrations were identified. High levels of BOD₅ and total P during the first 2 years of operation may have been the result of initial flushing of dissolved organic matter from the organic soil in the wetland. Spikes in outflow concentration of BOD₅ and TKN occurred during the winter of 1994–95, following a severe freeze. A series of freezes during February 1996 resulted in total loss of the water hyacinth cover in L1; the water hyacinths were not replaced following the freeze. As a result, nitrate concentration in the wetland, and especially in L1, fluctuated considerably, and frequently exceeded 20 mg N L⁻¹ in the outflow (Figure 11.4). This is primarily attributed to a sharp decline in denitrification activity in L1 associated with the loss of plant cover (see discussion). Prior to 1996, i.e., before the loss of water hyacinths in L1, the constructed wetland had shown no trends toward decreased contaminant removal capability during the operational period. In fact, the P and Fe removal efficiency had apparently increased with the age of the wetland system.

11.5 DISCUSSION

TOC represents a wide range of organic compounds with varying degrees of biodegradability. Humic acids, for example, are highly resistant to biodegradation, and are thus generally stable in the environment (Reddy and D'Angelo, 1994). In contrast, the BOD₅ assay is a measure of

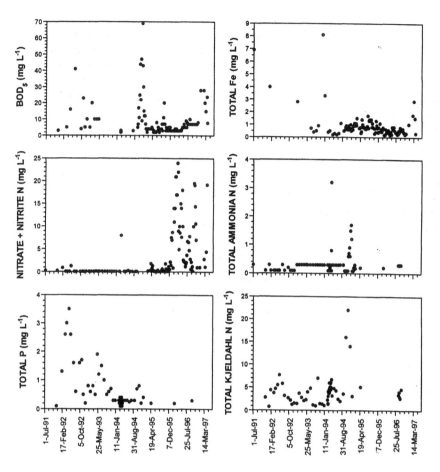

Figure 11.4   Temporal trends in wetland outflow (SF) concentrations of selected parameters.

biologically available carbon, although the test may lead to overestimation of C in the presence of reduced inorganic compounds, which may also be oxidized, either chemically or biologically (e.g., Fe^{2+} and NH_4^+). Net removal of organic C entering wetlands is limited by the fact that these systems produce and store vast quantities of organic C, due to high rates of photosynthesis (C fixation) and low rates of decomposition. Removal of the more-labile organic C probably occurred rapidly in the primary treatment pond and wetland. This is due to the fact that the most-labile organic C would be the preferred energy source for microbial decomposers. Between L1 and the wetland outflow (SF), the relative decrease in average BOD_5 was greater than the relative decrease in TOC concentration. Wetlands typically export dissolved and particulate organic C, much of it highly resistant to biological degradation (Mitsch and Gosselink, 1993). Therefore, more-recalcitrant native organic C was added to the surface water, while the more-labile portion of imported (by way of leachate) organic C was biologically oxidized.

TSS removal efficiency was typical for emergent macrophyte-dominated wetlands. Settling of particulate matter in the landfill treatment wetland was enhanced by the effect of vegetation to reduce wave action and flow velocity. Trapping of sediment particles occurs most readily in the plant litter layer in wetlands (Kadlec and Knight, 1996). In addition, the low flow rates characteristic of the treatment wetland result in laminar, rather than turbulent, flow, thus sediment resuspension is not likely to occur downstream of L1.

Iron and manganese were present at relatively high concentrations (compared with typical environmental levels) in the leachate, but were efficiently removed by the wetland treatment system.

Probable removal mechanisms are plant uptake and eventual burial, and probably more importantly, precipitation and sedimentation of Fe and Mn oxides (Ponnamperuma, 1972). Reduced, and consequently soluble, forms (Fe^{2+} and Mn^{2+}) in the leachate were probably oxidized to insoluble forms (Fe^{3+} and Mn^{4+}) by aeration in the L1 pond.

Inorganic nitrogen and phosphorus compounds are potential causes of eutrophication of natural water bodies. Addition of these compounds generally results in accelerated growth of algae and macrophytes, and may significantly alter species diversity and overall ecosystem structure and function. Leachate from the Perdido Landfill typically contained only moderate concentrations of total phosphorus (mean = 2.1 mg L^{-1}), but extremely high levels of nitrogen (mean TKN = 460 mg L^{-1}). Average molar ratio of TKN:total P in the leachate during the period 1991 to 1996 was 489, compared with 15 in the SF outflow.

Total P concentration was highly variable in the leachate inflow (L0), L1, and the wetland cells. Although the mean concentration of total P was higher in L1 than in L0, it was typically lower in L1 than in L0 for individual sampling events where both sites were sampled. The discrepancy between mean values and individual observations is an artifact of the sampling schedule; L0 was not sampled as frequently as L1. Phosphorus chemistry is complex, and may be controlled by a number of factors in (and outside of) the wetland treatment system. The high concentration of Fe in the leachate suggests that P availability may be significantly affected by Fe solubility. This is highly dependent on the redox status of the water and soil. Ferric phosphates may precipitate and settle out of the water column under aerated conditions (Ponnamperuma, 1972). However, under reduced conditions, these compounds may become soluble (ferrous), resulting in release of PO_4^{3-}. In addition, PO_4^{3-} may be tightly adsorbed to positively charged surfaces of clays. The potential for long-term PO_4^{3-} retention is substantial for soils containing appreciable amounts of clay. Organic forms of P may be broken down to inorganic forms (e.g., HPO_4^{2-}) through microbial enzyme activity. Finally, floating and emergent macrophytes, as well as algae, take up varying amounts of phosphorus. Plant uptake may result in net accumulation of P (and other nutrients) through accretion of resistant organic matter (such as peat). Based on available data, it is not possible to pinpoint the factors governing P cycling in the system. Nevertheless, data collected thus far indicate that substantial removal of P occurred in the wetland, particularly in the downstream portion (W5 through SF).

Although the leachate was substantially more enriched with N than with P, N removal (based on concentration) in the treatment system was enormously greater. Substantial removal and immobilization of N may occur through plant uptake and sediment or peat accretion. However, microbial metabolic processes were probably the major sink for N in the treatment system. Surface water concentration data (Table 11.4) indicated that most of the N present in the leachate was in the form of ammonia ($NH_3 + NH_4^+$; at circumneutral pH most would occur in the ionized form NH_4^+), and that nearly all of the N removal occurred in L1, before entering the wetland. The processes that apparently mediated N removal in L1 were sequential nitrification and denitrification of the incoming ammonia. The combination of aeration and a dense water hyacinth mat in L1 provided a suitable environment for these reactions. Nitrification of ammonium to nitrate took place under aerated conditions. Convective and diffusive transport resulted in movement of nitrate away from the aerated zone and into anoxic zones created by the dense mat of water hyacinth. Under conditions of depleted oxygen, high labile organic C (energy source) and neutral pH denitrification should proceed rapidly. These conditions were met in L1 prior to winter 1995–96, at which time the water hyacinth cover was lost.

Following the loss of water hyacinths in the primary treatment pond L1, a number of changes occurred in the wetland cells, as well as L1, related to spatial patterns of nutrient concentrations and overall removal efficiency. The predominant form of N entering the wetland cells from L1 changed from NH_4^+ (reduced) to NO_3^- oxidized) (Figure 11.5). Also, concentration of NO_3^- was significantly increased throughout the constructed wetland, while NH_4^+ concentration decreased. The bulk of total N removal in the treatment system shifted from L1 to the wetland cells

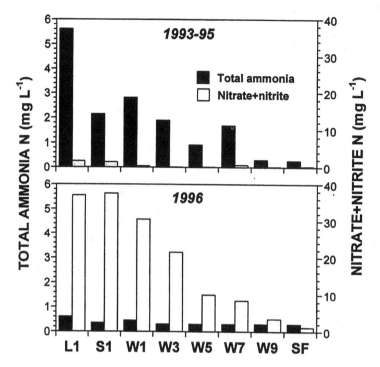

Figure 11.5　Comparison of total ammonia and nitrate + nitrite concentrations in the wetland treatment system prior to the loss of water hyacinths in L1 (1993–1995 data) and subsequent to the loss (1996 data).

(Figure 11.6), as total N concentration in the wetland inflow (from L1) increased. In contrast, total phosphorus concentration in L1 and the wetland cells was considerably lower after the loss of water hyacinths (Figure 11.6). Thus, most of the phosphorus removal in the treatment system was occurring at the front end, in the L1 primary treatment pond, rather than in the wetland cells.

Changes in N and P cycling in the wetland were also indicated by water chemistry data before and after the loss of hyacinths. The mean N:P mass ratio in surface water entering the wetland at S1 increased from 4.7 to 108, suggesting that P may have replaced N as the growth-limiting nutrient in the wetlands. The increase in N:P ratio resulted from a decrease in total P as well as an increase in total N in L1. The decrease in total P in L1 may have resulted from accelerated algal growth or coprecipitation of phosphate with Fe in the open-water, highly aerated pond.

Frequent periods of high NO_3^- concentration in the wetland outflow following hyacinth mortality in L1 may have resulted from a chain reaction of events catalyzed by the shift in surface water N:P ratio. Decomposition of plant litter may have been stimulated by increased N availability, especially during the spring months of 1996 and 1997 when detrital mass was high following winter freezes. Prior to 1996, the limited availability of N to microbial decomposers probably resulted in substantial immobilization of N in the plant litter. With the subsequent increase in the availability of N, the capacity of the wetlands to serve as an N sink has probably decreased because of the relative excess of N loading. It is unclear whether or not denitrification potential of the wetland litter and soil increased with increased loading of NO_3^-. Nevertheless, the increased throughput of inorganic N to the SF pond created the potential for significant accumulation of NO_3^- in the outflow through nitrification activity in SF, which may have been highly stimulated by algal blooms.

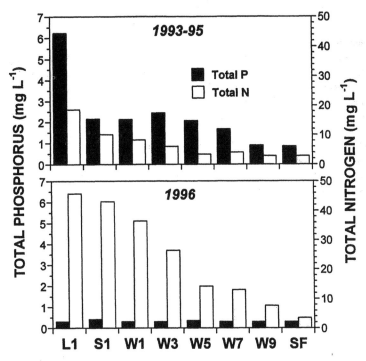

Figure 11.6 Comparison of total N and total P concentrations in the wetland treatment system prior to the loss of water hyacinths in L1 (1993–1995 data) and subsequent to the loss (1996 data).

11.6 SUMMARY AND CONCLUSIONS

The constructed wetland treatment system at the Perdido Landfill has achieved substantial removal of N, P, organic C, Fe, and other contaminants from the leachate. During the past 6 years of monitoring, BOD_5 was reduced by an average of 96%, while TSS, Fe, and TKN concentrations were decreased by >98%, on average.

Mean concentration of total P, a nutrient widely associated with eutrophication of natural water bodies, was reduced to 0.6 mg L^{-1} in the wetland outflow, compared with 2.1 mg L^{-1} in the landfill leachate. Nitrogen, another nutrient of environmental concern, is present at high concentrations in the landfill leachate (mean TKN = 460 mg L^{-1}), primarily in reduced inorganic form (ammonia-N). The primary treatment pond L1, utilizing a combination of aeration and a dense water hyacinth mat, served as an extremely efficient sink for N, through the microbial processes of nitrification and denitrification. The recent loss of the water hyacinth cover in L1 has diminished the N removal efficiency of the system. However, P removal efficiency of the wetland treatment system has been maintained, and possibly increased.

The Perdido Landfill constructed wetland has proved to be a sustainable means for removal of nutrients and metals from leachate during an operational period of 6 years. The performance of this system strongly supports the use of wetlands as efficient, low-cost alternatives to conventional treatment of landfill leachate.

REFERENCES

Kadlec, R. H. and Knight, R. L., 1996. *Treatment Wetlands*, Lewis Publishers, Boca Raton, FL, 893 pp.

Martin, C. D., Moshiri, G. A., and Miller, C. C., 1993. Mitigation of landfill leachate incorporating in-series constructed wetlands of a closed-loop design, in Moshiri, G. A., Ed., *Constructed Wetlands for Water Quality Improvement*, Lewis Publishers, Boca Raton, FL, 473–476.

Mitsch, W. J. and Gosselink, J. G., 1993. *Wetlands*, Van Nostrand Reinhold, New York.

Ponnamperuma, F. N., 1972. The chemistry of submerged soils, *Advances in Agronomy* (Academic Press) 24:29–96.

Reddy, K. R. and D'Angelo, E. M., 1994. Soil processes regulating water quality in wetlands, in Mitsch, W. J., Ed., *Global Wetlands: Old World and New*, Elsevier Science, Amsterdam.

U.S. Environmental Protection Agency (U.S. EPA), 1983. Methods for Chemical Analysis of Water and Wastes, EPA 600/4-79-020, USEPA, Cincinnati, OH.

CHAPTER 12

An Integrated Natural System for Leachate Treatment

Joseph Loer, Katrin Scholz-Barth, Robert Kadlec, Douglas Wetzstein, and Joseph Julik

CONTENTS

ABSTRACT: The Isanti–Chisago Sanitary Landfill, an unlined municipal solid waste facility located near Cambridge, MN was closed in 1992. Leaching of soluble wastes had contaminated the surficial and increasingly deeper aquifers with toxic organic compounds and heavy metals. The Minnesota Pollution Control Agency requested an innovative treatment system with operating and maintenance costs far below a conventional system. The selected approach was a natural systems engineering design that relies on existing topography for gravity flow (with the exception of groundwater pumping), solar and wind energy inputs rather than electrical, and natural biological, chemical, and physical interactions rather than petrochemical inputs.

Volatile organic compounds (VOCs) are removed by cascading the water down the side of the landfill in a polypropylene "step aerator", which is designed also to increase dissolved oxygen concentrations in the water to oxidize ferrous to ferric iron, thereby precipitating the hydroxide and other solids. Coprecipitation of heavy metals also occurs in this stage.

A sedimentation basin was selected as the second component of the treatment train, to continue aeration via natural surface agitation, and oxidation/degradation via ultraviolet mechanisms; and to allow settling of insoluble metals and other inorganic and organic solids following cascade aeration. The basin was sized for 6 days residence time at a pumping rate of 600 m³/day (110 gpm) and a liquid depth of 1.2 m (4 ft) (including settled sludge). Basin size and residence times were selected based on settling rate studies, anticipated sludge volume generation calculations, and land availability. The sedimentation basin was constructed of earthen materials and a soil-covered polypropylene liner to minimize infiltration through the base.

The next component of the treatment train is a 0.6 ha (1.5 acre) free-water surface constructed wetland. Three parallel-flow cells were seeded with cattails in the fall of 1995, and developed into a dense stand during the summer of 1996. The wetland provides 3 days of residence time at a pumping rate of 600 m³/day (110 gpm) and an average free-water depth of 30 cm (1 ft). Continued treatment occurs via aeration, sorption, biological storage and transformation, and trapping of solids. The wetland was constructed of earthen materials and polypropylene liner to minimize infiltration through the base. Water from the sedimentation basin enters the constructed wetland by means of gated inlet pipes that promote evenly distributed flow. A mid-cell, deep-water channel re-creates sheet flow to the second half should channeling occur through the vegetation. The water level is controlled by adjustable stoplogs at the cell outlets.

Discharge from the constructed wetland is to a borrow pit wetland (pond), modified to infiltrate treated water into the surficial aquifer from which the contaminated groundwater was initially removed. At the completion of the first season of treatment, results indicate that the system efficiency ranges from 85 to 100% for VOCs and from 95 to 99% for iron, with varying reductions in heavy metals. Operation and maintenance requirements of this simple, natural system are minimal, as was desired.

12.1 INTRODUCTION

The Isanti–Chisago Sanitary Landfill (Site) is located approximately 74 km (45 miles) north of the St. Paul/Minneapolis metropolitan area, near the city of Cambridge (Figure 12.1). It was permitted as a municipal solid waste (MSW) disposal facility and operated from 1973 to 1992. The Site covers 8.9 ha (22 acres) and contains an estimated 315,000 m³ (412,000 yd³) of waste. Like many closed landfills, this site has environmental problems concerning groundwater contamination.

Isanti and Chisago counties were issued a Request for Response Action (RFRA) under the Minnesota Environmental Response and Liability Act (state Superfund program) in June 1988, because of groundwater contamination in the downgradient monitoring system and in nearby residential wells. The Remedial Investigation (RI) conducted by the counties found that, like many landfills, the plume was wide (300 to 660 m) and shallow (1.5 to 6.0 m thick). However, quarterly monitoring had indicated contamination in lower aquifers as well, suggesting that they may be hydraulically connected. The plume has high concentrations of organic compounds of which only a small fraction are priority pollutants. There are high concentrations of inorganic compounds including some heavy metals. The counties conducted remedial activities under state oversight until reaching a joint political subdivision liability limit of $800,000 in 1993. Response actions completed by the counties included the RI, a 2-meter (6-ft) cover with 60 cm (2 ft) of compacted clay over the Site, and drilling deeper drinking water wells for affected downgradient residences.

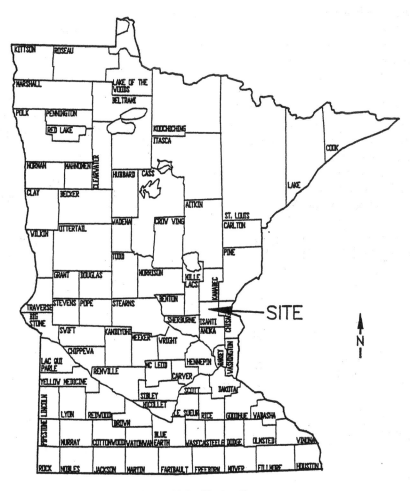

Figure 12.1 Site location map.

The State of Minnesota (State) took control of the Site under the state Closed Landfill Cleanup Program* in 1993. The State retained Delta Environmental Consultants, Inc. (Delta), to develop an innovative approach to groundwater remediation at the Site. The State directed Delta to design a groundwater treatment system for this site that would be both simple in design and easy to maintain and operate. Faced with the responsibility for more than 100 landfills, the State was not interested in complex, conventional treatment systems. Additionally, the State requested that capital costs of innovative approaches be comparable with, but annual operation and maintenance costs significantly below, conventional approaches.

Hauling of leachate or extracted groundwater to nearby municipal wastewater treatment plants was not possible because the plants would not accept the waste because of concerns over effluent limitations and long-term heavy metal residuals in sludge. Conventional on-site treatment methods for wastewater, such as aeration tanks or lagoons, pH adjustment, flocculent addition, clarification, and filtration were considered but determined to be unacceptable based on the cost and logistical concerns associated with the continual energy and chemical inputs required to drive these unit

* The site has been included within the Closed Landfill Program created in 1994 by the Minnesota Legislature effectively to remove many private and public landfills from state and federal Superfund programs. Closed, permitted MSW landfills that qualify for the program are taken over by the state once they have been closed according to the rules in effect when the landfill ceased taking waste. The state is then responsible for any further corrective action and the long-term care and maintenance of the landfill.

processes, as well as the relatively constant labor necessary to operate such systems. Additionally, current influent data may not be representative of the contaminant types and volumes leaching from a landfill over the decades-long life of treatment. Design requirements of conventional treatment lead to narrowly focused systems that are susceptible to failure if influent chemistry varies far from design criteria.

The State and Delta decided to pursue an innovative approach using natural systems engineering. The advantages to be gained were the use of existing topography to facilitate gravity flow, a minimization of energy inputs, elimination of chemical inputs, and a treatment system that with few "moving parts" would greatly reduce operation and maintenance requirements and also, by blending into the surrounding landscape, would not be a highly visible target for vandals. The selected design consisted of three extraction wells, a cascade aerator built into a landfill side slope, a lined sedimentation basin, and a lined constructed wetland, with surface discharge to a pond in an adjacent soil borrow area for infiltration back into the surficial aquifer (Figure 12.2).

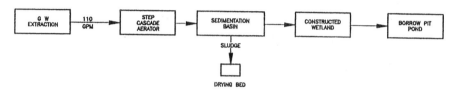

Figure 12.2   Process flow diagram.

12.2 DESIGN PROCESS

Treatability studies conducted on the extracted groundwater included laboratory-bench-scale and field evaluations of

- The aeration required to volatilize organics and oxidize metals to insoluble compounds,
- The settling rate of the insoluble compounds, and
- The composition of the supernatant and sludge produced.

A further objective of the treatability studies was to determine the minimum level of treatment that could still produce the desired effluent characteristics. These studies indicated that while discharge objectives for organics could be met with low-efficiency aeration processes, removal of heavy metals required near-saturation levels of dissolved oxygen over extended periods (e.g., 4 h) to oxidize the ferrous (Fe^{2+}) iron in the wastewater sufficiently to less-soluble ferric (Fe^{3+}) iron and precipitate the hydroxide, ($Fe[OH]_3$).* The studies also showed that extended periods of settling after the aeration period were required to achieve cosettling of the heavy metals of interest with the insoluble iron and calcium compounds presumably acting as flocculating agents (e.g., removal to discharge standards required from 12 to 48 h).

Following evaluation of treatability studies, it was felt that a sufficient level of aeration and settling could be achieved with cascade aeration if the aeration was followed by long residence in an oversized sedimentation basin that would also continue aeration. However, it was not felt that surface discharge effluent standards could be consistently achieved from the basin without accepting an undue level of risk. Specifically, mercury, zinc, copper, lead, and arsenic were not consistently removed to acceptable levels without aggressive aeration and long-term settling.

---

* It is assumed that oxidation, precipitation, and settling of iron, and to a lesser extent calcium and magnesium compounds, are the primary indicators of treatment effectiveness because of their high influent concentrations and tendency to induce cosettling of heavy metals.

Therefore, infiltration basins upgradient of or within the groundwater extraction zone were evaluated as discharge options. However, because infiltration basin operation is susceptible to clogging of the void spaces, especially with an iron-rich effluent stream, it was determined that discharge of the wastewater in this fashion introduced an unacceptable level of operation and maintenance requirements to the system, including the energy input necessary to either pump to upgradient infiltration basins or increase the capture zone to allow downgradient infiltration.

The decision to design a constructed wetland as a means to polish the effluent from the settling basin to surface discharge standards was based on

- The intention of seasonal (April through October) groundwater extraction and the published performance characteristics of constructed wetlands in warm-weather operation (primarily within the mining industry with regard to heavy metals removal);
- The high level of adsorptive surfaces with sediments and vegetation;
- Literature descriptions of the ability of sediments to adsorb and fix, via ion exchange, the heavy metals in the wastewater;
- The ability to create both aerobic and anaerobic zones;
- The diverse physical/chemical/biological processes that allow a broad spectrum and concentration of contaminants — which are likely over the operating lifetime of a leachate treatment system — to be effectively treated;
- The availability of land immediately adjacent to the Site; and
- The overall aesthetic appeal and improvement of such a "natural" system.

12.3 SYSTEM COMPONENTS

The groundwater treatment system consists of groundwater extraction followed by a treatment train composed of a cascade aerator, a sedimentation basin, and a constructed wetland, with surface discharge. The system operates via gravity flow from the top of the cascade aerator through the discharge outlet (Figure 12.3). Beginning with the sedimentation basin, all components are located off of the landfill itself to avoid problems associated with waste settlement. The following sections describe each component in more detail. Precedents for natural system treatment have been described at a number of other locations (Mæhlum, 1993; Martin and Moshiri, 1993; Martin et al., 1993).

12.3.1 Groundwater Extraction

Three recovery wells are located on the east-central portion of the landfill (Figure 12.4). The wells are constructed through the landfill cap and garbage, 16 to 23 m (50 to 70 ft), into the underlying alluvial sand aquifer. Each well is equipped with an electric submersible pump and discharge piping. The discharge piping is connected through a pitless adapter to a buried pipe, which transmits the extracted groundwater to a discharge sump at the top of the cascade aerator. The sump consists of a shallow polypropylene basin open at the top for easy access and cleaning. The sump serves as an equalization basin for the extracted groundwater, reduces water velocity prior to cascade treatment, and provides a representative sampling location for wastewater quality prior to treatment. The pumping system is designed to capture contaminated groundwater as it migrates from beneath the east portion of the landfill with minimal clean water capture. The submersible pumps are sized to extract a combined maximum of 600 m^3/day (110 gpm) of groundwater. Submersible pump operation can be monitored remotely via a programmable logic controller and phone/modem connection, which allows the operator to check system status without a site visit.

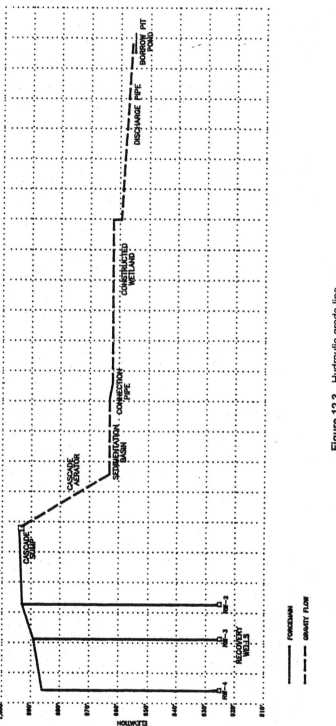

Figure 12.3   Hydraulic grade line.

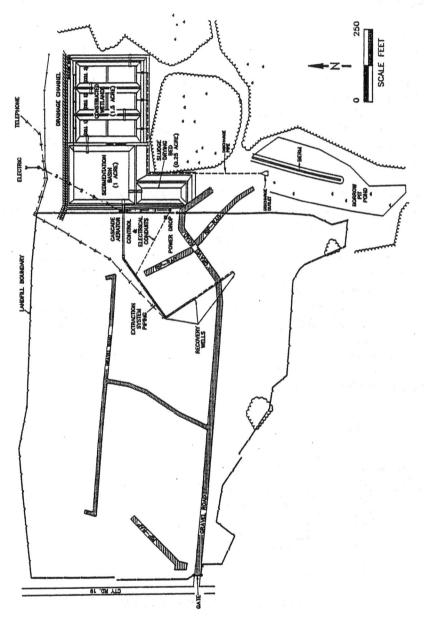

Figure 12.4 System layout.

12.3.2 Cascade Aeration

The first component of the treatment train is the cascade aerator (Figure 12.4). The cascade is a gravity-flow, "stepped" design, relying on flow in a thin layer over the steps. Oxygen transfer occurs across the air–water surface within each step, but primarily as the water free-falls (cascades) from one step to the next, thereby entraining air. Wastewater aeration within the cascade serves two primary purposes:

- To remove volatile organic compounds
- To increase the dissolved oxygen concentration to initiate metals precipitation

Besides aeration, the function of the cascade is to transport extracted groundwater from the top of the landfill to the sedimentation basin at the base of the landfill. Several alternative designs and materials were evaluated based on several criteria, including the expected degree of aeration, the ability to tolerate potential differential settlement, anticipated maintenance, and cost. A thermo-plastic, sectional design was chosen over materials such as fiberglass and concrete based on these criteria.

The cascade aerator is constructed from the top of the landfill to a spillway on the west side of the sedimentation basin, which is set east of the landfill boundary. Because the top of the cascade was constructed over 9 m (30 ft) of waste, and the base of the cascade was founded on more-competent natural soil, anticipated differential settlement could be as much as several feet over the length of the cascade and life of the system. Because of these poor foundation conditions, the cascade was designed to be flexible and essentially freestanding and to have individually replaceable steps. The lineal distance of the cascade is approximately 27 m (80 ft) and the vertical drop is approximately 10 m (30 ft). There are approximately 25 2.5-m-long (8-ft-long) individual sections, each providing a drop or step of 30 to 50 cm (1 to 1.5 ft). The sections are constructed with 60-mm (¼ in.) and 130-mm (½ in.) thick polypropylene sheets welded together to form a trapezoidal channel (Figure 12.5). Each section is independent of the others and is held horizontally by steel guideposts driven into the cap on both sides. Sections are free to move vertically within the confines of the steel guideposts, and are constructed primarily at or below grade to minimize potential erosion and weather effects (Figure 12.6).

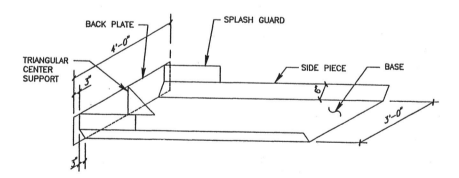

Figure 12.5 Cascade section fabrication.

The geometry of the cascade sections and predictions of flow parameters were developed using open-channel fluid flow relationships and calculations, including flow over an open weir, hydraulic jump calculations, critical depth calculations, and specific energy/velocity relationships. Based on these calculations, it was predicted that at a flow rate of 600 m³/day (110 gpm) the critical depth at the end of each 1-m-wide (3-ft-wide) section will be a little less than 2.54 cm (1 in.), and will

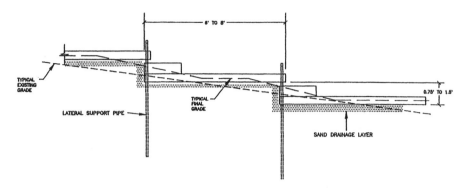

Figure 12.6 Cascade section installation.

be constant. The flow velocity at the end of each cascade section will be 0.5 m/s (1.4 fps). The average force exerted by the flow stream striking each downstream cascade will be about 1.8 kg (4 lb). Sidewalls, 15-cm (6-in.) high, would be sufficient to contain the maximum hydraulic jump; an additional 15-cm (6-in.) section was added to the sidewalls of the first 4 ft of each section to account for splashing and wind effects.

12.3.3 Sedimentation Basin

The sedimentation basin consists of a 0.49-ha (1 acre) surface impoundment located immediately northeast of the landfill (Figure 12.4). A sedimentation basin was selected as the second component of the treatment train to continue aeration via natural surface agitation, oxidation/degradation via ultraviolet mechanisms, and to allow settling of insoluble metals and other inorganic and organic solids following cascade aeration. The basin is sized to allow 6 days of residence time at a pumping rate of 600 m³/day (110 gpm) and a liquid depth of 1.2 m (4 ft) (including settled sludge). Basin size and residence time were selected based on settling rate studies, anticipated sludge volume generation calculations, and land availability. The sedimentation basin is constructed of earthen materials and a covered polypropylene liner to minimize infiltration through the base.

Sludge settling within the basin is dependent upon a number of factors, the most important of which are influent suspended solids concentrations and basin retention time. The influent suspended solids concentrations will be determined by the dissolved suspended solids levels in the extracted groundwater and the degree to which dissolved solids are precipitated out of solution by aeration. Flocculation and settling of the insoluble iron and calcium compounds is anticipated to be the primary removal mechanism of heavy metals, via surface adhesion.

12.3.4 Constructed Wetland

The last component of the treatment train is the constructed wetland, a 0.6-ha (1.5-acre) surface impoundment located immediately northeast of the landfill and east of the sedimentation basin (Figure 12.7). Design techniques are described in a number of publications (U.S. EPA, 1988; WPCF, 1990; Kadlec and Knight, 1996). However, these sources deal primarily with domestic wastewater, and consequently contribute only a framework for design of leachate treatment.

The constructed wetland is a free-water-surface (FWS) wetland consisting of three parallel-flow cells planted with emergent aquatic vegetation (cattails, *Typha* spp). The basin size is based on the available land, which allows approximately 3 days of residence time at a pumping rate 600 m³/day (110 gpm) and an average free-water depth 30 cm (1 ft) above the saturated rooting bed. The average liquid depth of 30 cm promotes continued aeration of the water while allowing sufficient residence time for continued treatment and settling of solids.

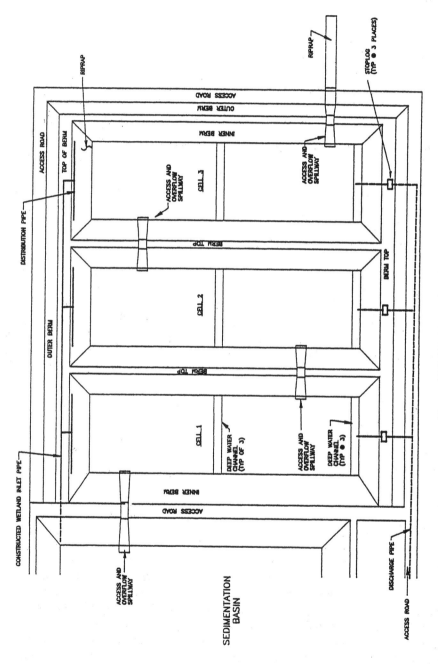

Figure 12.7 Constructed wetland.

The wetland is constructed of earthen materials and a polypropylene liner to minimize infiltration through the base. Water from the sedimentation basin flows into the constructed wetland by means of gated inlet pipes located on the north side of the cells. The inlet pipes at the head of each cell distribute the incoming water across the width of the cell to promote even, laminar flow. A mid-cell, deep-water channel will re-create sheet flow to the second half should channeling occur through the vegetation. The water level within and discharge from the constructed wetland is controlled by adjustable level stoplogs located on the south (effluent) side of the cells.

The primary treatment mechanisms within the constructed wetland are precipitation of metal hydroxides, due to continued oxidation within aerobic zones, and sulfate reduction to insoluble metal sulfides within anaerobic zones. The dissolved and precipitated metals contact adsorptive surfaces and ion exchange sites provided by the plant surfaces and sediments, allowing the metals to be removed from the water stream and to be fixed into the organic matter base. Full efficiency of this system is not expected for two to three growing seasons until the wetland is fully populated with emergent vegetation, and a litter layer becomes fully established. Treatment of residual organic compounds will occur via biodegradation and settling within the constructed wetlands. The metal and organic absorptive and loading capacities of the wetland are expected to be sufficient for the design life of the system, although it is possible that removal of organic matter will be necessary to maintain treatment efficiency.

12.3.5 Discharge

A water balance over the system indicates that only slight reductions in discharge volume relative to influent volume will occur, because combined evaporative and liner permeation losses only slightly outweigh rainfall. The treated groundwater is discharged to a borrow pit southeast of the landfill via gravity flow from the constructed wetland (Figure 12.4). The borrow pit also receives surface water runoff from the landfill. The discharge area was modified to contain a receiving half with shallow water to promote vegetative growth (i.e., a shallow emergent marsh wetland) and a discharging half with deep water to discourage emergent growth and allow infiltration of the surface water to the surficial aquifer.

12.3.6 Sludge Management

Based on the variability in sludge generation evidenced in treatability studies, the sludge generation rate will greatly depend upon the long-term influent chemistry from the extraction wells. Sludge buildup is likely to interfere at some point with settling and affect sedimentation basin operation. Once this point is determined based upon actual operation, the sludge will be removed, in the form of an estimated 1 to 5% solids content slurry, by pumping to a sludge drying bed. The sedimentation basin has been designed to facilitate sludge removal by sloping the bed to the eastern berm.

The sludge drying bed consists of an unlined, 0.1-ha (0.25-acre) basin located immediately south of the sedimentation basin along the eastern edge of the landfill (Figure 12.4). This drying bed will allow sludge dewatering through evaporation and infiltration of the water into the subsurface where it can be recaptured by the pumping system. Based on treatability studies of groundwater removed from landfill-monitoring wells, a nonhazardous sludge (predominantly iron and calcium compounds) will be generated from treatment of the wastewater. The dried sludge can be removed for land spreading, composting, or placement within an active landfill as solid waste. The sludge drying basin is sized to allow disposal of the sludge generated from a season of treatment system operation, based on treatability studies 1050 m^3 (37,000 ft^3). However, it is likely that sludge volume reduction within the sedimentation basin due to an increase in solids content from hydrostatic pressure will limit the frequency of sludge-pumping events to less than annual.

Table 12.1   First-Year Removal Characteristics

Parameter	Average System Removal Efficiency (%)	Average System Removal Rate (kg/ha/season)
Volatile organic compounds	97	0.081
Iron	97	3.8
Zinc	93	0.044
Manganese	91	0.36
Arsenic	89	0.0064
Lead	80	0.00021
Mercury	75	0.000037
Chromium	67	0.001
Cadmium	65	0.00056
Nickel	19	0.013
Copper	—	0.00056

12.4 FIRST-YEAR OPERATION AND MAINTENANCE

The system was constructed in fall 1995. The cattails were established in the wetlands in spring 1996 by manual dispersion of seeds from native plants on site. The rapid rate of cattail growth (fully established within 3 months) was higher than anticipated. Some experimentation of the effect of water level within the cells was conducted, by varying the free-water depth at 15, 30, and 45 cm (6, 12, and 18 in.) in the cells. However, no significant difference in cattail establishment was observed. The initial system startup was performed in late May 1996. Only two recovery wells (RW-2 and RW-3) were started. Recovery well RW-2 operated continuously at an average pumping rate of 71 m³/day (13 gpm) throughout the first season, whereas RW-3 operated at an average pumping rate of 245 m³/day (45 gpm), for a total average pumping rate of 316 m³/day (58 gpm). Table 12.1 below shows average removal efficiencies and mass removal rates for total VOCs and various inorganic/heavy metal parameters.

Figures 12.8 through 12.17 show how the dissolved concentrations of these parameters changed through the course of the treatment system. While removal from the solubilized form

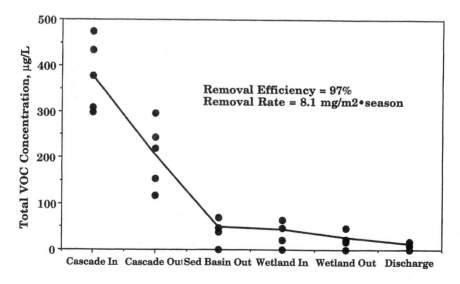

Figure 12.8   Changes in concentrations in volatile organic compounds.

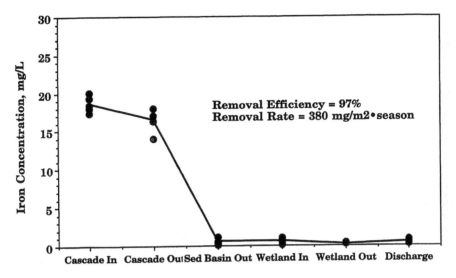

Figure 12.9 Changes in dissolved concentrations of iron.

was consistent for VOCs, iron, zinc, manganese, arsenic, and mercury, with primary removal occurring across the sedimentation basin (and cascade for VOCs), several heavy metals exhibited less-consistent patterns of removal. The removal efficiency for chromium ranged up to 90%, but during the August sampling event, an increase in chromium concentration was observed across the wetland. While high percent and absolute removal efficiency was achieved for lead in July and September, in August only 39% removal efficiency was measured. Cadmium removal ranged from 26 to 98%, although a significant increase occurred across the wetland in one sampling event. Nickel removal efficiencies were inconsistent, and no discernible pattern emerged for copper.

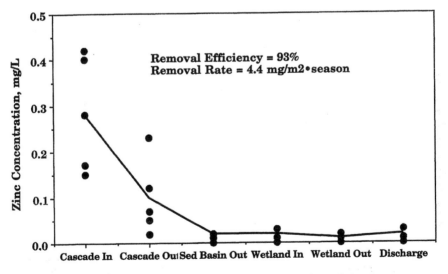

Figure 12.10 Changes in dissolved concentrations of zinc.

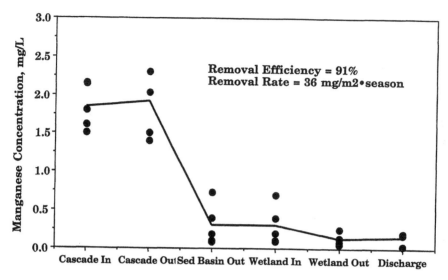

Figure 12.11   Changes in dissolved concentrations of manganese.

Overall, however, efficiency of the system was higher than anticipated for the first year of operation. Heavy metal removal anomalies, such as increases across the wetlands, are expected to attenuate over time as the wetlands become more established and develop an anaerobic zone, thereby encouraging sulfide reduction of metals into the sediments. It is also possible that some heavy metals were solubilized from the MSW-based compost used as rooting soil in the wetlands; again, long-term anaerobic processes should fix the metals in the sediments.

The cascade was cleaned from precipitation and sludge once per month with a stiff push broom, which was more than adequate. Power washing, as originally expected, was not necessary. Sludge accumulation was 15 to 30 cm (6 to 12 in.) near the base of the cascade, but sludge removal will not pose a maintenance issue for the next several years. No significant weather damage has occurred in either of the first two winters endured by the system (although both were harsh by even Minnesota standards). Maintenance of the system was as minimal as intended.

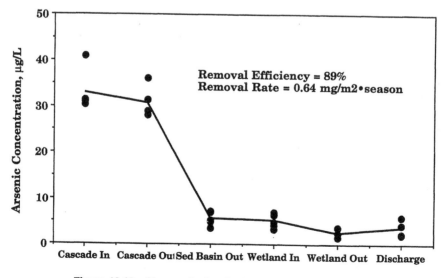

Figure 12.12   Changes in dissolved concentrations of arsenic.

12.5 COSTS

The costs associated with this system include approximately $40,000 for treatability testing, $95,000 for design, and $550,000 for bidding and construction (which is estimated to be approximately 75% of the construction cost of a conventional treatment system). The operation and maintenance costs of this system during the first (start-up) year were $20,000, significantly lower than a conventional treatment system, because operation and maintenance requirements are limited to

A. Energy costs of the pumping system
B. Monthly visits to
    1. check integrity of berms and flow structures
    2. record pumping information
    3. conduct monitoring
    4. mow grass
    5. repair cascade, berms, or pumping system, as necessary
    6. handle sludge, as necessary
    7. remove organic buildup, as necessary

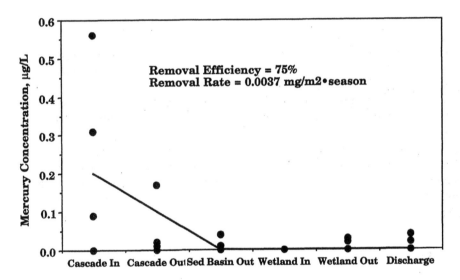

Figure 12.13   Changes in dissolved concentrations of mercury.

The cost of operating natural treatment systems is markedly less than a conventional system as a result of far fewer energy, chemical, and labor costs. Future annual operating costs in present dollars are estimated at $10,000 to $15,000.

It should be noted that treatability testing results and design knowledge gained would significantly reduce engineering costs to utilize this design at another site. In future work, treatability testing would be limited to confirming leachate characteristics to be amenable to natural systems treatment, and design efficiency would increase as the information learned from this innovative design was reapplied.

12.6 SUMMARY AND CONCLUSIONS

An innovative, cost-effective system for landfill leachate treatment has been completed. The design includes groundwater extraction to capture through-flow beneath the landfill coupled with

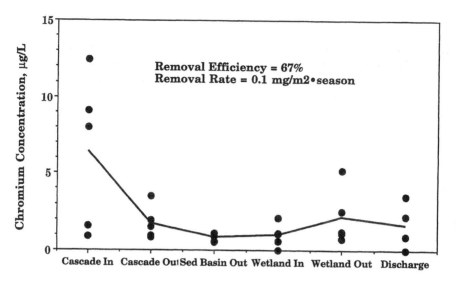

Figure 12.14   Changes in dissolved concentrations of chromium.

a gravity-driven treatment system. The treatment system consists of a cascade aerator, a sedimentation basin, and a FWS constructed wetland. Discharge is to a surface water pond in a borrow area adjacent to the landfill.

Treatment is based upon removal of volatile organics of concern via the cascade aerator, while inorganics will be removed through precipitation, adhesion, settling, and ion exchange in both the sedimentation basin and the constructed wetland. Sludge from the sedimentation basin will be removed as necessary and allowed to dry in a bermed sludge bed adjacent to the sedimentation basin. VOC removal efficiency during the first year of operation varied from 85% to 100%. Efficiency of iron removal varied from 95% to 99%. Heavy metal removal varied from insignificant to 100%.

Leachate control at landfills needs to occur with minimal cost and attention due to the wastewater volume, length of treatment, and often-isolated landfill settings. The high capital and operation and

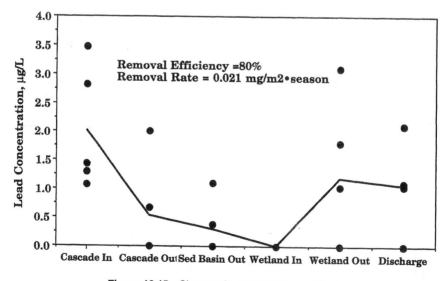

Figure 12.15   Changes in concentrations of lead.

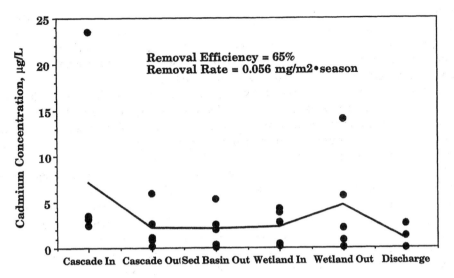

Figure 12.16 Changes in concentrations of cadmium.

maintenance costs of conventional treatment systems dependent upon energy, chemicals, and operator attention, or the cost and difficulty of acceptance associated with off-site hauling of leachate, require that alternative methods be sought to treat groundwater contaminated from landfill leachate. An alternative exists with the use of natural systems engineering to reduce costs while meeting treatment and discharge requirements.

Treatability studies at the Isanti–Chisago Sanitary Landfill indicated a high oxygen requirement to both volatilize and degrade organic compounds and oxidize inorganic compounds such as iron. Treatment efficiency is highly dependent upon the ability to transfer oxygen into the wastewater to obviate the need for chemical or energy inputs. Following aeration, residence times and sufficient storage to allow settling of insoluble compounds must be met. Once these mechanisms have occurred, the wastewater can be polished to meet discharge requirements with the biofiltration capabilities of a constructed wetland.

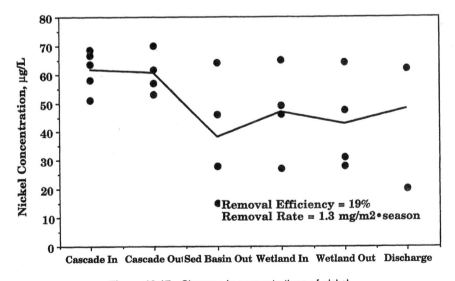

Figure 12.17 Changes in concentrations of nickel.

Advantages of a natural engineering treatment approach, besides reduced cost, include a more ecologically sound system because of far fewer chemical and energy inputs, a system that is fundamentally more capable of treating the varying influent chemistry from a nonuniform waste source due to the diversity of its biological base, and the ability to limit human involvement to monthly visitations and occasional sludge management events.

Disadvantages or uncertainties associated with these systems are the lack of existing data for design without accepting a residual level of risk, because such systems are relatively new and insufficient information is available to date regarding physicochemical processes and capabilities under varying conditions. Design decisions involving layout, length-to-width ratios, parallel vs. series operation, and loading limitations have to be based largely on professional judgment at present. Note, however, that this situation is rapidly changing with the advent of published design texts and increased activity in this field. Additionally, the ability of these systems to "fix" toxic metals within sediments (vs. re-entrainment of these compounds over time) is uncertain, and is dependent upon the influent wastes and operating conditions. The availability of land is also a significant factor in the cost of natural systems, which tend to be land intensive.

However, at locations such as landfills, buffer lands are often available and owned by the landfill owner to keep costs attractive relative to conventional treatment. Overall, the potential gains in cost and ecological worthiness outweigh the uncertainties.

REFERENCES

Kadlec, R. H. and Knight, R. L., 1996. *Treatment Wetlands*, CRC Press, Boca Raton, FL, 893 pp.

Mæhlum, T., 1993. Treatment of Landfill Leachate in On-Site Lagoons and Constructed Wetlands, Jordforsk, Centre for Soil and Environmental Research, N-1432 Ås, Norway, 553 pp.

Martin, C. D. and Moshiri, G.A., 1993. The Use of In-Series Surface-Flow Wetlands for Landfill Leachate Treatment, Wetlands Research Laboratory, The University of West Florida, 11000 University Parkway, Pensacola, FL 32514-5741, 513 pp.

Martin, J. P., Girts, M., and Koncewicz, F., 1993. Design and Construction of a Natural System for Combined Treatment of Mine Tailings and Industrial Landfill Leachates, in *Proceedings of Water Environment Federation, 66th Annual Conference and Exposition*, Anaheim, CA, Oct. 3–7, Vol. 9, pp. 37–45.

U. S. Environmental Protection Agency, 1988. Design Manual: Constructed Wetlands and Aquatic Plant Systems for Municipal Wastewater Treatment, U.S. EPA 625/1-88/022, September, 83 pp.

Water Pollution Control Federation (WPCF), 1990. Natural Systems for Wastewater Treatment, Manual of Practice FD-16, Chap. 9, Alexandria, VA.

CHAPTER 13

A Constructed Wetland System for Treatment of Landfill Leachate, Monroe County, New York

David A.V. Eckhardt, Jan M. Surface, and John H. Peverly

CONTENTS

ABSTRACT: A wetland system was constructed in 1995 for an evaluation of the treatment of landfill leachate. The system consisted of (1) a surface-flow (SF) bed of topsoil and (2) a subsurface-flow (SSF) bed of pea-sized gravel. Both contained *Phragmites australis* reeds and were designed for leachate application at a rate of 1.8 m^3/day. The SF bed served as a pretreatment site for dissolved metals through aeration, oxidation, and subsequent precipitation in a ponded-flow environment. The SSF bed, just downslope, provided an environment for uptake by the reeds and for microbial and geochemical processes after the pretreatment. Samples of leachate inflow and outflow were collected biweekly at the wetland site, from

June 1995 through August 1996 and analyzed for 33 organic and inorganic constituents. Water budgets, chemical loads, and the removal rates of each constituent were calculated for each sampling period. Extractions and analysis of selected constituents were also conducted on samples of the deposited residue and plant tissues from four locations in the wetland system at the end of the study. The data demonstrate that removal rates for the 14 constituents ranged from 49% to 100%. The removal rates were highest for metals, total phosphorus, biochemical oxygen demand (BOD), and volatile organic compounds (VOCs). Removal rates were lowest for the major inorganic ions and barium. Outflow from the wetland system met New York State discharge regulations for all constituents except ammonium, phenol, magnesium, nickel, sodium, and dissolved oxygen. The results indicate the (1) total iron removal was 98%, and load reduction for most metals species was facilitated by oxygen from rainfall, aeration, plant input, especially in the SF bed; (2) nitrogen removal (mostly as ammonium) was 91%; and (3) total phosphorus removal was 99% and was mainly through plant uptake and concentration in plant tissues, especially the rhizomes. An evaluation of the accumulation of solids through sedimentation, sorption, and precipitation reactions is a critical part of pedicting the long-term operation of a wetland system and the ultimate fate of the solid-phase residues.

13.1 INTRODUCTION

Landfill leachate in humid regions is a common source of contamination to surface and groundwater. Leachate collected from landfills typically is hauled to licensed sewage treatment plants for treatment, at substantial transportation and disposal cost, but large volumes of leachate can adversely affect the operation of conventional sewage treatment plants. Conventional on-site treatment entails the costs of (1) constructing a wastewater treatment plant and (2) operation and monitoring costs. Pretreatment of the leachate is often required, especially for metals, which increases the costs. Thus, other methods of leachate treatment are needed that are economically feasible and environmentally sound.

Wetlands have been shown to improve leachate and wastewater quality through processes that include microbially mediated transformations, biotic uptake of organic chemicals and nutrients, and precipitation, complexation, and adsorption reactions (Reddy and Smith, 1987; Brix, 1993a; Kadlec and Knight, 1996). Considerable work has been done on the diverse mechanisms of wastewater improvement within wetlands (Hammer, 1989; Cooper and Findlater, 1990; Water Pollution Control Federation, 1990; Moshiri, 1993). A thorough understanding of these mechanisms is required before a wetland system can be developed that can provide successful, long-term treatment of landfill leachate on a large scale.

The performance of wetland systems can be evaluated through water quality monitoring, where the operational goal is to produce outflow that meets state and federal discharge requirements. Environmental laws differ among different jurisdictions, and the quality and quantity of leachate vary among sites. Hence, each wetland system must be designed individually to provide the appropriate hydraulic and biochemical mechanisms because these mechanisms ultimately determine the success or failure of a system.

The environmental benefits of on-site treatment in a constructed wetland include (1) decreased potential for spills by eliminating the need for off-site transportation, (2) sharp reduction in use of transportation fuel, and (3) decreased energy consumption by using natural processes rather than conventional, electrically driven wastewater treatment processes. Studies of the long-term use of wetlands for leachate treatment have demonstrated significant economic advantages, mainly through lowered construction, transportation, and operation costs (Water Pollution Control Federation, 1990; Kadlec and Knight, 1996).

The City of Rochester, NY, receives about 900 mm of precipitation annually, and the surrounding area of Monroe County, in western New York, has several municipal landfills that generate leachate. Leachate from one of the county landfills, the Northeast-Quadrant Solid-Waste Facility (NQSWF), is trucked to the county Frank E. Van Lare wastewater treatment facility for disposal. The NQSWF landfill was operated from 1975 through 1980 in an unlined pit excavated in a drumlin (glacial till). A leachate collection system was installed after its closure and about 4.5 m³/d of leachate was collected during this study (1995–96) for disposal.

In 1992, the county, in cooperation with the U.S. Geological Survey (USGS) and the New York State Energy Research and Development Authority, began a 1-year pilot study to evaluate the feasibility of treating NQSWF leachate in a constructed wetland. The intent of the study was to assess the potential savings that might be realized through use of the wetland system as an alternative to trucking and conventional treatment.

The objectives of the study were to (1) design and build a demonstration-scale treatment wetland, (2) determine whether outflows from the wetland system meet New York State discharge standards (New York State, 1994), and (3) identify the major water-renovating processes within the wetland and its capacity for constituent removal. The study was based on results from previous work in New York by Staubitz et al. (1989), Sanford et al. (1990), and Surface et al. (1993). This paper presents information on the system design and the analytical results of inflow, outflow, substrate residue, and plant tissue samples. The paper also evaluates the system performance through its first year of operation. Two identical wetlands were constructed side by side, but only one (bed 1) is discussed here because the results from both were similar.

13.2 METHODS

The general approach was to (1) design and construct the wetland system and apply leachate, (2) measure rainfall and the inflow and outflow volumes for calculation of the water budget, (3) collect samples of the inflow and outflow and conduct chemical analyses for calculation of loads and load-reduction efficiencies, and (4) upon completion of the field study, analyze plant tissue and substrate residue from the wetland beds for selected constituents.

13.2.1 Design, Construction, and Operation of the Wetland System

The treatment system is a modification of a constructed wetland that was used to treat leachate from a municipal landfill in nearby Tompkins County (Surface et al., 1993; New York State Energy Research and Development Authority, 1993). It consists of two parallel sets of gently sloping wetland beds lined with an impermeable synthetic liner (Figure 13.1). Each set has two components — a surface-flow (SF) pretreatment wetland followed by a subsurface-flow (SSF) gravel and reed wetland. The SF beds are primary-treatment areas for load reduction of metals and biochemical oxygen demand (BOD) through oxidation and deposition of precipitates in a ponded environment containing emergent aquatic reeds. The adjacent SSF beds are areas for continued treatment as the leachate passes through saturated pea-sized gravel. The SSF beds were designed to provide adequate contact time with the gravel substrate and plant roots to allow biotic and chemical transformations, adsorption and cation exchange, and uptake of water and dissolved chemical constituents by the reeds.

Retention time, expressed as the effective fluid pore space in a bed divided by the flow rate, directly affects the efficiency of treatment within a constructed wetland because treatment typically involves rate-controlled processes. This theoretical approximation of retention time assumes that fluid moves by piston flow through the wetland, but the validity of the assumption was not tested during the operation of the wetlands. First-order, time-dependent rate reactions for BOD derived

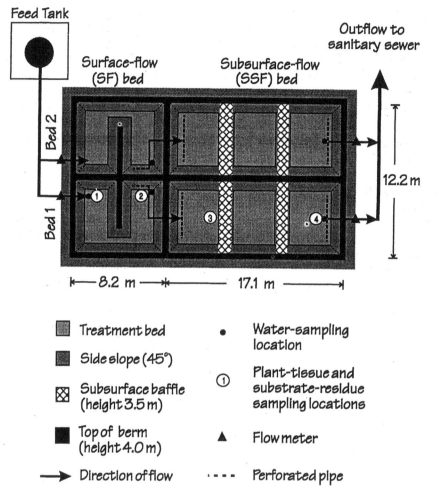

Figure 13.1 Plan view of constructed wetland plots at Van Lare Wastewater Treatment Facility in Rochester, NY.

from chemical reactor theory of treatment processes (Kadlec et al., 1993) were used in the design of the system dimensions and flow rates that control the retention times. Dimensions, slopes, inflow rates, and estimated retention times for the wetland beds are given in Table 13.1.

Each SF bed was filled with 15 cm of topsoil and planted with the reed *Phragmites australis*, although the common cattail *Typha latifolia* quickly invaded the SF beds during operation. A perforated drain pipe was placed at a fixed elevation at the outflow of each SF bed to maintain a 15-cm ponded depth above the topsoil and allow pretreated leachate to flow by gravity into the SSF beds. The SSF beds were constructed with a 60-cm layer of pea gravel substrate and planted with *P. australis*. A perforated outflow pipe was fixed at the bottom of the SSF beds to maintain a leachate level just below the gravel surface and to allow treated effluent to leave the beds. Outflow from the SSF beds was discharged by gravity to a sanitary sewer.

The SSF beds each contained two baffles perpendicular to the flow direction to prevent preferential flow, minimize density stratification, and maintain a fluid level just below the gravel surface throughout the beds. Preferential flow of leachate in a density-stratified region of the beds would reduce retention time and treatment efficiency (Sanford et al., 1990). The baffles also increased the exposure of the relatively dense leachate to the full wetted thickness of the gravel beds and roots near the surface and enhanced the vertical mixing of leachate with rain that entered the beds.

Table 13.1 Design Specifications for Constructed Wetland, Monroe County, NY

Property	Surface-Flow (SF) Bed	Subsurface-Flow (SSF) Bed
Length (m)	18.0	17.1
Width (m)	3.5	6.1
Depth (m)	0.3[a]	0.6[b]
Area (m²)	63	104
Slope (m/m)	0.002	0.005
Aspect ratio (length/width)	5.1	2.8
Inflow rate (m³/day)		
Design	0.9	0.9
Actual	0.6	0.6[c]
Retention time (day)		
Design	21	28
Actual	32	42[c]

[a] Mean ponded water depth.

[b] Saturated thickness of gravel bed.

[c] Value estimated; porosity = 0.4.

Planting of the reeds was completed in spring 1993, and leachate was first applied in July 1995, after the reeds were well established. Leachate was delivered from the landfill to a 26-m³ supply tank at the wetland site and applied at a rate of 0.9 m³/day to the top of each SF bed through separate metering pumps equipped with Teflon and stainless steel pump heads and backpressure valves. The quantity of leachate at the inflow and outflow of each system was measured continuously through in-line flowmeters (Figure 13.1).

13.2.2 Water Budget

A water budget of leachate inflow (In), the SSF outflow (Out), precipitation (P), and evapo-transpiration (ET) was computed at biweekly intervals. Precipitation was directly measured with a tipping-bucket raingauge at the site, and ET was computed as the residual of the other water budget components for each biweekly budget period:

$$ET = In + P - Out \tag{13.1}$$

where units are volumetric (m³) and changes of storage within the system are assumed to be negligible. The volumetric components can be expressed in linear units (m) by dividing the volume by the surface area of bed 1 (167 m²).

13.2.3 Leachate Analysis

Samples of leachate at the inflow to the SF beds and at the outflow of the SF and the SSF beds were collected every 2 weeks for inorganic constituent analysis, and every 4 weeks for organic analysis, from June 1995 through August 1996. In all, 19 sample sets were taken because problems, such as frozen pipes, lack of leachate delivery, and clogged pumps, sometimes prevented biweekly sample collection. Alkalinity was measured on-site on filtered samples by incremental titration with sulfuric acid directly after sample collection. Specific conductance, temperature, and pH were measured on-site by a portable environmental laboratory probe at the sampling points (ponded SF water or through piezometers in the SSF gravel). Dissolved oxygen concentration was measured at the site through Winkler titrations.

Samples for dissolved inorganic constituent analysis were filtered through a 0.45-µg pore-size cartridge, and samples for dissolved organic carbon analysis were filtered through 0.45-µg silver membranes. Samples for metals analysis were immediately preserved with nitric acid. Laboratory analyses were performed by the Monroe County Department of Health, Environmental Health Laboratory in Rochester, NY in accordance with standard analytical methods (American Public Health Association et al., 1985, 1988; Fishman and Friedman, 1989); the laboratory also is a quality-assurance participant in the USGS Standard Reference Sample Program. Cation concentrations were measured by flame atomic-absorption (AA) spectroscopy. Total and soluble-reactive phosphorus (as P) was measured by the ascorbic acid spectrophotometric method. Anion concentrations, except chloride, were measured colorimetrically through an autoanalyzer; chloride concentrations were measured through titrations. Ammonium concentrations were measured through the cadmium reduction and autoanalyzer method. Concentrations of organic nitrogen were not measured. Metal concentrations were measured through AA or graphite furnace methods. Volatile organic compounds (VOCs) were determined through gas chromatography. Dissolved and total organic carbon concentrations were measured through the persulfite ultraviolet oxidation method. BOD was measured through standard methods for a 5-day period.

Constituent loads in the inflow to SF bed 1 and the outflow from SSF bed 1 were calculated for each sample period, and load-reduction efficiencies were evaluated for each constituent over time. The constituent loads (L, in kg) at the inflow and the SSF outflow were calculated for each sample period as

$$L = 10^{-3} \, CV \tag{13.2}$$

where C is the liquid concentration (mg/L) and V is the flow volume (m³). Load-reduction efficiency (LR, in percent) was calculated for the entire monitoring period from the sum of the loads (ΣL)

$$\mathrm{LR} = 100 \, (\Sigma L_{\mathrm{inflow}} - \Sigma L_{\mathrm{outflow}}) \, / \, \Sigma L_{\mathrm{inflow}} \tag{13.3}$$

13.2.4   Plant Tissue and Substrate Residue Analysis

Plant tissue (from above- and belowground) was collected and analyzed for metals and total phosphorus content at the completion of the field study in August 1996, when the reeds had reached maximum biomass after 2 years of growth. The plant tissue samples were collected at four sites in bed 1 — sites 1 and 2 were in the SF bed and sites 3 and 4 were in the SSF bed (Figure 13.1). The reed tissue was separated into three components — leaf, rhizome, and root. The samples were dried at 75°C, and the dry weight of each component was measured. The samples were then ground to a fine powder (mesh size #40), and 1-g subsamples were ashed overnight in a muffle furnace at 450°C. The ashed samples were dissolved with 5N HNO₃, baked at 400°C, and then redissolved with concentrated HCl (Greweling, 1976, p. 21). The HCl solution and remaining residues were filtered, the residues were rinsed with 10% HCl, and the filtrate was diluted to 100 ml in acid-washed bottles. The extract solutions were sent to the laboratory for analysis as water samples. Determinations of analyte mass in the water samples were then normalized to the dry weight of the plant tissues.

The partitioning of metals and phosphorus between the liquid and the solid phase was studied through chemical extractions of substrate residue (Berndt, 1987; Fishman and Friedman, 1989). Substrate material was collected at the same four sites at which plant tissue was sampled. Substrate material at sites 1 and 2 (SF bed) was residue that was deposited during operation of the bed. Substrate material at sites 3 and 4 (SSF bed) was pea gravel with the biofilms and solid-phase residues that resulted from the wetland operation. An extraction of clean (unexposed) gravel that was reserved during construction of the SSF beds provided baseline concentrations for comparison with the samples of SSF gravel that were exposed to leachate.

13.3 RESULTS AND DISCUSSION

13.3.1 Water Budget

A total inflow volume of 249 m³ of leachate was applied to bed 1 during the 410-day monitoring period from July 1995 through August 1996. The volume of treated effluent that was discharged from the SSF outflow during this period was 229 m³, and the loss of water is attributed to ET. Total precipitation during the period was 1110 mm, which amounted to a volume of 184 m³ for the 167 m² area of bed 1. Precipitation was considerably greater than the long-term average monthly rainfall in October 1995 and from April through July 1996. The ET from bed 1 was estimated through Equation 13.1 to be 1.22 m (204 m³). Daily mean ET was 4.6 mm/day in summer (June through September) and 2.3 mm/day in winter (December through March). ET during winter months, when vegetation is dormant, occurs mainly as evaporation, although broken, hollow reed stems may have enhanced water vapor exchange from the wetland to the atmosphere (Brix, 1990).

The monthly mean ET measured in this study (90 mm/month) is comparable to pan-evaporation rates for this region (Brutsaert, 1974). Despite the above-average rainfall, ET caused an 8% decrease in outflow relative to inflow. High rates of ET during two summer sample collection periods coincided with dry weather, and the SSF outflow volumes (and the chemical loads) were zero. The high ET rates were due in part to (1) a constant supply of available water, (2) relatively high leachate temperatures, nearly 23°C, in the SF beds during the summer, and (3) interception of rainfall and uptake by the reeds, especially in the SSF beds, where the reeds grew to a density of about 250 plants/m² and a height of nearly 3 m. The high rate of ET also is attributed to an oasis effect (Brutsaert, 1982), where evaporation is enhanced when a wet area is surrounded by a dry soil area.

13.3.2 Reductions in Concentrations and Loads

Data from the 410-day field study demonstrate that load reductions for the selected constituents ranged from 49 to 100% (Table 13.2). Box plots showing loads of four representative constituents in inflow and outflow are shown in Figure 13.2, and their concentrations are shown in Figure 13.3. The highest removal efficiencies (greater than 90%) were for metals, total phosphorus, phenol, VOCs, BOD, and ammonium, all of which exceeded 90% removal efficiency. The relatively soluble inorganic constituents, such as calcium, magnesium, sodium, chloride, and sulfate, had the lowest removal efficiencies, typically less than 80% (Table 13.2). Constituents whose total loads exceeded 50 kg at the inflow included bicarbonate, chloride, sodium, potassium, BOD, total organic carbon, and ammonium. Concentrations at the SSF outflow met New York State regulations for all constituents except ammonium, phenol, magnesium, nickel, sodium, iron, and dissolved oxygen.

Reductions in concentrations for several constituents within the SF bed differed from those in the SSF bed. For example, ammonium and phosphorus concentrations (Figure 13.3) decreased more in the SSF bed than in the SF bed, where the nutrients had less direct contact with plant roots. In contrast, iron and BOD concentrations decreased mostly in the SF bed, where the slow-moving ponded flow provided direct exposure to the atmosphere and enhanced oxidation. Concentrations of some constituents, such as sodium and chloride, actually increased in transit through the SF beds, apparently through evaporative concentration (Table 13.2). In general, the relatively soluble inorganic ions underwent greater removal in the SSF bed than in the SF bed, where uptake by plants and cation exchange were less active. Magnesium and sodium concentrations in the SSF outflow consistently exceeded the state limits of 35 and 20 mg/L, respectively, however.

13.3.2.1 Dissolved Oxygen

Dissolved oxygen (DO) in the inflow samples was consistently zero, but DO was detected in 1 of 19 samples at the SF outflow and 6 of 19 samples at the SSF outflow. The DO in these samples

Table 13.2　Median Concentrations, Total Loads, and Load Reduction for Leachate Constituents Applied to Constructed Wetland, Monroe County, NY, 1995–96.

Constituent	Median Concentration, mg/L except as noted			Load, kg		Load Reduction, %
	Inflow	SF outflow	SSF outflow	Inflow	SSF outflow	
pH (standard units)	6.9	7.4	6.9	—	—	—
Specific conductance (µS/cm)	4740	3900	968	—	—	—
Oxygen, dissolved	0.0	0.0	0.0	—	—	—
Alkalinity, as $CaCO_3$	2440	1780	680	605	124	79
Bicarbonate, as $CaCO_3$	2970	2170	830	737	152	79
Ammonium, dissolved, as N	230	133	18	53	4.8	91
Nitrite, dissolved, as N	< 0.05	< 0.05	< 0.05	0.031	0.006	80
Nitrate, dissolved, as N	< 0.05	< 0.05	< 0.05	0.004	0.210	—
Phosphorus, total, as P	1.9	1.5	0.18	5.9	0.042	99
Phosphorus, soluble-reactive, as P	0.36	0.36	0.09	0.150	0.019	88
Calcium, total	180	70	77	42	16	63
Magnesium, total	160	150	50	37	10	72
Potassium, total	269	196	64	70	12	83
Sodium, total	410	430	130	110	23	79
Chloride, dissolved	490	510	160	125	28	78
Sulfate, total, as SO_4	23	20	10	8.0	3.0	62
Organic carbon, total	160	120	33	52	6.7	87
Organic carbon, dissolved	140	115	34	40	6.4	84
Biochemical oxygen demand, 5-day	70	27	10	51	2.4	95
Aluminum, total	0.21	0.08	< 0.05	0.475	0.005	99
Aluminum, dissolved	< 0.05	< 0.05	< 0.05	0.009	Nd	100
Barium, total	0.310	0.140	0.380	0.151	0.078	49
Barium, dissolved	0.195	0.080	0.380	0.059	0.074	—

						%
Cadmium, total	< 0.001	< 0.001	< 0.001	0.712	0.028	96
Cadmium, dissolved	< 0.001	< 0.001	< 0.001	Nd	Nd	—
Chromium, total	0.013	0.009	< 0.005	0.00660	0.00015	98
Chromium, dissolved	0.010	0.006	< 0.005	0.00310	0.00003	99
Copper, total	0.030	< 0.010	< 0.010	0.0260	0.0032	89
Copper, dissolved	< 0.010	< 0.010	< 0.010	0.0048	0.0003	93
Iron, total	51	15	1.3	15	0.27	98
Iron, dissolved	26	5.4	1.1	6.2	0.17	97
Lead, total	0.013	< 0.005	< 0.005	0.0960	0.0002	100
Lead, dissolved	< 0.005	< 0.005	< 0.005	0.0008	Nd	100
Mercury, total	< 0.0002	0.0002	< 0.0002	0.000012	Nd	100
Nickel, total	0.065	0.060	0.030	0.028	0.005	82
Nickel, dissolved	0.060	0.040	0.030	0.020	0.004	81
Selenium, total	< 0.002	< 0.002	< 0.002	Nd	Nd	—
Silver, total	< 0.002	< 0.002	< 0.002	0.0013	Nd	100
Zinc, total	0.227	0.080	0.025	0.750	0.005	99
Zinc, dissolved	0.048	0.025	0.010	0.020	0.003	86
Phenol, total	56	19	3.0	15	0.60	96
Benzene	0.0055	0.0006	<0.0005	0.00150	0.00008	94
Ethylbenzene	0.0190	0.0007	0.0004	0.0063	0.0001	99
Toluene	0.0220	0.0013	<0.0005	0.0180	0.0001	99
Xylene	0.045	0.002	0.001	0.0140	0.0002	98

Note: Locations shown in Figure 13.1. SF, surface-flow bed; SSF, subsurface-flow bed; Nd, not detected; dashes indicate no data.

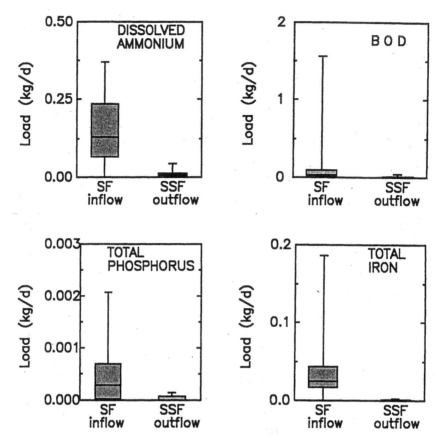

Figure 13.2 Median and range in values of loads in inflow to surface-flow (SF) bed and in discharge from subsurface-flow (SSF) bed at constructed wetland, 1995–96. (Central bar indicates median; box encompasses 25th to 75th percentiles; whiskers indicate minimum and maximum.)

was less than 5 mg/L and was present only after heavy rainfall and therefore is attributed to dissolved oxygen in the rainfall. Values for pH ranged from 6.0 to 7.8, but no consistent differences were observed among the inflow, SF outflow, and SSF outflow samples. A pE value of about –3 at the inflow was estimated through a geochemical speciation model (Parkhurst et al., 1980) on the basis of a sulfide–sulfate redox couple measured in one sample. The low pE and the lack of oxygen in the leachate indicate reducing conditions and the presence of iron and nitrogen species mainly as ferrous iron and ammonium.

13.3.2.2 Nitrogen Species

Ammonium is often the most persistent constituent encountered in leachate remediation. Ammonium concentrations in wastewater discharges are limited by the state to prevent excessive DO demand and toxicity to fish and invertebrates in the receiving streams. Ammonium concentrations in discharge from wastewater treatment wetlands in North America have ranged from 0.01 to 23 mg/L (Knight et al., 1993). Treatment efficiencies in these systems typically are about 90% for ammonium, as was observed in this study (Table 13.2). Nonetheless, the ammonium concentrations in the SSF outflow in this study were unacceptably high and consistently exceeded the state limit (2 mg/L) by about a factor of 10 or more (Figure 13.4). Knight et al. (1993) suggest that the efficiency of ammonium removal declines as inflow concentrations increase or when inflow-loading rates exceed 20 kg/ha/day. The loading rate for ammonium in this study was 7.7 kg/ha/day, and

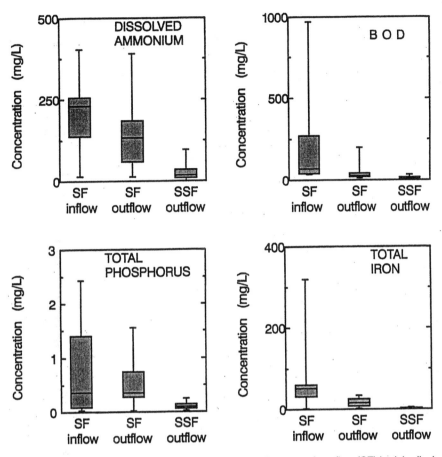

Figure 13.3 Median and range in values of concentrations in inflow to surface-flow (SF) bed, in discharge from SF bed, and in discharge from subsurface-flow (SSF) bed at constructed wetland, 1995–96.

inflow concentrations ranged from 14 to 400 mg/L (Figure 13.3). However, mineralization of organic nitrogen, which was not measured but inferred to be present in the leachate (Pearsall and Aufterheide, 1995), likely occurred in the wetland. The mineralization of organic nitrogen to ammonium would effectively increase the loading rate of ammonium to the wetland beds and decrease its treatment efficiency.

Nitrification and plant uptake are primary removal mechanisms for the ammonium (Kadlec and Knight, 1996). Some ammonium is nitrified where oxygen enters the system, and a net increase in nitrate load was measured between the inflow and SSF outflow (Table 13.2). In addition to nitrification, atmospheric deposition contributes some nitrate, but nitrogen from this source was not directly measured. Nitrite and nitrate were detected in only 1 of 18 inflow samples, but both were detected at concentrations less than 7 mg/L in outflow samples — nitrite was in 5 of 18 samples and nitrate was in 11 of 18 samples at the SF outflow, and nitrite was in 4 of 18 samples and nitrate was in 7 of 18 samples at the SFF outflow. Nitrate concentrations exceeded nitrite concentrations in all samples, but the difference was greater at the SSF outflow than at the SF outflow. The presence of nitrite in an anaerobic environment indicates some nitrogen removal through denitrification and outgassing. The loss of ammonia through volatilization is assumed to be nearly negligible, however, because dissolved ammonia (gas) concentrations are insignificant at the nearly neutral pH of the leachate (Stumm and Morgan, 1981, p. 450). The sum concentration of nitrite and nitrate in outflows was always less than the state limit of 10 mg/L.

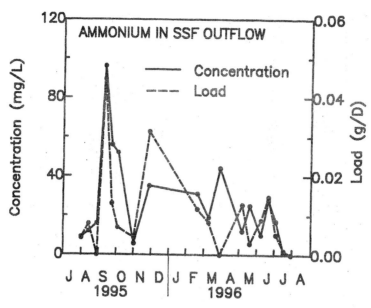

Figure 13.4 Dissolved ammonium concentrations and loads in outflow from subsurface-flow (SSF) bed, July 1995 through August 1996.

13.3.2.3 Alkalinity and Metals

Leachate from the NQSWF is highly alkaline and provided buffering that maintained a nearly neutral pH of the leachate within the wetlands (Table 13.2). The alkalinity in the inflow has been shown to be a major factor in the kinetics of contaminant removal, notably for iron and manganese (Hedin and Nairn, 1993). Although manganese concentrations were not monitored in this study, the load of total iron in the inflow was 15 kg, much of which was deposited as precipitated sludge in the SF beds. In this system, treatment efficiency for the total iron load was 98% (Table 13.2, Figure 13.2), but iron concentrations in the SSF outflow consistently exceeded the New York State limit (0.3 mg/L) by at least a factor of 3. Much of the iron removal was through oxidation and hydrolysis reactions that produce oxyhydroxides, such as goethite (FeOOH), which precipitate and settle. Hydrolysis also produces a net gain in the hydrogen ion activity. Without bicarbonate alkalinity to neutralize the hydrogen ion activity, the pH would rapidly decrease, which would increase iron solubility and reduce the treatment efficiency for iron. The bicarbonate in the leachate reacts with the hydrogen ion, however, to form water and carbon dioxide, and thereby decreases the alkalinity load. Thus, the load-reduction processes for metals and bicarbonate alkalinity are favorably linked.

Dissolved iron and alkalinity concentrations in the SSF outflow were inversely correlated with rainfall volume ($r = -0.52$, $p = 0.02$ for iron; and $r = -0.47$, $p = 0.05$ for alkalinity) during the 19 sampling periods; this is attributed to the presence of oxygen that enters the wetlands through meteoric water. Mixing of the meteoric water and its oxygen load with the denser leachate is enhanced by the SSF bed baffles (Figure 13.1). Diffusive inputs of oxygen at the ponded surface of the SF beds also is important in the treatment of iron, as shown by its concentration decrease in the SF beds (Figure 13.3). Weider (1989) found that treatment efficiency for iron was directly correlated to bed area and inversely correlated to bed depth; clearly, surface area and ponded depth affect oxygen transfer from the atmosphere. Additional oxygen input is provided at root surfaces and is discussed later.

In general, metals in the inflow were efficiently removed (Table 13.2), although the treatment for nickel was notably less efficient than for others. This result is consistent with findings from other studies (Sinicrope et al., 1992; Eger et al., 1993) and may be due to the relatively slow kinetics for nickel hydrolysis in the presence of other metals. Kadlec and Knight (1996) indicate that the efficiency of nickel removal may be directly correlated with its inflow concentration. In this study, nickel in the SSF outflow consistently exceeded the state limit (5.6 μg/L) by a factor of more than 5.

13.3.2.4 Barium

Barium, which can be present as colloidal hydroxides and oxides (Hem, 1985), was not efficiently removed, apparently because barium from witherite (barium carbonate) in the SSF beds was being released into solution. As a result, the median dissolved and median total barium concentrations at the SSF outflow exceeded the respective median values at the SF inflow. The total barium load in the inflow decreased only 49%, but the dissolved barium load actually increased (Table 13.2). The median dissolved barium concentration in the SSF outflow was 0.38 mg/L, which exceeds by about a factor of 3 the solubility equilibrium for barite (barium sulfate). In the presence of 10 mg/L of sulfate, which is the median sulfate concentration in the SSF-bed outflow, the equilibrium barium concentration is 0.14 mg/L (Sillen and Martell, 1964). Thus, barium was probably released from witherite, and possibly barite, in the SSF substrate. Barium concentrations in the outflow did not exceed the state limit of 1 mg/L, however.

13.3.2.5 Organic Compounds

The leachate contained relatively high concentrations of total (TOC) and dissolved organic carbon (DOC), which were decreased more effectively in the SSF beds than the SF beds. The median DOC concentration in the SSF outflow was 33 mg/L, which is typical of DOC concentrations in natural wetland outflows (Kadlec and Knight, 1996). Phenols and the VOCs that include benzene, ethylbenzene, toluene, and xylene were also efficiently removed, as indicated by load reductions of at least 90%. Most of the VOC load was effectively reduced in the SF beds (Table 13.2). The load-reduction efficiency for phenol was 96%, but phenol concentrations in the SSF outflow frequently exceeded the state standard of 1 mg/L.

13.3.3 Plant Tissue and Substrate Residue Analyses

Concentrations of constituents in plant tissue and substrate residue samples from the four sampling sites at the completion of the study are shown in Table 13.3. The total biomass of the plant tissue from the SF bed at the time of sampling was about 400 kg, and that in the SSF bed was about 1000 kg. Concentrations in the sludgelike, solid-phase residue that was sampled in the SF bed (sites 1, 2) show that the SF bed was the main sink for the metals, notably, iron and manganese. The total mass of the SF bed residue, estimated from its bulk density and thickness measurements, exceeded the total plant tissue biomass by a factor of at least 10. Thus, the significant load reductions of the metal species occurred in the SF beds mostly through precipitation reactions and deposition of the solid phases. Oxide, hydroxide, and carbonate precipitates probably represent the greatest removal of metals, especially iron, from the leachate; the metals may also react to form phosphate, sulfate, and sulfide precipitates. Metals also are removed in the wetlands through (1) bioaccumulation in the plants, (2) adsorption onto solid-phase surfaces, and (3) chelation or complexation by organic material and biofilms (Ross, 1994).

The accumulation of solids through sedimentation, sorption, and precipitation reactions is a critical part of predicting the long-term operation of a wetland system and the ultimate fate of the solid-phase residues. Hydraulic conductivity of SSF substrates can significantly decrease as pores become blocked by bacterial films and organic and mineral precipitates. At the end of the 410 days of operation in this study, the SF outlet pipes were partially clogged with sludgelike residue and would require regular maintenance to enable long-term operation. Clogging was not observed in

Table 13.3 Chemical Content of Substrate Residue Extracts and Plant Tissue Digests from Four Locations within Constructed Wetland, Monroe County, NY, 1996

Material and Site	Constituent (mg/kg)												
	P	K	Na	Ca	Mg	Fe	Mn	Cd	Cu	Pb	Zn	Ba	Cr
Substrate													
1	375	786	786	4150	1290	2720	307	0.2	5.0	19.6	13.6	50.8	1.3
2	418	378	679	3250	1080	2560	298	0.3	4.9	16.9	13.8	58.1	1.4
3	25.5	30.6	50.0	40400	25900	644	99.9	<0.1	0.5	1.1	2.0	20.2	0.3
4	23.4	21.9	57.6	45900	21600	504	126	<0.1	0.8	1.1	3.2	17.1	0.3
Unexposed	41.2	23.6	42.9	35000	21900	578	105	<0.1	0.4	0.9	3.2	14.9	0.3
Leaves													
1	883	4120	10600	14600	3530	618	264	0.1	2.9	0.8	16.7	15.7	0.5
2	994	6800	8690	9390	3800	799	478	0.3	3.0	2.0	16.0	13.0	1.4
3	897	11000	706	1280	883	108	66.7	<0.1	2.0	<0.5	12.7	10.8	0.8
4	701	9120	637	3240	883	265	265	0.5	1.0	<0.5	33.3	26.5	0.7
Rhizomes													
1	2600	6100	4400	4400	4000	1800	142	0.2	2.0	1.0	50.0	24.0	0.5
2	2140	6770	4860	3720	3910	2190	168	<0.1	4.8	1.2	47.7	29.6	1.5
3	979	9180	1110	404	807	172	33.3	<0.1	2.0	<0.5	11.6	5.0	0.9
4	206	13900	663	316	384	125	49.9	<0.1	1.0	1.0	11.5	3.8	0.9
Roots													
1	795	3200	6080	13900	3200	9280	227	1.5	6.4	14.0	28.8	11.7	3.4
2	1090	290	2490	10100	2680	7060	—	0.5	4.0	5.0	37.8	72.6	5.0
3	866	14300	14300	11600	2600	2500	195	0.3	10.0	2.1	24.0	25.0	3.2
4	395	7040	7040	35100	16000	4570	567	<0.1	23.8	2.9	41.9	21.9	27.3

Note: Sampling locations shown in Figure 13.1. Sites 1 and 2 are in SF bed; Sites 3 and 4 are in SSF bed; dashes indicate no data.

the gravel substrate at the two sampling sites in the SSF bed (sites 3, 4), probably because most of the precipitates were deposited in the SF wetland and did not reach the SSF bed. The absence of significant deposition of mineral precipitates in the SSF beds is clearly supported by the similarity in concentrations in the substrate extracts at sites 3 and 4 relative to those in the unexposed substrate sample (Table 13.3).

The highest concentrations of iron were extracted from the surfaces of roots and, to a lesser degree, from rhizomes, especially in the SF bed (Table 13.3 and Figure 13.5). The mass transfer of oxygen through the vascular systems of the reeds is well documented (Armstrong et al., 1990; Brix, 1993b), and iron deposition occurs where the oxygen is exuded from the plant tissues in contact with the anaerobic pore water in the root zone. In turn, carbon dioxide gas can pass upward and leave the wetland through the reeds. Concentrations of iron in the leaves (Figure 13.5) show that little iron is assimilated into plant tissue.

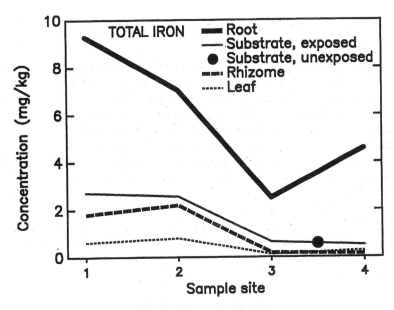

Figure 13.5 Total iron concentrations in substrate residue extracts and plant tissue digests. (Sample locations shown in Figure 13.1.)

Total phosphorus removal (99%) occurred mainly through plant uptake and assimilation in plant tissues, especially the rhizomes. The data in Table 13.3 show that phosphorus was concentrated in all parts of reed tissues, but most notably in rhizomes in the SF beds. Relatively low concentrations in the roots and rhizomes at site 4, in the SSF bed, indicate that plant uptake of phosphorus had not yet reached its full potential. Relatively high concentrations of phosphorus were seen in the solid-phase residue in the SF beds, and this is attributed to organic complexes, decay of biomass that is entrapped in the deposits, and precipitation of phosphate minerals. In contrast, negligible deposition of phosphorus was observed in the SSF substrate.

At the conclusion of the study, plants in the SF beds were notably less vigorous and healthy than those in the SSF beds, although the invasive *Typha* grew better than the *Phragmites* in the SF beds. Restricted root growth and loss of plant vigor in the SF beds are attributed mainly to sludge deposition. Accordingly, leachate treatment by the reeds was less effective in the SF beds than in the SSF beds. The importance of plant uptake in the SSF bed is supported by the results of potassium extractions from plant tissues (Table 13.3), which show higher concentrations in the SSF samples (sites 3, 4) than in SF samples (sites 1, 2). Potassium, a macronutrient, is less reactive than phosphorus and ammonium and shows a lower treatment efficiency. Some wetlands actually release potassium through mineral dissolution (Richardson, 1989), although the clean, unexposed gravel sample appears to be an insignificant source (Table 13.3). Thus, a primary removal mechanism for the nutrients (nitrogen, phosphorus, and potassium) is plant uptake, although, in the absence of periodic plant harvesting, decaying plant litter could readily release the nutrients back into solution.

Excavation of plant roots indicated that root growth extended through only half of the saturated thickness of the gravel substrate in the SSF beds. Shallow rooting depths (about 30 cm) limited nutrient uptake and also reduced the contact of leachate with oxygen at the roots. As the beds mature and root density and depth increase during reed growth over several years, treatment efficiency in the SSF beds may increase, especially for iron and ammonium. An improved treatment efficiency for these constituents may also be realized through expansion of the SF and SSF bed area and additional aeration of the wetland outflow.

13.4 SUMMARY AND CONCLUSIONS

A constructed wetland system in Monroe County, NY, was operated for 410 days in 1995–96 for an evaluation of the treatment of municipal landfill leachate. The system consisted of (1) a surface-flow (SF) bed of topsoil and (2) a subsurface-flow (SSF) bed of pea-sized gravel. Both contained *Phragmites australis* reeds and were designed for leachate application at a rate of 1.8 m³/day. Results indicate that load reductions for the 33 constituents of interest ranged between 49 and 100%. Removal efficiencies exceeded 90% for most metal species, total phosphorus, phenol, VOCs, BOD, and ammonium. Water discharged at the wetland outflow met New York State discharge regulations for all constituents except ammonium, phenol, magnesium, nickel, iron, sodium, and dissolved oxygen. An improved treatment efficiency for these constituents, however, may be realized through expansion of the SF and SSF bed area and additional aeration of the wetland outflow. The treatment efficiencies measured in this initial phase of operation are expected to increase with time as the biological activity develops in the beds and as root density and depth increase, especially in the SSF beds.

ACKNOWLEDGMENTS

Thanks are extended to Richard Burton, Anna Madden, and Charles Knauf of the Monroe County Environmental Health Laboratory, who provided project reviews and coordinated the analytical laboratory services; to Michael Schifano, John Pitts, and Edward Harding of the Monroe County Department of Environmental Services, who assisted in project reviews and coordinated leachate deliveries at the field site; to John Kierecki and Andy Scarboro of the Frank E. Van Lare Wastewater Treatment Facility, who assisted in the construction and maintenance of the wetland system; and James Reis of the New York State Energy Research and Development Authority, which provided funding support.

REFERENCES

American Public Health Association, American Water Works Association, and Water Pollution Control Federation, 1985. *Standard Methods for the Examination of Water and Wastewater*, 16th ed., American Public Health Association, Washington, D.C., 1268 pp.

American Public Health Association, American Water Works Association, and Water Pollution Control Federation, 1988. *Standard Methods for the Examination of Water and Wastewater*, suppl. to 16th ed., American Public Health Association, Washington, D.C., 161 pp.

Armstrong, W., Armstrong, J., and Beckett, P. M., 1990. Measurement and modeling of oxygen release from roots of *Phragmites australis*, in Cooper, P. F. and Findlater, B. C., Eds., *Constructed Wetlands in Water Pollution Control*, Pergamon Press, Oxford, U.K., 41–52.

Berndt, M. P., 1987. Metal Partitioning in a Sand and Gravel Aquifer Contaminated by Crude Petroleum. Master's thesis, Syracuse University, Syracuse, NY, 63 pp.

Brix, H., 1990. Gas exchange through the soil-atmosphere interphase and through dead culms of *Phragmites australis* in a constructed wetland bed receiving domestic sewage, *Water Research* 24:259–266.

Brix, H., 1993a. Wastewater treatment in constructed wetlands — system design, removal processes, and treatment performance, in Moshiri, G. A., Ed., *Constructed Wetlands for Water Quality Improvement*, Lewis Publishers, Boca Raton, FL, 391–398.

Brix, H., 1993b. Macrophyte-mediated oxygen transfer in wetlands — Transport mechanisms and rates, in Moshiri, G. A., Ed., *Constructed Wetlands for Water Quality Improvement*, Lewis Publishers, Boca Raton, FL, 391–398.

Brutsaert, W., 1974. Evaporative Water Loss from Large Land Areas in the Finger Lakes, Ithaca, NY, Cornell University Water Resources and Marine Sciences Center, Technical Report No. 86, 5 pp.

Brutsaert, W., 1982. *Evaporation into the Atmosphere — Theory, History, and Applications*, Kluwer Press, Boston, 299 pp.

Cooper P. F., and Findlater, B. C., Eds., 1990. *Constructed Wetlands in Water Pollution Control*, Pergamon Press, Oxford, U.K., 605 pp.

Eger, P., Melchert, G., Antonson, D., and Wagner, J., 1993. The use of wetland treatment to remove trace metals from mine drainage, in G. A. Moshiri, Ed., *Constructed Wetlands for Water Quality Improvement*, Lewis Publishers, Boca Raton, FL, 171–178.

Fishman, M. J. and Friedman, L. C., Eds., 1989. *Methods for Determination of Inorganic Substances in Water and Fluvial Sediments: Techniques of Water Resources Investigations of the United States Geological Survey*, Book 5, Chapter A1, 545 pp. U.S. Geological Survey, Reston, VA.

Greweling, T., 1976. Chemical analysis of plant tissue, *Search Agriculture, Cornell University Agronomy*, Ithaca, NY, 6:1–35.

Hammer, D. A., Ed., 1989. Constructed Wetlands for Wastewater Treatment, in *Proceedings from the First International Conference on Constructed Wetlands for Wastewater Treatment*, Chattanooga, TN, June 13–17, 1988, Lewis Publishers, Ann Arbor, MI, 831 pp.

Hedin, R. S. and Nairn, R. W., 1993. Contaminant removal capabilities of wetlands constructed to treat coal mine drainage, in G. A. Moshiri, Ed., *Constructed Wetlands for Water Quality Improvement*, Lewis Publishers, Boca Raton, FL, 187–195.

Hem, J. D., 1985. *Study and Interpretation of the Chemical Characteristics of Natural Water*, 3rd ed., U.S. Geological Survey Water-Supply Paper 2254, 264 pp.

Kadlec, R. H. and Knight, R. L., 1996. *Treatment Wetlands*, Lewis Publishers, Boca Raton, FL, 893 pp.

Kadlec, R. H., Bastiaens, W., and Urban, D. J., 1993. Hydrological design of free-water surface treatment wetlands, in G. A. Moshiri, Ed., *Constructed Wetlands for Water Quality Improvement*, Lewis Publishers, Boca Raton, FL, pp. 77–86.

Knight, R. L., Ruble, R. W., Kadlec, R. H., and Reed, S. C., 1993. Wetlands for wastewater treatment — performance database, in Moshiri, G. A., Ed., *Constructed Wetlands for Water Quality Improvement*, Lewis Publishers, Boca Raton, FL, 35–58.

Moshiri, G. A., Ed., 1993. *Constructed Wetlands for Water Quality Improvement*, Lewis Publishers, Boca Raton, FL, 632 pp.

New York State, 1994. Water quality regulations — surface water and groundwater classifications and standards, in *New York State Codes, Rules, and Regulations*, Albany, Title 6, Chapter X, parts 700–705.

New York State Energy Research and Development Authority (NYSERDA), 1993. *Constructed Wetlands for Municipal Solid Waste Landfill Leachate Treatment*, NYSERDA Report 94-1, Albany, 258 pp.

Parkhurst, D. L., Thorstenson, D. C., and Plummer, L. N., 1980. *PHREEQE - A Computer Program for Geochemical Calculations*, U.S. Geological Survey, Water Resources Investigation Report, 80-96, 193 pp.

Pearsall, K. A. and Aufterheide, M. J., 1995. *Groundwater Quality and Geochemical Processes at a Municipal Landfill, Town of Brookhaven, Long Island, New York*, U.S. Geological Survey, Water Resources Investigation Report 912-4154, 45 pp.

Reddy, K. R. and Smith W. H., Eds., 1987. *Aquatic Plants for Water Treatment and Resource Recovery*, Magnolia Publishing, Orlando, FL, 1032 pp.

Richardson, C. J., 1989. Freshwater wetlands — transformers, filters, or sinks? in Sharitz, R. R., and Gibbons, J. W., Eds., *Freshwater Wetlands and Wildlife*, Oak Ridge, TN, U.S. Department of Energy, 25–46.

Ross, S. M., Ed., 1994. *Toxic Metals in Soil-Plant Systems*, John Wiley & Sons, New York, 469 pp.

Sanford, W. E., Kopka, R. J., Surface, J. M., and Lavine, M. J., 1990. Rock-Reed Filters for Treating Landfill Leachate, American Society of Agricultural Engineers, December 1990, Chicago, Paper No. 90-2537, 14 p.

Sillen, L. G., and Martell, A. E., 1964. *Stability Constants of Metal–Ion Complexes*, Chemical Society, Special Publication 17, London, 754 pp.

Sinicrope, T. L., Langis, R., Gersberg, R. M., Busanardo, M. J., and Zedler, J. B., 1992. Metal removal by wetland mesocosms subjected to different hydroperiods, *Ecological Engineering* 1:309–322.

Staubitz, W. W., Surface, J. M., Steenhuis, T. S., Peverly, J. H., Lavine, M. J., Weeks, N. C., Sanford, W. E., and Kopka, R. J., 1989. Potential use of constructed wetlands to treat landfill leachate, in Hammer, D A., Ed., *Constructed Wetlands for Wastewater Treatment*, Lewis Publishers, Boca Raton, FL, 735–742.

Stumm, W. and Morgan, J. J., 1981. *Aquatic Geochemistry*, 2nd ed., John Wiley & Sons, New York, 780 pp.

Surface, J. M., Peverly, J. H., Steenhuis, T. S., and Sanford, W. E., 1993. Effect of season, substrate composition, and plant growth on landfill leachate treatment in a constructed wetland, in Moshiri, G. A., Ed., *Constructed Wetlands for Water Quality Improvement*, Lewis Publishers, Boca Raton, FL, 461–472.

Water Pollution Control Federation, 1990. *Natural Systems for Wastewater Treatment*, Manual of Practice FD-16, Alexandria, VA, 270 pp.

Weider, R. K., 1989. A survey of constructed wetlands for acid-coal mine drainage treatment in the eastern United States, *Wetlands* 9:299–315.

CHAPTER 14

Seasonal Growth Patterns
in Wetland Plants Growing in Landfill Leachate

John M. Bernard

CONTENTS

ABSTRACT: Large wetland plant species have been shown to be successful in a variety of constructed wetland treatment systems. They are generally easy to propagate and tolerate a range of climates, water levels, and chemical conditions. Shoots of these plants, separated by horizontal rhizomes, grow rapidly in early spring by photosynthesis and translocation of materials from belowground. They reach large size and some flower by approximately July 1, then begin to translocate carbohydrates belowground for growth of new rhizomes. In late summer and autumn, rhizomes produce new shoots, many of which overwinter and form the next year's population. In constructed wetlands, shoots filter leachate, which helps to prevent clogging of the soil and may remove quantities of some chemicals if harvested during the growing season.

14.1 INTRODUCTION

Constructed wetland treatment systems have become increasingly common in recent years and have been used as filters for a wide variety of pollutants. While large wetland macrophytes are an important part of these systems, relatively few studies have focused on the seasonal growth patterns, production, and nutrient uptake capabilities of the plants used, and this is particularly true for those growing in landfill leachate. Thus, predictions about plant behavior in landfill leachate must come in large part from consideration of growth patterns in both natural wetland systems and other, different types of polluted systems.

Species life history is an important attribute of plants, and studies devoted to seasonal growth patterns are important to a full understanding of ecosystem function. A knowledge of life history has become even more important in recent years because of the development and use of constructed wetlands to reduce pollutants (Richardson and Davis, 1987; Hammer, 1989; Cooper and Findlater, 1990; Huttunen et al., 1996). These constructed wetlands have been used in a variety of projects including those to treat sewage (Brix and Schierup, 1989), agricultural (Findlayson and Chick, 1983) and urban runoff (Meiorin, 1989) and, more recently, landfill leachate (Trautman et al., 1989; Dobberteen and Nickerson, 1991; Peverly et al., 1993; Bernard and Lauve, 1995). While almost all projects have included plantings of aquatic or wetland species in the constructed beds, most studies have focused on input–output relationships of pollutants with relatively little attention paid to the plants in the beds (Guntenspergen et al., 1989). This is changing because it is now realized that different species may have markedly different success depending on such factors as type and toxicity of individual pollutants, water level, temperature, and soil type.

Different projects have used different types of plants and while Guntenspergen et al. (1989) listed 32 different species tested for use in constructed wetlands, most projects done in the north temperate zone have used large perennial wetland species, and this paper will focus on such plants. Probably the majority of projects have used reed (*Phragmites australis*), but reed canary grass (*Phalaris arundinacea*), mannagrass (*Glyceria* sp.), cattail (*Typha* sp.), and various species of sedges such as *Carex* sp. and *Scirpus* sp. have been used. These species and others of the same type fulfill the most important requirements for use in treatment systems, including ecological acceptability, tolerance to a range of climates, resistance to disease, and ability to tolerate a wide range of pollutants (Tanner, 1996). These species are also easy to propagate and generally grow very well although, as Daniels (1991) and Tanner (1996) point out, *Phragmites* may not perform as well in constructed wetlands as in natural wetlands.

The life history of many large wetland species has been studied in recent years. For example, large wetland *Carex* species have been the focus of many studies (see reviews by Bernard, 1990 and Bernard et al., 1988), while Haslam (1970; 1973) focused on *Phragmites* and Grace and Wetzel (1981; 1982) studied *Typha* sp. Fiala (1976; 1978) studied seasonal development in roots and rhizomes of *Phragmites* and *Typha* sp. Thus, although much is known of seasonal life history, production, and nutrient requirements of these species in natural wetlands, little is known of their growth in constructed beds, particularly those receiving landfill leachate. This paper will rely importantly on results and observations made in a project in central New York at the Town of Fenton landfill (Bouldin et al., 1993; Bernard and Lauve, 1995) which used *Phalaris* in two overland flow beds and the cattail *T. glauca* in two root-zone beds, set up in series. A somewhat similar project in the Town of Ithaca landfill used *Phragmites* (Staubitz et al., 1989; Peverly et al., 1993) in four individual beds in parallel. The purpose of this paper is to outline important aspects of seasonal development and, where possible, compare species behavior in natural and constructed wetlands receiving landfill leachate. Although some of the conclusions are made from data on plants receiving landfill leachate, some are based on information from beds receiving sewage or other types of pollutants.

14.2 PLANT STRUCTURE

Emergent wetland species all have a similar structure as illustrated in Figure 14.1. The above-ground shoots, also termed *modules* or *ramets*, arise from underground horizontal stems, the rhizomes. Shoots differ morphologically in the different species; some such as the grasses *Phragmites* and *Phalaris* have shoots that are true stems with leaves arising along the length of the stem.

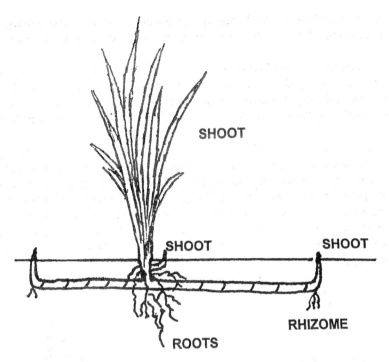

Figure 14.1   Structure of wetland macrophytes.

In other species, such as cattails and most sedges, as shown in Figure 14.1, the shoot is composed of a series of overlapping leaves, the stem tip being located at just above the soil surface. Either of these shoot types can be annual or perennial; most of the perennial species shoots in temperate regions living for 2 years and more southern species such as sawgrass (*Cladium jamaicense*) in the Florida Everglades living for 8 to 9 years (Stewart and Ornes, 1975; Davis, 1989). Species life spans may vary depending on the environment. *Carex rostrata*, for example, has biennial shoots in temperate climates but may live 3 to 4 years in northern Sweden (Solander, 1983; Hultgren, 1988).

Individual shoots are separated by horizontal rhizomes, which are sometimes termed *spacers*. Rhizome growth patterns differ; some species produce only long rhizomes, some only short clumping rhizomes, and some a combination of both as illustrated in Figure 14.1. The species discussed in this paper are of the latter type, producing both long and short rhizomes. Such a growth pattern produces a loose clump of aboveground shoots formed from short rhizomes or axillary buds and termed a *tiller clump* separated by open spaces into which individual shoots on the end of long rhizomes invade. This type of growth pattern leads to the species producing a pure stand of dominant plants by growing over an area and preventing other species from invading the site.

Rhizomes are important storage organs for nutrients and carbohydrates during winter and are the principal reproductive structure of the plants. New rhizomes begin growth in midsummer and produce new shoots either at their ends or along their length beginning often in late July and continuing until winter dieback. During this time they also accumulate carbohydrates and some nutrient elements by translocation from aboveground tissues.

Root systems of aquatic plants are of two types. The first have been termed *soil roots* and are generally long, sparsely branched, and grow deep into the soil. Such roots have been noted for *C. gracilis* and *C. vescaria* (Koncalova, 1990), *C. rostrata* (Bernard and Gorham, 1978), *Cladium mariscus* (Conway, 1936), *Phragmites australis* (Sculthorpe, 1967; Haslam, 1973) and *Typha angustifolia* (Lukina and Smurova, 1988, in Koncalova, 1990). These develop typically where the soil is anaerobic and apparently serve primarily to anchor the plants. The second type has been termed the *aquatic root*. These are finely branched, typically grow more horizontally, and develop best in moist aerated soil. These are the roots principally involved in uptake of water and minerals from the soil.

Root systems vary in their distribution on the plants. Some species such as *Phragmites* produce roots only at the bases of the shoots, while others such as *Phalaris* have roots arising along the whole length of the rhizome. These different growth patterns have an obvious effect on root distribution in the soil; the former growth pattern often causes areas of the wetlands to be without a large root mass.

An internal structural feature of these plants is that tissues contain large open spaces called *aerenchyma*. Plants such as *Phragmites* and cattails that grow typically in somewhat deep water typically have more aerenchyma than species such as *Phalaris* growing in more shallow water. Figure 14.2 illustrates aerenchyma tissue in *T. angustifolia*, a species found typically in deeper water sites (Grace and Wetzel, 1981; 1982). Both the leaves and roots have large air spaces while the rhizome has a more spongy consistency. This aerenchyma is important because it allows oxygen to diffuse from the atmosphere into stems and leaves, then into the belowgound roots and rhizomes (Brix and Schierup, 1990).

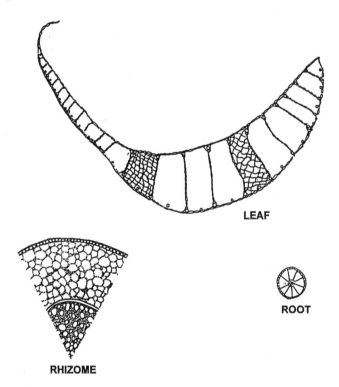

Figure 14.2 Aerenchyma tissue in leaf, rhizome, and root systems of *Typha angustifolia*.

The typical structural characteristics of these plants can be modified by such environmental factors as water depth and nutrient supply, both important considerations in constructed wetlands. High water levels may have a number of effects on plant structure. Studies have found that plants in deep water modify their morphology and allocate biomass differently than the same species in shallow water. Generally, high water levels cause stems to grow taller and heavier as in the cattails *T. latifolia* and *T. domingensis* (Grace, 1989), *T. glauca* (Waters and Shay, 1990), *Scirpus maritimus* (Lieffers and Shay, 1981), and *Carex lacustris* (Bernard, 1975). More recently, Coops et al. (1996)

in a study of four species found maximum shoot length and aboveground biomass at a water depth of 30 cm for *Phalaris*, 55 cm for *Phragmites* and *S. maritimus*, but the growth of all three species declined significantly when water depth increased to 80 cm. They found that in *S. lacustris*, the fourth species they studied, shoots were longer and biomass was not significantly lower at 80 cm depth than at 55 cm depth.

Belowground, high water tends to inhibit growth of roots and rhizomes. Coops et al. (1996) found for the four species they studied that rhizome growth was inhibited for *Phalaris*, *Phragmites*, and *S. maritimus* but not for *S. lacustris*. In addition, they found that *Phragmites* and *Phalaris* showed no increase in aerenchyma at deeper water levels. In summary, they found that morphology and biomass allocation patterns change along water depth gradients and these changes are related to a species position on the gradient, *Phalaris* being a typically shallow water species, the other species mentioned being more characteristic of deeper water sites.

Excessive flooding that covers the shoots may lead to anoxic conditions in the root zone, killing most of the plants. This is important where harvesting is undertaken because water levels high enough to cover the cut ends of the shoots would result in great damage to plant populations in the beds. High water levels are also a consideration when planting in the beds; Bedish (1967) found that over half the planting of what he called the cattail hybrid (here termed *T. glauca*) did not grow when rhizome fragments were planted under water.

In addition to high water effects, the amount and availability of soil nutrients may also have an effect on root and rhizome structure and function. Koncalova (1990), for example, found that under high nitrogen and organic matter concentrations, or additions of piggery sewage, the amount of root gas space in *C. vescaria* declined by approximately 50%, enough to have a significant negative effect on plant function. She also found that the total storage carbohydrates in rhizomes in both *C. vescaria* and *C. gracilis* declined by approximately 40% from control plants growing in normal nutrient solution.

Another important consideration in constructed wetlands, especially where harvesting is planned, will be the distribution of nutrient and metal elements in the plants. Some metals such as zinc are not transported abovegound to any degree (Larsen and Schierup, 1981; Schierup and Larsen, 1981; Auclair, 1982; Bernard et al., 1988; Bernard and Lauve, 1995). Table 14.1 compares metal contents of *T. glauca* in above- and belowground tissues at the Fenton landfill site and the wetland in Irondiquoit Creek in western New York. Harvesting will not remove large quatities of these substances. However, harvesting will be much more successful for removing nitrogen, phosphorus, and potassium, all of which are in high concentrations in shoots in summer (see discussion).

Table 14.1 Average Chemical Concentrations in Above and Belowground Tissues of *T. glauca* from Input and Output Ends of Bed 3 Constructed Wetlands at Fenton Landfill and Irondequoit Creek Wetlands

	Bed 3 Input		Bed 3 Output		Irondequoit	
	Above	Below	Above	Below	Above	Below
Copper	4.1	20.6	2.9	44.6	5.3	15.6
Nickel	1.8	11.5	1.3	10.5	1.6	9.6
Zinc	13.3	27.0	9.1	57.3	17.5	59.8
Aluminum	61	1300	56	2330	17	1875
Iron	368	19439	292	10745	67	18006
Lead	1.6	8.6	1.0	14.6	0.6	18.9
Boron	27.7	37.8	14.0	26.0	12.7	132.0

Note: All values are in mg/kg.

14.3 SEASONAL GROWTH PATTERNS

Seasonal growth and production studies have been done for some wetland species, such as *Phragmites* (Haslam, 1973), *T. glauca* (Garver et al., 1988), and *Carex* sp. (Bernard and Gorham, 1978; Bernard, 1990) growing in the north temperate zone. In winter, the shoots from the last season turn brown and most, such as those of *Phragmites*, *Phalaris*, *Typha*, and many sedges, die. Even species with biennial or perennial shoots die back and will produce new leaves from the stem tip the next summer. Rhizome biomass is high in winter and has the highest carbohydrate and nutrient levels of the year (Fiala, 1971; 1978; Roseff and Bernard, 1979).

During spring, growth of the already developed shoots begins. This early growth is fueled by photosynthesis, but a large part may be fueled by translocation of carbohydrates and nutrient elements from the rhizomes. The amount of this translocation has been estimated to provide from 25% of early shoot growth (Bernard, 1990) in *Carex* sp. to almost 100% in *Phragmites* and *T. latifolia* (Fiala, 1976; 1978). In any case, shoots in early spring have the season's high percentages of nutrient elements and carbohydrates, most of them probably translocated from belowground tissues. This growth can be very rapid; shoots of *Phalaris* may grow 2 cm/day during late April and early May (Bernard, unpublished).

After spring shoot emergence and growth, emergence of new shoots slows during late May and fewer still emerge during June. By July 1 approximately, some shoots have flowered and vegetative shoots have reached maximum size and are producing excess carbohydrates for transport below-ground (Fiala, 1971; Roseff and Bernard, 1979). New rhizome growth begins during July, and new shoots and associated root systems begin to develop. This development and growth of new shoots from the rhizomes continues into late autumn. Some of the new shoots may grow aboveground before dieback; Bernard (1975) found some shoot emergence in *C. lacustris* in early December even after all large abovegound shoots had turned brown.

These seasonal growth patterns can be modified by environmental conditions such as elevated temperature or high water levels. At the Fenton constructed wetlands, leachate temperatures were high and even in midwinter water temperature averaged between 5 and 8°C. This warm water had a number of effects on growth of *Phalaris*. First, dieback of shoots in autumn was delayed by approximately 2 weeks with the result that more new small shoots emerged during autumn; the green standing crop was approximately 30% higher at Fenton than the control marsh during winter. Second, spring growth began approximately 2 weeks earlier than the control marsh; the result was an approximately 200 g/m^3 higher standing crop abovegound in summer. This is an important outcome where harvesting of plant material is planned. It is also important because it indicates the possibility that some typically northern species may grow year-around in locations a bit farther south than at Fenton and stay green and productive all year.

High water levels may also have an effect on growth patterns. Bernard (1975) in a study of *C. lacustris* found that high water levels prevented rhizome growth and new shoot emergence for up to 2 months when flooding of the marsh occurred. Lieffers and Shay (1981) found a slowing of phenological development in *S. maritimus* with increasing water depth. This finding is important in constructed wetlands where water levels may be high.

14.4 SHOOT LIFE SPANS AND MORTALITY

Life spans of individual shoots of aquatic macrophytes vary widely. Most temperate zone species such as *Phragmites*, *Phalaris*, *Typha*, and *Carex* have annual shoots that live for 12 months or less, commonly from the autumn of one year to summer or autumn of the next, but others such as *Cladium mariscoides* and some *Carex* have shoots that may live for approximately 24 months. In more southern areas, such as the Florida Everglades, Davis (1989) found that saw grass (*Cladium jamaicense*) shoots lived for up to 8 to 9 years and cattails (*T. domingensis*) shoots lived for over

6 years. Even in the temperate zone, shoot life span can be modified by environmental factors. For example, work with *Carex rostrata* has shown that shoots in temperate areas live for up to 24 months (Bernard, 1990) whereas in Sweden they live for 36 to 48 months (Solander, 1983; Hultgren, 1988), apparently not because of the rigorous climate but because of a lack of nutrients in oligotrophic lakes.

Few shoots of species in the north temperate zone attain their maximum life span, and mortality rates of 80 to 90% have been found for some species in the genus *Carex* (Bernard, 1990). Apparently, mortality is somewhat lower in other species, Dickerman and Wetzel (1985) found a mortality rate of approximately 10% in cattails within 3 months of shoot emergence. Southern plants live longer than temperate zone plants, and mortality is death of individual leaves instead of the whole shoot system. Davis (1989) found growth of new leaves and death of old leaves to be a continuous process, and turnover in saw grass leaves was estimated at 1.7 to 6.2 times average biomass and in cattail 3.4 to 11.3 times average shoot biomass during the long life of the whole shoot system. Although mortality in wetland species may be high, it is balanced by emergence of new shoots so that the population numbers do not change greatly throughout the year. It appears that all species, either temperate or tropical, attempt to maximize the growth of young, photosynthetically active tissues while older tissues, either the whole shoot or individual leaves, are subject to senescence.

Mortality may be caused by a number of factors including the time of emergence of the shoots, grazing by animals such as the muskrat, and prolonged flooding. High mortality rates may also result from large amounts of nutrients. Callaghan (1976) found that shoots growing in favorable sites produced more new shoots during the season, but later, when conditions changed or crowding was intense, higher than the normal number of shoots died. Others doing experiments with fertilizer additions have found this same pattern of more emergence of new shoots under optimum conditions followed by greater mortality.

14.5 BIOMASS AND NUTRIENT RELATIONSHIPS

Wetland macrophytes are among the most productive of plants, and it appears from the data available that those species growing in waste water rich in nutrients attain higher biomass and production values than those growing in natural wetlands. Table 14.2 gives data on four temperate zone species growing in constructed wetlands treating landfill leachate, sewage treatment wetlands, or in natural wetlands receiving large amounts of runoff from urban areas. Biomass values range from 1130 g/m² in constructed sewage treatment wetlands planted with *Phalaris* and harvested twice to 2608 g/m² in a *T. glauca* wetland receiving urban runoff. The average for these sites is approximately 1625 g/m², although, as pointed out by Boar (1996), biomass varied by as much as 350 g/m² in *Phragmites* stands over a 4-year period. The warm leachate waters may also increase biomass production and, as pointed out by Gorham (1974), a correlation exists between warm summer temperatures and biomass in *Carex* wetlands.

Seasonal nutrient patterns follow a similar course in wetland species. Shoots typically have high percentages of nutrients during winter and early spring, but as the season progresses the percentages decline; by midsummer perhaps only 50% of nitrogen, phosphorus, and potassium are left, and by late autumn only about 33% is left (Garver et al., 1988; Bernard et al., 1988; Hurry and Bellinger, 1990).

Although percentages of nutrients are less in summer and autumn, the standing stocks in biomass are much higher than in winter. Table 14.2 also gives maximum summer nutrient standing stocks for the same species and sites. Nitrogen and potassium are present in high amounts, averaging for all sites 34.9 g/m² and 21.4 g/m², respectively. Phosphorus, another important and metabolically mobile element, averages just 2.4 g/m² in midsummer.

Harvesting for nutrient removal has been done in some wetland treatment systems, and at least two harvests are recommended to remove the maximum amount of nutrients. Braxton (1981), for

Table 14.2 Maximum Biomass and Macronutrients in Four Species Growing in Constructed Wetlands or Wetlands Receiving Large Chemical Inputs

Species	Location	Biomass	N	P	K	Ca	Mg	Source
Typha glauca	Landfill, NY	1446	29	1.6	18.7	8.9	4.3	Bernard & Lauve, 1995
	Urban runoff, NY	2608	58.6	6.5	37.1	22.6	4.6	Bernard & Seischab, 1997
Phalaris arundinacea	Landfill, NY	1713	37.6	—	32.5	1.1	1.1	Bernard & Lauve, 1995
	Sewage, Long Island	1130	47.7	5.4	—	—	—	Braxton, 1981
Phragmites australis	Landfill, NY	1520	26.9	0.7	12.6	1.9	—	Peverly et al., 1993
	Fishpond, Czech Rep.	1800	26.7	1.9	14.8	2.7	1.9	Kvet, 1973
Carex lacustris	Marsh, NY	1145	18.3	1.6	12.7	4.1	1.4	Bernard & Solsky, 1977

Note: All Values are in g/m².

example, used three harvests of *Phalaris* to remove 5.4 g/m² of phosphorus. The first harvest in May removed 61% of the total, the second in late June 27%, and the third in late September removed an additional 12%. Suzuki et al. (1989) found for *Phragmites* that two harvests increased the yearly total biomass production by 14% over that of one harvest and increased nitrogen and phosphorus removal by 22% and 10%, respectively. Harvesting may be particularly important in nutrient-rich constructed wetlands because, as Shaver and Melillo (1984) found, plants in rich sites do not conserve nutrients. Thus, a considerable amount of nutrients would be lost from the beds as plants age in autumn.

14.6 SUMMARY AND CONCLUSIONS

Large wetland macrophytes have potential for growth in constructed wetlands. They are as a group easy to propagate and grow well in polluted waters, where they help to filter leachate and perhaps prevent clogging of the soil. They are plastic organisms and can modify their structural characteristics and seasonal growth patterns when exposed to high water and nutrient levels and inputs of warm leachate from landfills. Most species can be harvested during the season without harm so long as water levels do not rise above the cut ends of the plants.

Landfills vary greatly in their location, leachate type, amount and concentration of different chemicals, and treatment objectives. Also, wetland species that may be used in treatment projects differ as to their site preferences, and therefore recommendations for use of a particular species in any one constructed wetland is not possible at this time. Much more work with due attention to life history attributes of plants will be necessary to specify the best plant for each individual situation.

REFERENCES

Auclair, A. N. D., 1982. Seasonal dynamics of nutrients in a *Carex* meadow, *Canadian Journal of Botany* 60:1671–1678.

Bedish, J. W., 1967. Cattail moisture requirements and their significance to marsh management, *American Midland Naturalist* 78:288–300.

Bernard, J. M., 1975. The life history of shoots of *Carex lacustris*, *Canadian Journal of Botany* 53:256–260.

Bernard, J. M., 1990. Life history and vegetative reproduction in *Carex*, *Canadian Journal of Botany* 68:1441–1448.

Bernard, J. M. and Gorham, E., 1978. Life history aspects of primary production in sedge wetlands, in Good, R. E., Whigham, D. F. and Simpson, R. L., Eds., *Freshwater Wetlands: Ecological Processes and Management Potential*, Academic Press, New York, 39–51.

Bernard, J. M. and Lauve, T. E., 1995. A comparison of growth and nutrient uptake in *Phalaris arundinacea* L. growing in a wetland and a constructed bed receiving landfill leachate, *Wetlands* 15:176–182.

Bernard, J. M. and Seischab, F. K., 1997. Vegetation of the Wetlands Near the Mouth of Irondequoit Creek, 1996. Report to U.S. Geological Survey.

Bernard, J. M. and Solsky, B. A., 1977. Nutrient cycling in a *Carex lacustris* wetland, *Canadian Journal of Botany* 55:630–638.

Bernard, J. M., Solander, D., and Kvet, J., 1988. Production and nutrient dynamics in *Carex*, *Aquatic Botany* 30:125–147.

Boar, R. R., 1996. Temporal variations in the nitrogen content of *Phragmites australis (Cavanilles) Trinius* ex *Steudel* from a shallow fertile lake, *Aquatic Botany* 55:171–181.

Bouldin, D. R., Bernard, J. M., and Lauve, T. E., 1993. Leachate Treatment System Using Constructed Wetlands, Town of Fenton Sanitary Landfill, Broome County, New York, Final Report to New York Energy Research and Development Authority No. 94-3.

Braxton, J. W., 1981. Nitrogen and Phosphorus Accumulation and Biomass Production in a Meadow-Marsh-Pond Sewage Treatment System, Ph.D. thesis, Rutgers University, 107 pp.

Brix, H. and Schierup, H. H., 1989. The use of aquatic macrophytes in water-pollution control, *Ambio* 18:100–107.

Brix, H. and Schierup, H. H., 1990. Soil oxygenation in constructed reed beds: the role of macrophyte and atmosphere interface oxygen transport, in Cooper, P. F. and Findlater, B. C., Eds., *Constructed Wetlands in Water Pollution Control, Advances in Water Pollution Control*, Pergamon Press, Oxford, U.K., 100–107.

Callaghan, T. V., 1976. Growth and population dynamics of *Carex bigelowii* in an alpine environment, *Oikos* 27:402–413.

Conway, V. M., 1936. Studies in the autecology of *Cladium mariscus R. BR*. I. Structure and development, *New Phytologist* 35:177–205.

Cooper, P. F., and Findlater, B. C., Eds., 1990. Constructed wetlands in water pollution control, in *Proceedings of the International Congress on Use of Constructed Wetlands in Water Pollution Control*, Cambridge, U.K., Pergamon Press, Oxford, U.K.

Coops, H., van den Brink, F. W. B., and van der Velde, G., 1996. Growth and morphological responses of four helophyte species in an experimental water-depth gradient, *Aquatic Botany* 54:11–24.

Daniels, R. E., 1991. Variation in the performance of *Phragmites australis* in experimental culture, *Aquatic Botany* 42:41–48.

Davis, S. M., 1989. Sawgrass and cattail production in relation to nutrient supply in the Everglades, in Sharitz, R. R. and Gibbons, J. W., Eds., *Freshwater Wetlands and Wildlife*, CONF-8603101, DOE Symposium Series No. 61. U.S. DOE Office of Scientific and Technical Information, Oak Ridge, TN, 325–341.

Dickerman, J. A. and Wetzel, R. G., 1985. Clonal growth in *Typha latifolia*: population dynamics and demography of the ramets, *Journal of Ecology* 73:535–552.

Dobberteen, R. A. and Nickerson, N. H., 1991. Use of created cattail (*Typha*) wetlands in mitigation strategies, *Environmental Management* 15:797–808.

Fiala, K., 1971. Comparison of seasonal changes in the growth of underground organs of *Typha latifolia* L. and *Typha angustifolia* L., *Hydrobiologica* 12:235–240.

Fiala, K., 1976. Underground organs of *Phragmites communis*, their growth, biomass and net production, *Folia Geobotanica Phytotaxonomica, Praha* 11:225–259.

Fiala, K., 1978. Underground organs of *Typha angustifolia* and *Typha latifolia*, their growth, propagation and production, *Acta Scientiarum Naturalium* (Brno) 12:1–43.

Finlayson, C. M. and Chick, A. J., 1983. Testing the potential of aquatic plants to treat abattoir effluent, *Water Research* 17:415–422.

Garver, E. G., Dubbe, D. R., and Pratt, D. C., 1988. Seasonal patterns in accumulation and partitioning of biomass and macronutrients in *Typha* sp., *Aquatic Botany* 32:115–127.

Gorham, E., 1974. The relationship between standing crop in sedge meadows and summer temperature, *Journal of Ecology* 62:487–491.

Grace, J. B., 1989. Effects of water depth on *Typha latifolia* and *Typha domingensis, American Journal of Botany* 76:762–768.

Grace, J. B. and Wetzel, R. G., 1981. Phenotypic and genotypic components of growth and reproduction in *Typha latifolia*: experimental studies in marshes of differing successional maturity, *Ecology* 62:789–801.

Grace, J. B. and Wetzel, R. G., 1982. Niche differentiation between two rhizomatous plant species: *Typha latifolia* and *Typha angustifolia, Canadian Journal of Botany* 60:46–57.

Guntenspergen, G. R., Stearns, F., and Kadlec, R. A., 1989. Wetland vegetation, in Hammer, D. A., Ed., *Constructed Wetlands for Wastewater Treatment: Municipal, Industrial and Agricultural,* Lewis Publishers, Chelsea, MI.

Hammer, D. A., Ed., 1989. *Constructed Wetlands for Wastewater Treatment: Municipal, Industrial and Agricultural, Proceedings 1st International Conference on Constructed Wetlands for Wastewater Treatment,* Chattanooga, Tennessee, Lewis Publishing, Chelsea, MI.

Haslam, S. M., 1970. The development of the annual population in *Phragmites communis, Annuals of Botany* (London) 34:571–591.

Haslam, S. M., 1973. Some aspects of the life history and autecology of *Phragmites communis,* a review, *Polskie Archiwum Hydrogiologii* 20:79–100.

Hultgren, A., 1988. A demographic study of aerial shoots of *Carex rostrata* in relation to water level, *Aquatic Botany* 29:81–93

Hurry, R. J. and Bellinger, E. G., 1990. Potential yield and nutrient removal by harvesting of *Phalaris arundinacea* in a wetland treatment system, in Cooper, P. F. and Findlater, B. C., Eds., *Constructed Wetlands in Water Pollution Control, Proceedings International Conference on Use of Constructed Wetlands in Water Pollution Control,* Cambridge, U.K., Pergamon Press, Oxford, U.K., 543–546.

Huttunen, A., Heikkinen, K., and Ihme, R, 1996. Nutrient retention in the vegetation of an overland flow treatment system in northern Finland, *Aquatic Botany* 55:61–73.

Koncalova, H., 1990. Anatomical adaptations to waterlogging in roots of wetland graminoids: limitations and drawbacks, *Aquatic Botany* 38:127–134.

Kvet, J., 1973. Mineral nutrients in shoots of reed (*Phragmites communis Trin.*), *Polskie Archiwum Hydrobiologii* 20:137–147.

Larsen, V. J., and Schierup, H. H., 1981. Macrophyte cycling of zinc, copper, lead and cadmium in the littoral zone of a polluted and a non-polluted lake. II. Seasonal changes in heavy metal content of above-ground biomass and decomposing leaves of *Phragmites australis (Cav.) Trin., Aquatic Botany* 11:211–230.

Lieffers, V. J. and Shay, J. M., 1981. The effects of water level on the growth and reproduction of *Scirpus maritimus* var. *paludosus, Canadian Journal of Botany* 59:118–121.

Lukina, L. F. and Smirnova, N. N., 1988. *The Physiology of Higher Aquatic Plants,* Naukovaya Dumka, Kiev (In Russian).

Meiorin, E. C., 1989. Urban runoff treatment in a fresh/brackish water marsh in Fremont, California, in Hammer, D. A., Ed., *Constructed Wetlands for Wastewater Treatment: Municipal, Industrial and Agricultural,* Lewis Publishers, Chelsea, MI, 677–685.

Peverly, J., Sanford, W. E., Steenhuis, T. S., and Surface, J., 1993. Constructed Wetlands for Municipal Solid Waste Landfill Leachate Treatment. Final Report to the New York State Energy Research and Development Authority No. 94-1.

Richardson, C. J. and Davis, J. A., 1987. Natural and artificial wetlands ecosystems: ecological opportunities and limitations, in Reddy, K. R. and Smith, H., Eds., *Aquatic Plants for Water Treatment and Resource Recovery,* Magnolia Publishing, Orlando, FL, 819–854.

Roseff, S. and Bernard, J. M., 1979. Seasonal changes in carbohydrates in tissues of *Carex lacustris, Canadian Journal of Botany* 57:2140–2144.

Schierup, H.-H. and Larsen, V. J., 1981. Macrophyte cycling of zinc, copper, lead and cadmium in the littoral zone of a polluted and a non-polluted lake. I. Availability, uptake and translocation of heavy metals in *Phragmites australis (Cavanilles) Trinius, Aquatic Botany* 11:197–210.

Sculthorpe, C. D., 1967. *The Biology of Aquatic Vascular Plants,* Edward Arnold, London.

Shaver, G. R. and Melillo, J. M., 1984. Nutrient budgets of marsh plants: efficiency concepts and relation to availability, *Ecology* 65:1491–1510.

Solander, D., 1983. Biomass and shoot production of *Carex rostrata* and *Equisetum fluviatile* in unfertilized and fertilized subarctic lakes, *Aquatic Botany* 15:349–366.

Staubitz, W. W., Surface, J. M., Steenhuis, T. S., Peverly, J. H., Lavine, M. J., Weeks, N. C., Sanford, W. E., and Kopka, R. J., 1989. Potential use of constructed wetlands to treat landfill leachate, in Hammer, D. A., Ed., *Constructed Wetlands for Wastewater Treatment: Municipal, Industrial and Agricultural*, Lewis Publishers, Chelsea, MI, 735–742.

Stewart, K. K. and Ornes, H., 1975. The autecology of sawgrass in the Florida Everglades, *Ecology* 56:162–171.

Suzuki, T., Ariyawathie Nissanka, W. G., and Kurihara, Y., 1989. Amplification of total dry matter, nitrogen and phosphorus removal from stands of *Phragmites australis* by harvesting and reharvesting regenerated shoots, in Hammer, D. A., Ed., *Constructed Wetlands for Wastewater Treatment: Municipal, Industrial and Agricultural*, Lewis Publishers, Chelsea, MI, 530–535.

Tanner, C. C., 1996. Plants for constructed wetlands treatment systems — a comparison of the growth and nutrient uptake of eight emergent species, *Ecological Engineering* 7:59–83.

Trautman, N. M., Martin, J. H., Jr., Porter, K. S., and Hawk, K. C., Jr., 1989. Use of artificial wetlands for treatment of municipal solid waste landfill leachate, in Hammer, D. A., Ed., *Constructed Wetlands for Wastewater Treatment: Municipal, Industrial and Agricultural*, Lewis Publishers, Chelsea, MI, 245–251.

Waters, I. and Shay, J. M., 1990. A field study of the morphometric response of *Typha glauca* shoots to a water depth gradient, *Canadian Journal of Botany* 68:2339–2343.

CHAPTER 15

The Use of a Constructed Wetland to Treat Landfarm Leachate at the Sunoco Refinery in Sarnia, Ontario

James Higgins and Tom Brown

CONTENTS

ABSTRACT: A petroleum refinery operates a landfarm to biodegrade liquid and solid industrial wastes. It is underlain by a tile bed, and treated leachate from it is collected in a retention pond. The refinery wanted to consider options other than trucking, but determined that piping the leachate back to the refinery would be prohibitively expensive. Evaluations of wetland and related treatment options as alternates were carried out. This paper presents analyses and related treatment options for treating the leachate, details how one was selected for recommendation, and presents a preliminary sizing, suggested layout, and costing of that recommendation. A small free-water-surface (FWS) constructed wetland was recommended as suitable for treating the leachate, and was compared with an equivalent subsurface-flow (SSF) wetland.

1-56670-342-5/99/$0.00+$.50
© 1999 by CRC Press LLC

15.1 INTRODUCTION

Sunoco, Inc. (Suncor) operates a landfarm to biodegrade liquid and solid industrial wastes from its Sarnia refinery. The landfarm is underlain by a tile bed, and treated leachate from it is collected in a small retention pond. This leachate is currently periodically collected and trucked to the refinery wastewater treatment plant for treatment. Suncor wishes to consider options other than trucking, but has determined that piping the leachate back to the refinery would be prohibitively expensive. Of interest is the potential to use a wetland for this treatment. An evaluation of wetland and related treatment options was carried out, and it was determined that a small free-water-surface (FWS) constructed wetland would be suitable for treating the leachate.

15.2 SOIL BIOREMEDIATION

Bioremediation is defined by the U.S. EPA (U.S. EPA, 1993) as a method of "enhancing the development of large populations of micro-organisms" and a way of "bringing these organisms into intimate contact with pollutants." Where bioremediation involves the handling or use of soil, it is called soil *bioremediation*. Some contaminants are relatively easy to bioremediate, especially if they are biogenic (i.e., occur in nature or are made from something natural). Others are much more difficult to biodegrade (degrade by biological means) and these include xenogenic ones (i.e., ones that do not occur naturally) such as many chlorinated compounds (e.g., polychlorinated biphenyls, PCBs).

Biostimulation refers to techniques that stimulate naturally occurring microbes already existing in the soil by adding water, air, nutrients, and other materials to it to promote their growth. *Bioaugmentation* involves adding to the soil specifically grown contaminant-degrading microorganisms. These may be ones already found naturally at the site and identified as having particular biodegradation properties or ones from other sites where they have proved effective for the type of contaminants being bioremediated. Landfarming generally involves biostimulation, although the land application of biosolids from activated sludge plants (largely microbial biomass) has bioaugmentation aspects.

Nutrients are often added during bioremediation and the major ones are nitrogen (usually in the form of ammonium nitrate or urea) and phosphorus (in the form of orthophosphates, pyrophosphates, or phosphoric acid).

Most soil bioremediation projects involve biostimulation. In 1993, the U.S. EPA monitored 124 ongoing bioremediation projects and reported 77% as biostimulation projects, 14% as bioaugmentation, and 9% as a combination. There are two basic kinds of soil bioremediation processes: cleanup processes and disposal processes. Cleanup processes involve the bioremediation of contaminated soils, while disposal processes involve converting/degrading a contaminated material (e.g., an organic sludge) using a soil-based system. The biodegradation processes are the same, but in the former the soil is already contaminated, whereas in the latter the contamination is added to the soil. Cleanup bioremediation processes may involve treating contaminated soil either on site or off, and if the former, either *in situ* or after moving it (*ex situ*). Disposal processes are always *ex situ* and may be carried out either on or off site. A major application of a disposal process is the spreading of municipal sewage treatment biosolids on farmland. Landfarming is also a disposal-type bioremediation process.

15.3 LANDFARMING

Landfarms are forms of rapid infiltration systems that involve the biodegradation of organic contaminants in open fields or on prepared beds. In them, microorganisms are "managed" into using the added contaminants as food sources, thereby breaking them down. Biological processes predominate in landfarms, but physical–chemical ones also play a role.

Landfarms are used to dispose of various types of organic sludges and biosolids including municipal and industrial biological wastewater treatment plant sludges, lagoon sludges, septage, bioreactor and biofilter sludges, oily sludges, and other kinds of raw and dewatered biosolids. They find their widest applicability in handling contaminated materials from petroleum refining and wood-preserving operations. Landfarming is neither labor nor energy intensive and is most often used where low costs and liabilities are the principal considerations and rapid cleanup is not a factor (O'Malley, 1996).

Industrial landfarming is generally carried out with 1 to 20 acre, well-drained field plots (soil beds containing adequate communities of microorganisms) onto or into which appropriate waste materials are spread, sprayed, and/or injected. These beds are then tilled with ordinary farm equipment to enhance oxygen infiltration into the soil and mix the contaminants with soil microorganisms (bacteria, fungi, etc.). The types of contaminants disposed of in a landfarm must be ones that are both biodegradable and of kinds that bind well with soil. Since many of the biosolids produced in petroleum-refining operations bind very well with soil, landfarms are often associated with refineries. Often, a crop is periodically grown on a landfarm plot, using phytoremediation to assist with biological activity.

Landfarming is restricted by seasons, requires lots of space (three to ten times as much as for ordinary bioremediation), and operators must control precipitation runoff and percolation. Treatment efficiencies vary widely and at maximum usually are only one half as fast as those for ordinary biostimulation.

Most of the soil microorganisms that predominate in landfarming are aerobic and require an adaptation period to become acclimatized to particular types of contaminants. Where the sludge being treated on a landfarm contains metal contaminants, these usually become strongly bonded to the soil. Metals are not transformed in a landfarm in the same sense as are degradable organics, but their valence states may be changed and chemical bonds involving them may be broken in a manner that changes both their toxicity and mobility for the better. Strongly bonded heavy organics and metals may be removed from landfarm soil by soil washing or phytoremediation. Another option is to dispose of some of the landfarm soil periodically in a landfill.

The environmental impact of landfarms is generally low, as the types of sludges selected for landfarming are of kinds that do not have appreciable amounts of volatile or especially water-soluble components. In them, contaminants are reduced to levels/conditions where they are no longer a danger to human health and/or the environment.

15.4 LANDFARM LEACHATE

Leachates have long been a source of both surface water and groundwater contamination and authorities are now increasing efforts to limit or prevent the migration of them into receiving waters. Leachates are produced when rainfall and/or groundwater percolate through solid wastes and combine with inorganic and degraded organic materials in them. Some leachates (i.e., those from hazardous waste landfills and acid mine drainage from mine tailings areas) can be very complex and highly contaminated. These require sophisticated treatment methods, whereas those from municipal and nonhazardous industrial waste landfills require less treatment.

Industrial landfarm beds are usually underlain with tile drains to collect leachate that results from any contained moisture in the sludges being treated, any infiltration of groundwater into the landfarm (negligible in a well-designed and operated landfarm as they are usually located in areas of low permeability to prevent infiltration into groundwater), and rainwater and snowmelt that percolate through it (the bulk of the flow). Since the landfarm operates as a terrestrial wastewater treatment system as well, contaminants in the leachate are usually relatively low. They generally consist of very small amounts of water-soluble, low-molecular-weight organic degradation products, such as acids, alcohols, and phenols, which will be measured as biological oxygen demand (BOD) and chemical oxygen demand (COD).

If inorganic fertilizers are used on the landfarm to stimulate microbial activity, traces of them may appear in the leachate as well, although, since landfarms are aerobic systems, most of the oxidizable compounds will not get through. For example, little of the ammonia nitrogen from ammonium nitrate fertilizer would get through as it would be microbially transformed to nitrate nitrogen. Any heavy metals in landfarmed sludges are unlikely to appear in the leachate as they are strongly bound in the soil system.

Biodegradation in a landfarm will not reduce contaminants in leachate from a landfarm to zero, and some type of further treatment of it may be required. Many petroleum refineries dispose of their landfarm in their wastewater treatment plants. This is an adequate but relatively expensive way to handle a stream that is largely precipitation. The types of contaminants found in landfarm leachate (BOD, low-molecular-weight water-soluble organics, nitrates) are particularly amenable to treatment in a natural wastewater treatment system.

There is almost no data available in the literature on the degree of uptake that occurs with precipitation and other water passing through a landfarm, or of the composition of landfarm leachate. However, since landfarms are themselves natural treatment systems, it was anticipated that the degree of contamination in landfarm leachates would be relatively low (compared with other forms of leachates) and the polishing required to bring them to discharge quality would be minimal.

15.5 OBJECTIVES OF THE EVALUATION

The objective of the evaluation was to examine the leachate from the Suncor landfarm, to assess potential wetland and related treatment options for it, to select one to recommend, and to carry out a preliminary sizing, suggested layout, and costing of that recommendation.

As part of achieving this goal, there were several related subobjectives, including

1. Reviewing the condition and operations of the landfarm.
2. Evaluating wetland (and related natural) treatment options for it and assessing relevant design criteria.
3. Defining acceptable wetland effluent discharge criteria, defining expected permitting and regulatory requirements, and attempting to obtain conceptual agreement on the reasonableness of the recommended option.

In carrying out the evaluation, there were a number of relevant issues that had to be considered, including

1. The leachate is currently trucked to the refinery's wastewater treatment plant, a procedure which, although a nuisance, is not very costly, so any alternative could not be expensive.
2. Further onsite treatment of the leachate would represent a net environmental benefit, but not if it proved time-consuming, expensive or contentious.
3. Only limited data on leachate flow rate or quality were available.
4. Landfarming operations are carried out seasonally but leachate flow can occur year round.
5. Effluent from any new leachate polishing system would enter a surface water ditch system which discharges into a local stream via an Indian Reservation.
6. Space near the landfarm is limited and for that which is available, topography was a major consideration.

15.6 THE SARNIA REFINERY LANDFARM

The Suncor Sarnia refinery landfarm is 5 acres in size and is located at its tank farm in a noncontiguous area west of the main process area. It consists of four 1-acre, oblong plots, as may be seen from Figure 15.1. Sludges are applied to the landfarm from mid-April each year to mid-October.

As part of the evaluation, a soil investigation was undertaken. This soil investigation included hand augering to 1 m at two locations (with samples retained in tubes for further laboratory

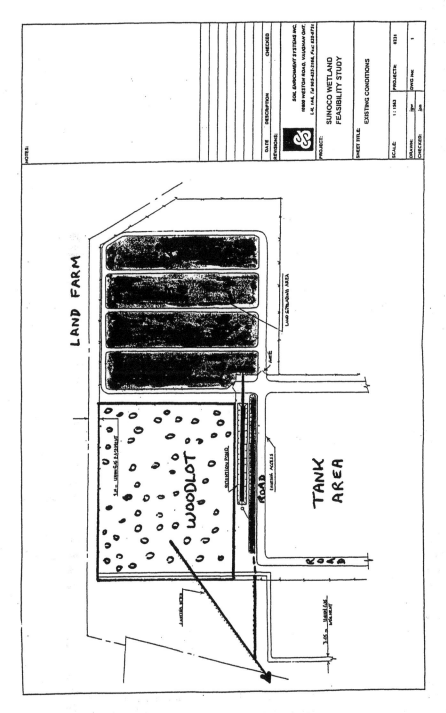

Figure 15.1 Existing conditions at the Suncor landfarm in Sarnia, Ontario.

evaluation); collecting an undisturbed core using a core tube (to a depth of 0.4 m); and digging an observation pit (0.5 m²) to 0.6 m. Soil profile observations were recorded from the pit and two bulk density cores taken from the surface horizon. One of the augered cores was taken in the center (width) of plot 2 of the landfarm, approximately 25 m from its southwest end and the other from the center (width) of plot 3, approximately 25 m from the northeast end. Table 15.2 summarizes the soil core observations.

Table 15.2 Landfarm Soil Core Observations — Plots 2 and 3

Horizon	Depth (cm)	Texture	Color	Structure	Consistency
			Plot 2		
A1	0–32	Clay loam	Very bark brown	Granular-fine SAB[a]	Friable
A2	32–45	Clay	Dark grayish brown	Fine SAB	Firm
Bt	45–63	Clay	Grayish brown	Massive	Firm
BC	63–73	Clay	Brown	Massive	Firm
Ck	73–100	Clay	Pale brown	Massive	Firm
			Plot 3		
A1	0–13	Clay loam	Very bark brown	Granular-fine SAB[a]	Friable
A2	35–43	Clay loam	Dark grayish brown	Fine SAB	Firm
Bt	43–56	Clay	Grayish brown	Massive	Firm
BC	56–67	Clay	Brown	Massive	Firm
Ck	67–100	Clay	Pale brown	Massive	Firm

[a] SAB = subangular blocky.

The surface structure (0 to 11 cm) in the A1 horizon is granular because of the tillage effects, while below this the soil becomes somewhat more compacted. The A2 horizon is the transition zone between the native soil and disturbed soil. According to available landfarm design blueprints, to build the landfarm, 48 to 65 cm of native soil was excavated and 20 to 30 cm of topsoil was placed back on at the excavated depth. From the coloring and structure it appears that the A2 horizon has been subjected to some mixing between the topsoil layer and the native B horizon. This probably is the result of deep cultivation at the time of topsoil placement. Hydrocarbon odors were apparent in the A1 and A2 horizons and some staining and lensing of hydrocarbon material was noticed in the A1 horizon.

The soil profile at this location is very similar to that of plot 2. According to the Soil Survey of Lambton County (1957), the soils at the landfarm location belong to the poorly drained Clyde Series. However, field examinations place the soils in the adjacent Perth Series. These imperfectly drained clay soils develop on undulating topography with natural vegetation growths of maple, hickory, and oak. Monthly (during the months of landfarm operation) soil sampling and analyses are carried out. pH and total salts tests on the A, B, and C horizons are carried out to determine if salts from fertilizer applications were accumulating in the soil.

Suncor uses the landfarm to bioremediate dewatered solids, wastewater treatment plant biosolids, and oily sludges. Dewatered solids come from a refinery belt filter press. The wastewater treatment plant biosolids are largely biomass from its activated sludge plant and are applied in undewatered form. Oily sludges are largely high-molecular-weight still bottoms and residues from various places in the refinery. During 1995, 21,000 kg of oily sludges were applied to the landfarm each week during its operating season, and for the entire season these and the other sludges led to the deposition of 133 t of solids, 33 t of oil, and 1781 t of water on the landfarm. The landfarm is fertilized three times a year to promote microbial growth. Almost 200 kg of phosphoric acid and 1700 kg of urea were used in 1995, amounting to a total phosphorus addition of over 1000 kg and a total nitrogen addition of over 9000 kg. Monthly soil samples are taken and analyzed, with 1995 results indicated in Table 15.3.

Table 15.3 Monthly Landfarm Soil Analyses, 1995

Parameter	Range	Average
pH	6.5–7.3	6.9
TKN (mg/kg)	7,300–17,000	9,900
Nitrate (mg/kg)	<1–498	100
Oil and grease (mg/kg)	18,700–48,700	37,000
Total phosphorus (mg/kg)	1,510–3,800	2,250
TOC (wt %)	7.96–10.90	9.4
Water (wt %)	15.0–31.9	23.1

As shown in Table 15.4, reported heavy metal analyses showed only a slightly high copper level (up to 384 mg/kg vs. a guideline level of 300 mg/kg) when compared with the Ministry of Environment and Energy (MOEE) cleanup guidelines for soils (commercial/industrial usage and medium/fine soil texture categories). These guidelines also recommend that total Kjeldahl nitrogen (TKN) not exceed 0.6% and values for it were reported at 0.7 to 1.7%, but this appears to be a function of nitrogen fertilizer application. (NO_3–N levels also peak after fertilizer application.) Oil levels were high, as expected, with the levels ranging between 1.9 and 4.9% (MOEE guidelines recommend 1 to 2%). Total phosphorus, and particularly the available phosphorus levels, appeared high, but again this probably reflects fertilizer application. Metals analyses (total salts) are as follows:

A horizon	3.01 mS/cm
B horizon	2.96 mS/cm
C horizon	2.46 mS/cm

These levels of salts in the soils are such that there should not be any phytotoxicity concerns for plants in contact with leachate from the landfarm.

Two undisturbed bulk density landfarm soil cores were removed from the surface horizon (taken parallel to the surface at a depth of 8 cm). The resulting bulk densities of 1.25 and 1.27 g/cm³ indicate slight to no compaction on the surface. Profile investigations from the pits did not reveal any hardpan layer beneath the cultivation zone. Infiltration of surface material to the leachate tile does not appear to be impeded, but it is expected to be slow because of low hydraulic conductivities associated with clay soils.

In general, the soil in the landfarm was found to be in good condition and appeared to be adequate for treating the biosolids from the refinery.

Table 15.4 Landfarm Soil Metals Analyses, 1995 (mg/kg)

Metal	Range	Average
Cadmium	<0.05–4	3
Cobalt	6–12	10
Chromium	24–48	39
Copper	220–384	268
Iron	17,000–29,200	24,740
Molybdenum	3–7	5
Sodium	130–430	270
Nickel	38–72	61
Lead	140–256	221
Vanadium	40–75	59
Zinc	260–432	354

15.7 THE LANDFARM RETENTION POND

Leachate from the landfarm collects in a tile bed under it and flows under an adjacent roadway into a retention pond (9 × 90 m). As may be seen from Figure 15.1, this pond is long and narrow with the tile bed outflow entering it at its southeast end. Pump-out is carried out by hose at the northwest end. Cattails (*Typha* sp.) have become established at the base of the bermed walls of the retention pond. A substantial growth of reeds (*Phragmites* sp.) is located around influent point at the southeast end of the pond. The retention pond is shallow (about 1 m deep on average) with an irregular outline due to clumps of cattails that intersect it. Surface runoff into the pond appears to be limited to a very small portion of the undisturbed area north of it. Most of this area, however, slopes northward into a wooded area and does not permit storm water runoff to enter the pond. Road runoff is collected in an adjacent ditch that is not connected to the retention pond.

Directly north of the retention pond, the elevation increases, possibly a result of the excavated material from the construction of the pond having been dumped there. Past this disturbed area there is a northward fall in elevation (sloping away from the pond). Hardwood trees are present throughout this area but tend to decrease in size farther down the slope. At the toe of the slope an open area occurs in which some wetland vegetation was observed. A ditch initiates at this toe area and flows westward into a ditch system paralleling a nearby road.

15.8 LEACHATE FLOW RATE

Leachate from the landfarm that is piped to the retention pond consists of contained water in the applied biosolids applied to the landfarm surface during the year, precipitation that falls on the landfarm and percolates through it, and any groundwater that might infiltrate into it. Because of the tight clay soils in the area, this latter flow (groundwater infiltration) appears to be negligible, so the bulk of the leachate collected in the tile drains under the landfarm should be the result of precipitation plus any contained water from the applied biosolids less evaporation and retained moisture in the soil.

Precipitation data for the Sarnia area were available from Environment Canada and indicate that average annual precipitation is 825 mm/year. By assuming that all of this infiltrates into the landfarm (as rain or melting snow), this indicates a yearly volume of over 17,300 m³ of precipitation. During 1995, 1947 t of biosolids were applied and these contained 1781 t of water (~1800 m³). By assuming that these 1995 numbers are indicative, the volume of the contained water in the applied biosolids averages about 10% of that of the precipitation. The total amount of leachate generated was about 11,400 m³ (3MM USG/264), indicating an annual evaporation rate of about 7700 m³ (17,300 + 1,800 − 11,400).

Leachate collected in the tile drain system would be joined by any groundwater infiltration into the retention pond (again assumed to be negligible), any leakage into the retention pond feed pipe (also assumed to be negligible), and any precipitation that enters the retention pond directly. This precipitation can enter the retention pond at any time during the year but due to the small surface area of this bermed structure and its related catchment area (<900 m²) relative to that of the landfarm (>20,000 m²), its amounts are not deemed to be significant. In any case, samples of leachate are taken from the retention pond, so for all practical purposes the water in the retention pond is the leachate.

During 1996, approximately 3 million U.S. gal of leachate was collected from the retention pond and trucked to the wastewater treatment plant at the refinery. The landfarm operates only from April to October and significant leachate flows can be expected only during frost-free periods. By assuming that relevant flows of leachate occur only for 200 days/year, an average daily flow rate of leachate for design purposes was calculated to be 60 m³/day (3MM/264 × 200) and this number was used as the basis for the wetland evaluations.

15.9 LEACHATE WATER ANALYSES

Water from the retention pond was analyzed at various times during 1996. The results of these analyses are summarized in Table 15.5. As may be seen from the analyses, based on the limited information available, the water in the retention pond does not appear to be highly contaminated, about what might be expected from a stream that is largely rainwater that has already passed through a soil bioremediation system (the landfarm). Small amounts of oil and grease and traces of phenolics do get through, again as might be expected, but these are very amenable to degradation in a wetland polishing system. The levels of nitrate, and probably some or all of the phosphorus, represent residual contamination from the fertilizers applied to the landfarm to stimulate microbial activity there.

Table 15.5 Retention Pond Water Analyses

Contaminant	Range	Average Value
TOC	55–89	72
Oil and Grease	2.4–38.0	14.8
Phenolics (µg/L)	6–29	15
NH_3–N	1.4–24.4	4.1
NO_2–N	0.18–0.24	0.21
NO_3–N	12–44	28
TKN	5–6	5.5
Sulfide	0.03–4.1	1.3
Total phosphorus	0–0.17	0.08
ortho-Phosphorus	0–0.03	0.02
pH (units)	7.1–8.5	7.5

Note: Values are mg/L except as noted.

No analyses for heavy metals contaminants in the leachate were available. As was discussed earlier, it is unlikely that any significant levels of them would get through the landfarm (and a wetland would quickly remove any that did). The analyses of the leachate are notable only from the fact that contaminants in it are in such low concentrations, again reflecting the high levels of treatment afforded by the soil system in the landfarm. In 1997, two leachate samples were collected, one from the outlet of the leachate pipe from the landfarm just prior to its entering the retention pond and the second near the outlet of the retention pond.

No BOD analyses were available for the landfarm leachate, but it is known that the levels of this parameter, which is used in wetland design rather than total organic carbon (TOC) are always less than those of TOC. (TOC is a rough, instrument-based measure of COD, and BOD is only a part of this value as there are materials in wastewaters that are not amenable to biological degradation). The design basis value for BOD was conservatively assumed to be 100 mg/L (Reed, 1997). Evaluation of the data showed that BOD (and hence TOC) reduction was the controlling design criterion, not nitrates. This will have to be verified when further analytical data are available at any subsequent preliminary or final design stage. Table 15.6 presents suggested design bases.

15.10 EVALUATION OF ALTERNATIVES AND DEFINITION OF DESIGN CRITERIA

Various alternative terrestrial and aquatic methods of treating the landfarm leachate were evaluated. Two systems were deemed possible options to use in polishing it, an FWS constructed wetland or an SSF constructed wetland. Over the past few years, the design and engineering of constructed wetlands has become systemized based on engineering models and increasing amounts of empirical data from successfully operating constructed wetland systems. Since constructed

Table 15.6   Landfarm Leachate — Suggested
Design Bases

Contaminant	Design Basis Value, mg/L
Biochemical oxygen demand	100
Oil and Grease	15
NH_3–N	2.5
NO_2–N	0.2
NO_3–N	44
Total Kjeldahl nitrogen	6
Total phosphorus	0.2

wetlands are more economic to build and operate than competitive systems, this has led to rapidly growing use of these systems in places where their use is applicable. For this evaluation, the design methods of Reed et al. (1995) were used. These methods allow the sizing of a wetland based on first-order areal kinetics for contaminant reduction using a plug-flow model.

Reaction kinetics for wetland design contain a volumetric rate constant term k (units/of day), which is available from the literature and design manuals for the various wastewater contaminants at base values for 20°C. For some contaminants (e.g., nitrogen compounds), these rate constants will vary with temperature, and, if they are used in designs of wetlands for cold-weather operation, they must be adjusted to values reflecting ambient temperatures using the Arrhenius equation. The Suncor landfarm operates largely in warmer weather, and, hence, most of the leachate would be treated in a wetland during warmer periods. Thus, only values of the rate constants at 20°C were used. More-detailed evaluations using rate constants reflecting ambient temperatures will be required, and these will be carried out at the subsequent detailed design stage. As may be seen from Table 15.5, except for nitrates and TOC, most of the contaminants in the leachate are of relatively low levels. It is apparent from them that wetland design will have to be based on either TOC or nitrate reduction targets.

The landfarm only operates from April to October. A wetland need not operate for much longer than the landfarm. Practically, there is no reason that the wetland could not continue operating until freeze up, even long after activity at the landfarm has ceased, drawing down water in the retention pond until it is almost all removed. Once the soil in the landfarm freezes in fall, there will be little more leachate from it to the retention pond until the thaw next spring. Precipitation into the retention pond during the winter season (both rain and snow) will lead to some water accumulation. This will dilute any residual leachate in it and can be either held (if there is sufficient storage capacity) or drained through the wetland during warmer days. If desired, treatment wetland can be designed to allow operation throughout the winter. Such a design would lead to a slightly more expensive wetland.

Both FWS and SSF wetlands are viable options to consider for polishing leachate from the refinery landfarm, but, after careful consideration, it was felt that the best selection for treating leachate would be an FWS wetland involving four cells of about 300 m² each, arranged in two parallel trains. After allowing for berms and other ancillaries, the wetland system will have a "footprint" of about 2500 m². Figure 15.2 shows proposed locations for the wetland. Figure 15.3 shows details on its layout.

15.11 FREE-WATER-SURFACE WETLAND EVALUATION

For an FWS wetland treating 60 m³/day of landfarm leachate with an influent concentration of 100 mg/L of BOD, conservative plug-flow, first-order calculations indicate a required treatment surface area of about 1250 m² with an estimated effluent BOD concentration of 22 mg/L.

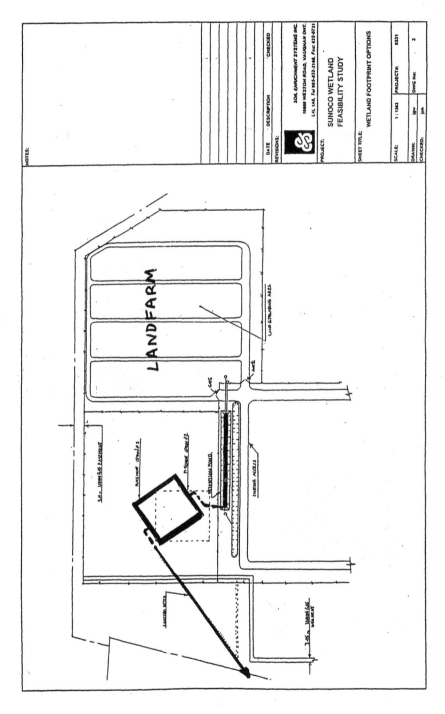

Figure 15.2 Proposed wetland location in Suncor landfarm, Sarnia, Ontario.

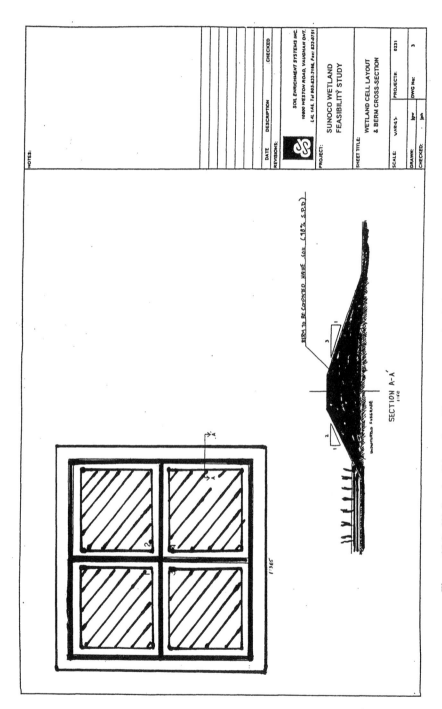

Figure 15.3   Wetland cell layout and berm cross section in Suncor landfarm, Sarnia, Ontario.

For flexibility, ease of operation, and other reasons, at least two treatment trains are recommended. Either two parallel cells, each of half the area, might be used, or four cells, each of one quarter the area, can be considered. The latter was recommended as it will provide more flexibility without significantly increasing costs.

In the past, treatment wetlands were designed with relatively high aspect ratios (wetland cell length-to-width ratios). This proved unnecessary and current wetland design practice is to minimize aspect ratio. Accordingly, the size of the wetland cells for this evaluation was set at 18 × 17 m, giving a cell area of 306 m² and a total wetland surface area of 1224 m², close to the initial estimate of 1250 m².

Inlet and outlet flows will be via slotted distributor pipes perpendicular to flow at each end of each wetland cell connected to culverts. Level in the wetland will be maintained by an adjustable riser connected to the outlets of the two final cells. Normal water depth in the wetland cells will be 0.3 m and floor slope 2%.

Each of the cells will be surrounded by earthen berms, 1 m in height and 1 m across at the top. Inside slopes for the berms would be 2:1 and outside ones 3:1.

Allowing for wetland surface and berms, the wetland will occupy an area of about 49 × 51 m (~2500 m²). It is not felt that a liner will be required under the wetland cells as an earlier hydrogeological study of the area indicated that the area is underlain by a very low permeability clay.

Hydric soil for the wetland will be blended onsite using local soils with some amendments. Flow into and out of the wetland will be by gravity.

FWS wetland cells will be vegetated with cattails and bulrushes native to the area. It is also suggested that duckweed be used in the open-water areas near the cell entrances and exits and that some be added to the retention pond as well to improve transformations in it.

15.12 SUBSURFACE-FLOW WETLAND EVALUATION

For an SSF wetland treating 60 m³/day of landfarm leachate with an influent concentration of 100 mg/L of BOD, conservative plug-flow, first-order kinetics calculations indicate a required treatment surface area of 415 m² with an estimated effluent BOD concentration of 12 mg/L.

As with FWS wetlands, at least two treatment trains are recommended for the SSF case, but here the smaller area calculated probably dictates that only two parallel cells, each of half the area, be used. As with the FWS case, current wetland design practice is to minimize aspect ratio. Accordingly, the size of the wetland cells for this evaluation was set at 15 × 15 m, giving a cell area of 225 m² and a total wetland surface area of 450 m², close enough to the initial estimate of 415 m².

With SSF wetlands, wastewater flows underground in a substrate. For this case the thickness of this substrate would be 0.6 m and normal water depth in it would be 0.5 m. The substrate would be imported pea gravel, which would be screened and sized on-site. Floor slope again would be 2%. The wetland surface would be vegetated with reeds. Berms for the SSF case would be similar to those for the FWS case.

Allowing for wetland surface and berms, the wetland would occupy an area of about 26 × 45 m (~1200 m²).

15.13 ESTIMATED COSTS

Although more-detailed cost calculations will be determined during the subsequent design phase, based on the sizings discussed above, implementation costs for the two cases can be estimated and these are presented in Table 15.7. The costs do not include the costs of an outlet pond or bay

**Table 15.7   Treatment of Landfarm
Leachate Estimated Costs ($K)**

	FWS	SSF
Mobilization	5	5
Site work	30	40
Hydrology/hydrogeology	6	6
Flow distribution	22	25
Vegetation	15	8
Design and engineering	18	40
Client/regulatory meetings	8	8
Site supervision	7	7
Manuals, specification, etc.	9	10
Contingency	21	36
Total	163	197

(estimated at $3K to 10K) or any costs to clean the existing outlet ditch, into which wetland effluent would be discharged. Final design and engineering costs are included in the costs in Table 15.7.

Operating costs for the new system (other than sampling) will be very small (<$2K/year for maintenance and repairs and, when compared with current Suncor staff costs associated with arranging for the trucking of the leachate, might even result in net savings. A summer student is employed to operate the landfarm, and this individual could carry out most of the operations and maintenance activities necessary for the wetland at no increased labor costs.

15.14 RECOMMENDED DESIGN

It is recommended that the FWS case be selected. There are a number of reasons for this recommendation besides the fact that it is estimated to be more economic. These include that an FWS is simpler and easier to operate, that a size which allows four cells would provide a more versatile and flexible system, that an FWS would not require the importation of substrate gravel media, and that FWS wetlands respond better to flow variations.

15.15 PERMITTING AND REGULATORY CONSIDERATIONS

The proposed wetland was reviewed with an official from the local office of the MOEE to discuss its possible regulatory implications. It was pointed out that currently leachate from the landfarm does not involve a Municipal Industrial Strategy for Abatement (MISA) discharge point and that the proposed wetland system will be a simple one, with few or no mechanical devices, one which would not justify a composite sampler or MISA sampling. It was noted that the leachate is already being treated in the landfarm and that the proposed wetland system would only be for polishing it further. It was also noted that the ditch system into which the effluent from the proposed wetland would flow was already extensively vegetated and undoubtably this aquatic vegetation was already cleaning up runoff passing through it. It would do the same for wetland effluent.

It was determined that the proposed wetland would require an application as sewer works under the Ontario Water Resources Act and have its own Certificate of Approval. The Ontario Environmental Bill of Rights (EBR) would require 30 days notification, but a public hearing would likely not be required.

REFERENCES

Kadlec, R. and Knight, R., 1996. *Treatment Wetlands*, Lewis Publishers, Chelsea, MI, 893 pp.

Matthews, B. C., Richards, N. R., and Wicklund, R. E., 1957. Soil Survey of Lambton County, Report No. 22, Ontario Soil Survey.

O'Malley, P., 1996. Another arrow in the quiver, *Remediation Management* May/June:35.

Reed, S., 1997. Personal communication.

Reed, S., Crites, R., and Middlebrooks, E., 1995. *Natural Systems for Waste Management and Treatment*, McGraw-Hill; New York.

U.S. Environment Protection Agency, 1993. Bioremediation Using the Land Treatment Concept, Office of Research and Development, EPA/600-R/164.

CHAPTER 16

A Review of Iron
Accumulation in Leachate Treatment
Wetlands: Toxicity to Benthic Invertebrates

Francine Kelly-Hooper

CONTENTS

ABSTRACT: Iron is a common constituent of municipal landfill leachate, which, when exposed to oxygenated aquatic environments, settles out of the water column to form a thick floc on the underlying sediment layer. This transformation can occur in treatment wetlands, and attention should be paid to the effects of chronically high iron concentrations on treatment efficiency and sustainability. The main focus of this review is benthic invertebrates as important components of treatment wetlands, and how they are affected by excessively high iron concentrations. The significance of ferrous iron (Fe^{2+}) is briefly discussed, with an emphasis on the effects of ferric hydroxide ($Fe(OH)_3$) flocs on individual species and ecosystem health in general.

16.1 INTRODUCTION

Municipal landfills are designed to protect human populations and the environment by collecting and confining wastes within a managed facility. The standard landfilling method involves the excavation of a large depression, into which refuse is deposited and covered with soil. Although this practice effectively confines solid waste, it also produces a highly polluted liquid by-product known as landfill leachate, which is created as precipitation percolates through the decomposing waste.

Typical municipal landfill leachate constituents include biochemical oxygen demand (BOD_5), chemical oxygen demand (COD), total suspended solids (TSS), total nitrogen (TN), metals, organics, and salts (U.S. Department of Commerce, National Technical Information Service, 1984; Ontario Ministry of the Environment and Energy, 1993. When left uncontrolled, these constituents can leach into and seriously degrade groundwater aquifers and aquatic habitats. Consequently, leachate collection and treatment systems are essential components of responsible landfill management.

Conventional leachate treatment systems are typically based on concrete and steel infrastructures, which rely on intensive inputs of chemicals and nuclear and/or fossil fuel energy. In recent years, however, leachate treatment wetlands have been considered as a natural alternative to these conventional technologies. As a chemical-free ecotechnology, it is aesthetically pleasing and less expensive to operate and maintain. It is therefore not surprising to see a growing interest in the potential use of constructed wetlands for leachate treatment.

16.1.1 Iron as a Primary Leachate Contaminant

Like all wastewater treatment facilities, wetlands have a finite capacity for water quality improvement. Inadequate planning and design can result in contaminant saturation of sediments, leading to possible system failure. It is therefore essential to anticipate and regulate those constituents that present the greatest probability of interfering with the treatment process. One such constituent is iron, which typically exceeds the Ontario Drinking Water Guideline of 0.3 mg/L and the Ontario Provincial Water Quality Objective of 0.3 mg/L, by an average of 94 mg/L (Ontario Ministry of Environment and Energy, 1993; 1994). Consequently, it is important to understand and predict the dynamics of iron within leachate treatment wetlands.

16.2 LEACHATE TREATMENT AND WETLAND DYNAMICS

One of the most commonly used treatment wetland designs is the surface-flow wetland (SFW), which typically includes basins or channels with a subsurface barrier to prevent off-site contaminant migration (U.S. Environmental Protection Agency, 1988). Treatment efficiency is highly dependent upon the interactions that occur among the following SFW structural and biological components: water, sediment, vegetation, and decomposer organisms including bacteria, fungi, and aquatic and benthic invertebrates.

Water quality improvement begins as incoming leachate is slowed by dense stands of emergent plants, allowing suspended sediments, organic debris, and trace metals to settle into the sediment layer. With regard to iron, the average removal rate is known to be 63% (Surface et al., 1992;

Kadlec and Knight, 1995). The next stage of the treatment process begins as aquatic and benthic microorganisms, such as bacteria, actively transform and mineralize pollutants into forms that enter the food chain and/or are sequestered within the sediment layer. A key component of this process is the release of essential nutrients, which are utilized by the wetland vegetation and microorganisms. Wetland plants support microorganisms by delivering oxygen to the sediments and by providing tissues upon which microorganisms can grow (Daukas et al., 1989; Fey, 1995). This mutually beneficial relationship allows plant and microbial communities to regenerate and to continue the water quality improvement process (Weller, 1987).

16.2.1 Benthic Invertebrates and Leachate Treatment Efficiency

Invertebrates include the 95% of species belonging to the Animal Kingdom that do not possess backbones (Barnes, 1980). Benthic invertebrates inhabit the sediments of aquatic environments including natural and treatment wetlands. They include many different species, with some of the most common being worms, clams, snails, mayfly nymphs, stone fly nymphs, and crayfish. Some benthic invertebrates feed on live and/or dead plant tissue, while others prey on bacteria and smaller invertebrates. Benthic invertebrates provide a critical link in the food chain by supporting many higher organisms including fish, amphibians, reptiles, birds, and mammals. Consequently, they can act as a pathway for contaminant uptake in food chains including those that exist within treatment wetlands (Godfrey et al., 1985; Feierabend, 1989; Fey, 1995).

Benthic invertebrate feeding activities (e.g., shredding and filtering) enhance water quality improvement by physically reducing detritic organic matter to a size that is more easily transformed by microorganisms (Murkin et al., 1988; Feierabend, 1989). For example, shredding activities could accelerate the rate at which leachate constituents are broken down and incorporated into the wetland sediment layer. It should also be noted, however, that benthic invertebrate foraging and burrowing activities can hinder the burial process by resuspending sediments (bioturbation) and associated constituents such as iron (Krantzberg, 1985). As a localized activity, however, bioturbation is unlikely to affect overall treatment efficiency.

In addition to their roles in food chain support and water quality improvement, benthic invertebrates can provide the early warning signs of treatment wetland oversaturation. As sediment dwellers, they eat, breathe, and reproduce within the most-concentrated areas of contamination. Under certain conditions, excessive iron concentrations can impede essential activities resulting in a combination of lethal and/or sublethal consequences. This could be followed by community structural changes such as replacement of sensitive species by more tolerant species and reduced species diversity. Benthic invertebrate–monitoring programs can therefore be used to detect the oversaturation of iron in treatment wetlands. They can also be used as indicators of potential bioaccumulation within aquatic food chains (Newman and McIntosh, 1991; Pinel-Alloul et al., 1995). Although this function is not typically applicable to iron, which does not tend to bioaccumulate in food chains, it can be highly useful in monitoring for many other types of contaminants.

16.2.2 Iron as an Essential Element and as a Toxin

As an essential element, iron is present in all plant and animal tissues and acts as a mediator of oxygen and energy transport. It is also necessary for photosynthesis and enzyme production in plants, and oxygen storage and transportation in animals. Iron is a key component of hemoglobin and can be highly toxic when present in high concentrations, certain chemical forms, and over certain periods of time (Venugopal and Luckey, 1978; Jacobs and Worwood, 1980; Berman, 1991). In benthic invertebrates, the toxicological effects of excessive iron exposure may be measured in terms of lethal and sublethal impacts as evidenced by increased mortality rates and impaired feeding, respiration, movement, and reproductive abilities (Koryak et al., 1972; Brenner et al., 1976; Osborne and Davies, 1979; McKnight and Feder, 1984; Rasmussen and Lindegaard, 1988; Hare, 1992; Gerhardt, 1992a; 1992b; 1994; Maltby and Crane, 1994; Wellnitz et al., 1994; Pynnone, 1996).

16.2.3 Iron Speciation

Iron species can be grouped into the following phases: (1) the aqueous phase (free ionic species and soluble complexes), (2) the solid phase (colloids and particles), and (3) the biological phase (adsorbed to biological surfaces or incorporated into cells) (Gerhardt, 1992b). It can exist in a wide variety of compounds, but for the purposes of this paper, only ferric iron Fe^{3+}, in the form of ferric hydroxide $Fe(OH)_3$, and ferrous iron Fe^{2+} species will be discussed.

16.2.3.1 Fe²⁺ and Fe³⁺ Phases and Characteristics

Fe^{2+} is a highly soluble iron species that exists in the aqueous phase as free ions or as complexed organic molecules in the biological phase. It is very unstable and readily oxidizes to form $Fe(OH)_3$ (Hem, 1985). As a soluble species, Fe^{2+} is easily absorbed into biological tissues and is considered to be the most acutely toxic form of iron (Venugopal and Luckey, 1978; Jacobs and Worwood, 1980; Berman, 1991; Gerhardt, 1992b). In contrast, Fe^{3+} is a highly insoluble iron species that exists in the solid phase. It forms stable complexes with a variety of ligands and readily joins with hydroxide ions to form $Fe(OH)_3$. The $Fe(OH)_3$ form of iron exists as an insoluble brown floc which settles to the sediment layer and/or remains in suspension adsorbed to organic matter (Hem, 1985). Many studies indicate that $Fe(OH)_3$ can exert toxic effects on individual aquatic organisms, decrease species diversity and disrupt community structures (Koryak et al., 1972; Sykora et al., 1972; 1975; Smith and Sykors, 1976; Brenner et al., 1976; Osborne and Davies, 1979; McKnight and Feder, 1984; Rasmussen and Lindegaard, 1988; Hare, 1992; Gerhardt, 1992a; Wellnitz et al., 1994).

16.2.3.2 Dissolved Oxygen Effects on Iron Speciation

The chemical behavior and solubility of iron is strongly dependent upon dissolved oxygen concentrations that exist within the water column and sediment layer. In the presence of oxygen, soluble ferrous hydroxides $(Fe(OH)_2)$ are oxidized from the aqueous Fe^{2+} phase to the insoluble $Fe(OH)_3$ solid phase (Sykora et al., 1972; Hem, 1985). Consequently, anaerobic (reduced) waters and sediments would likely be dominated by Fe^{2+} rather than by $Fe(OH)_3$. In contrast, oxygenated environments would likely be dominated by $Fe(OH)_3$ rather than by Fe^{2+}. Treatment wetlands tend to maintain low oxygen levels of approximately 1 mg/L, and may fluctuate between periods of oxia and anoxia (Kadlec and Knight, 1995). It would therefore be logical to expect both forms of iron to be present, with $Fe(OH)_3$ being the dominant species.

16.2.3.3 pH Effects on Iron Speciation

Acidic conditions affect iron by changing the complexation equilibrium to favor Fe^{2+} in the soluble aqueous phase. Under neutral conditions, however, Fe^{2+} is converted to Fe^{3+} by changing the complexation equilibria and competition between hydrogen and metal ions at the binding sites of organic and inorganic ligands (Hem, 1985; Gerhardt, 1992). Fe^{2+} would be expected to dominate under acidic pH conditions of less than 5.3, while $Fe(OH)_3$ would dominate under circumneutral to alkaline conditions of greater than 6.5 pH (Osborne and Davies, 1979; McKnight and Feder, 1984; Rasmussen and Lindegaard, 1988; Gerhardt, 1994). Leachate treatment wetlands would be expected to be slightly acidic to circumneutral pH ranges of 5.2 to 7.6 (Ontario Ministry of the Environment and Energy, 1993; Kadlec and Knight, 1995). It would therefore be logical to expect both forms of iron to be present under such pH conditions, with $Fe(OH)_3$ being the dominant species.

16.2.3.4 Iron Speciation within Leachate Treatment Wetlands

Within treatment wetlands, iron speciation is mainly affected by fluctuating dissolved oxygen concentrations as compared with the more stable and predictable pH conditions. Iron speciation varies according to the varying oxygen levels that occur within the oxic water layer, the oxic sediment layer, and the anoxic sediment layer. The oxic water layer is located within the water column above the sediment surface. This layer typically has an oxygen level of at least 1 mg/L and would likely be dominated by $Fe(OH)_3$. The oxic sediment layer is located below the oxic water layer, measures only a few millimeters in depth, and is rich in organic matter (e.g., detritus, live roots, and benthic organisms). This layer typically holds at least 1 mg/L oxygen and would also be dominated by $Fe(OH)_3$. In contrast, the underlying anoxic sediment layer is void of oxygen and would likely be dominated by Fe^{2+}. This harsh environment supports only the most tolerant organisms such as anaerobic bacteria (Solomens et al., 1987; Tessier et al., 1987).

16.2.4 Iron Toxicity

The toxicity of iron in surface water bodies has generally been considered as being low because as an essential element it is regulated and eliminated through bodily wastes. Although high doses of iron will bioaccumulate within organisms, constant elimination through bodily wastes prevents it from biomagnifying through to higher levels of the food chain. However, iron will bioaccumulate within individual organisms if the rate of consumption exceeds the rate of elimination (Venugopal and Luckey, 1978; Jacobs and Worwood, 1980; Berman, 1991).

16.2.4.1 Drinking Water Objectives

From a drinking water perspective, iron is considered to be more of an aesthetic nuisance than a toxic health hazard. The Ontario Drinking Water Objective of 0.3 mg/L, is based upon the assumption that iron-contaminated water will be consumed as $Fe(OH)_3$ rather than as Fe^{2+} (Ontario Ministry of the Environment and Energy, 1979). Since $Fe(OH)_3$ is insoluble and is not absorbed into organic tissue, it does not cause intracellular damage to biological organisms (Jacobs and Worwood, 1980). Consequently, the drinking water objective for iron is intended to address aesthetic issues such as unpalatable water and staining of laundry and plumbing fixtures.

16.2.4.2 Toxicity to Aquatic Life

Iron toxicity can disrupt aquatic communities in several ways. For example, concentrated pulses of Fe^{2+} can cause increased mortality rates. In contrast, long-term exposures to relatively low $Fe(OH)_3$ concentrations can cause sublethal impacts, such as decreased feeding activity, growth rates, and reproductive success. Consequently, the Provincial Water Quality Objective (PWQO) of 0.3 mg/L total iron is designed to protect aquatic organisms from the lethal and sublethal effects of Fe^{2+} and $Fe(OH)_3$ (Ontario Ministry of the Environment and Energy, 1979).

16.2.5 Fe²⁺ Toxicity

Gerhardt (1992a) refers to Fe^{2+} as a carcinogen that creates oxidative stress by inducing the formation of oxygen radicals resulting in membrane and DNA damage. Chronic Fe^{2+} exposure is also known to cause liver cirrhosis and ulceration of the gastrointestinal mucosa in mammals (Gerhardt, 1992a). In addition, Fe^{2+} can create neurotoxic effects that may cause motor inactivation leading to decreased feeding activity. These effects were observed by Gerhardt (1992a) during experimental studies of *Leptophlebia marginata* larvae that were exposed to Fe^{2+}. This study implicates Fe^{2+} as having impaired the muscular system, which led to peristalsis of the gut membrane. The resulting gut impermeability likely caused dehydration and hardening of ingested food. The impaired transport of essential nutrients through the gut wall is thought to have caused a significant decrease in larval motility. These results are confirmed by Pynnonen (1996) who observed the same toxicological effects in the baltic isopod, *Saduria (Mesidotea) entomon* L.

16.2.5.1 Fe(OH)₃ Toxicity

16.2.5.1.1 Habitat Impacts

— $Fe(OH)_3$ blankets the sediment with a highly turbid brown floc that reduces light penetration and primary productivity. The resulting reduction of benthic algae and detritus leads to decreased availability of food for grazers such as crayfish, snails, mayfly nymphs, and stone fly nymphs (Koryak et al., 1972; McKnight and Feder, 1984; Rasmussen and Lindeggard, 1988; Gerhardt, 1992). $Fe(OH)_3$ flocs can also cover essential food sources as was observed during a 6-day study of *Gammarus pulex* (Amphipoda, Crustacea) responses to iron-coated alder leaves (Maltby and Crane, 1994). The feeding activity of the test animals decreased as the body concentrations of iron increased, resulting in a significant increase in mortality. Increased turbidity also impairs sight perception, causing significant changes in visually cued reproductive, feeding, and shelter-seeking activities. Although this impairment of visual perception has not been studied in benthic invertebrates, it has been observed in brook trout and fathead minnows (Sykora et al., 1972; 1975).

On the sediment surface, $Fe(OH)_3$ can form hardened layers that act as a barrier to burrowing organisms. Such a barrier could inhibit essential foraging, shelter-seeking, and predator-avoidance activities (Osborne and Davies, 1979). This barrier effect was observed by Werner (1976) who documented the absence of macroinvertebrates in areas where $Fe(OH)_3$ precipitates had formed a cement-like coating over the substrate. The absence was attributed to habitat degradation and loss of food resources.

16.2.5.1.2 Direct Biological Impacts

— In addition to habitat degradation, $Fe(OH)_3$ can directly affect epidermal tissues of affected organisms. Chronic exposures to $Fe(OH)_3$ precipitates can encrust exterior tissues, preventing free movement (Osborne and Davies, 1979). Some species, such as *L. marginata* (mayfly), have been observed to attempt removing these bothersome coatings with their mouthparts. Over time, layers of iron hydroxides will become firmly attached to the epidermal tissues, which can actually be used to estimate the ages of individual organisms (Gerhardt, 1992).

$Fe(OH)_3$ can also coat benthic invertebrate mouth parts, which could inhibit feeding abilities leading to decreased growth rates and possible starvation. In addition, clogged gills can increase susceptibility to disease and cause respiratory distress leading to suffocation. The combined effects of $Fe(OH)_3$ are known to occur among a wide range of benthic invertebrate species (Sykora et al., 1972; Hare, 1992; Gerhardt, 1992; Wellnitz et al., 1994).

16.2.5.1.3 Habitat and Biological Impacts of Fe²⁺ and Fe³⁺

— As indicated earlier, the oxidation of Fe^{2+} to Fe^{3+} under acidic conditions occurs through the biotic processes of iron-depositing bacteria. Wellnitz et al. (1994) observed that the combined presence of both iron and manganese creates blooms of *Leptothrix ochracea* bacteria. The presence of these bacteria blooms negatively affect existing aquatic communities because they displace existing diatom populations as the primary food source for benthic invertebrates. In addition to the food-loss issue, the bacterial sheath material is known to clog the mouthparts and tracheal gills of species such as the mayfly *Stenonema fuscum*. The impacts of the bacteria blooms were demonstrated by the differences in caddisfly (*Glossosoma* spp., *Neophylax* spp., and Hydropsychidae) populations, which were present above and below the bloom but absent within the bloom itself. This absence was attributed to the inadequacy of the bacteria as a food source, starvation caused by clogged mouthparts, suffocation caused by clogged gills, and the toxic effects of iron and manganese ions and/or bacteria (Wellnitz et al., 1994).

16.2.6 Iron Toxicity Thresholds

Iron toxicity thresholds are dependent upon abiotic (e.g., dissolved oxygen, pH) and biotic (e.g., age, sex, species) factors. Dissolved oxygen and pH conditions will determine the presence of Fe^{2+} or $Fe(OH)_3$, with Fe^{2+} being more toxic at lower concentrations. For example, a 96-h bioassay of the mayfly *Leptophlebia marginata* found that 50% of the population lost their escape behavior when exposed to 63.9 mg/L of Fe^{2+} as compared with 70 mg/L of Fe^{3+}. The same study found 50% mortality among mayfly populations exposed to 89.5 mg/L of Fe^{2+} and 106.3 mg/L of Fe^{3+} (Gerhardt, 1994).

The age factor was observed in a study of the sow bug *Asellus aquaticus* which showed an increased sensitivity in younger as opposed to older individuals. With sex as a factor, iron-stressed females invested more energy in reproduction and produced smaller sized offspring (Gerhardt, 1992). Many benthic invertebrate species are highly sensitive to iron toxicity, whereas others are more tolerant. For example, a 96-h study of the mayfly *Ephemerella subvaria* found that 50% of the population died when exposed to 0.32 mg/L of Fe^{3+} (Maltby and Crane, 1994). In comparison, the less sensitive mayfly *Leptophlebia marginata* had a 50% mortality rate at an Fe^{3+} concentration of 106.3 mg/L (Gerhardt, 1992). The more tolerant species are usually considered to include members of Diptera and Tubificidae as compared with more sensitive species such as those of Ephemeroptera, Coleoptera, and Plecoptera (Koryak et al., 1972; Rasmussen and Lindegaard, 1988).

According to Rasmussen and Lindegaard (1988), ecosystem health can be maintained where Fe^{2+} and Fe^{3+} concentrations do not exceed 0.2 and 0.3 mg/L, respectively. Total iron concentrations ranging from 0.2 to 1.0 mg/L may allow for the survival of normal numbers of individuals with significant reduction in species diversity. A range of 1 to 10 mg/L total iron may result in a reduction in both number of individuals and taxa, while 10 mg/L or greater would likely support only a few species in very small numbers.

16.3 CONCLUDING COMMENTS

As a developing technology, leachate treatment wetlands hold both promise and some uncertainty. It is not known whether some leachate contaminants can be effectively transformed and eliminated while others, such as iron, have some portion retained within the wetland sediments (see Robinson et al., Chapter 6, and DeBusk, Chapter 11) which accumulate over time. In order to prevent the oversaturation of wetland sediments and toxic risks to benthic invertebrates, leachate should be pretreated before it is channeled to a treatment wetland. For example, sedimentation ponds could be incorporated into site designs as a means of minimizing iron levels before leachate is discharged to a treatment wetland. If iron concentrations reach toxic concentrations within the pond sediments, they can be dredged without having to disrupt the established community of wetland flora and fauna.

Comprehensive monitoring programs should also become standard protocol for all wetland treatment systems, including water quality monitoring. In addition, sediment monitoring would allow resource managers to assess the contamination history of the constructed wetland. Monitoring for changes in benthic invertebrate species diversity and community structures will signal the threshold at which the wetland has reached treatment capacity.

REFERENCES

Barnes, B. M., 1980. *Invertebrate Zoology*, 4th ed., W. B. Saunders, Toronto.

Berman, E., 1991. *Toxic Metals and Their Analysis*, U-M-I Out-of-Print Books on Demand, Ann Arbor, MI.

Brenner, F. J., Corbett, S., and Shertzer, R., 1976. Effect of ferric hydroxide suspension on blood chemistry in the common shiner, *Notropus cornutus*, Transactions American Fish. Society, 3:450–454.

Daukas, P., Lowry, D., and Walker, W. W., Jr., 1989. Design of wet detention basins and constructed wetlands for treatment of storm water runoff from a regional shopping mall in Massachusetts, in Hammer, D. A., Ed., *Constructed Wetlands for Wastewater Treatment: Municipal, Industrial, and Agricultural*, Lewis Publishers, Chelsea, MI, 104–118.

Feierabend, J. S., 1989. Wetlands: the lifeblood of wildlife, in Hammer, D. A., Ed., *Constructed Wetlands for Wastewater Treatment: Municipal, Industrial and Agricultural*, Lewis Publishers, Chelsea, MI, 109–117.

Fey, R. T., 1995. *Creating and Using Wetlands for Wastewater Disposal and Water Quality Improvement*, Van Nostrand Reinhold, Toronto.

Gerhardt, A., 1992a. Effects of subacute doses of iron (Fe) on *Leptophlebia marginata* (Insecta: Ephemeroptera), *Freshwater Biology* 27:79–84.

Gerhardt, A., 1992b. Review of impact of heavy metals on stream invertebrates with special emphasis on acid conditions, *Water, Air and Soil Pollution* 66:289–314.

Gerhardt, A., 1994. Short term toxicity of iron (Fe) and lead (Pb) to the mayfly *Leptophlebia marginata* (L.) (Insecta) in relation to freshwater acidification, *Hydrobiologia* 284:157–168.

Godfrey, P. J., Kaynor, E. R., Pelczarski, S., and Benforado, J., 1985. *Ecological Considerations in Wetlands Treatment of Municipal Wastewaters*, Van Nostrand Reinhold, New York.

Hare, L., 1992. Aquatic insects and trace metals: bioavailability, bioaccumulation, and toxicity, *Critical Reviews in Toxicology* 22:327–369.

Hem, J. D., 1985. Study and Interpretation of the Chemical Characteristics of Natural Water, U.S. Geological Survey Water-Supply Paper 2254, 3rd ed.

Jacobs, A. and Worwood, M., Eds., 1980. *Iron in Biochemistry and Medicine II*, Academic Press, London.

Kadlec, R. H. and Knight, R. L., 1995. *Treatment Wetlands*, Lewis Press, Boca Raton, FL, 893 pp.

Koryak, M., Shapiro, M. A., and Sykora, J. L., 1972. Riffle zoobenthos in streams receiving acid mine drainage, *Water Research* 6:1239–1247.

Maltby, L. and Crane, M., 1994. Responses of *Gammarus pulex* (Amphipoda, Crustacea) to metalliferous effluents: identification of toxic components and the importance of interpopulation variation, *Environmental Pollution* 84:45–52.

McKnight, D. M. and Feder, G. L, 1984. The ecological effect of acid conditions and precipitation of hydrous metal oxides in a Rocky Mountain stream, *Hydrobiologia* 119:129–138.

Newman, M. C. and McIntosh, A. W., Eds., 1991. *Metal Ecotoxicology: Concepts and Applications*, Lewis Publishers, Chelsea, MI.

Ontario Ministry of the Environment and Energy, 1979. *Rationale for the Establishment of Ontario's Provincial Water Quality Objectives*, Queen's Printer for Ontario.

Ontario Ministry of the Environment and Energy, 1993. *Guidance Manual for Landfill Sites Receiving Municipal Waste*, Queen's Printer for Ontario, PIBS 2741.

Ontario Ministry of the Environment and Energy, 1994. *Water Management: Policies Guidelines, Provincial Water Quality Objectives*, Queen's Printer for Ontario, PIBS 3303E.

Osborne, L. L. and Davies, R. W., 1979. Effects of limestone strip mining on benthic macroinvertebrate communities, *Water Research* 13:1285–1290.

Pinel-Allaul, B., Methot, G., Lapierre, L., and Willsie, A., 1996. Macroinvertebrate community as a biological indicator of ecological and toxicological factors in Lake Saint-Francois (Quebec), *Environmental Pollution* 91:65–87.

Pynnonen, K., 1996. Heavy metal-induced changes in the feeding and burrowing behaviour of a baltic isopod, *Saduria* (Mesidotea) *entomon* L., *Marine Environmental Research* 41:145–156.

Rasmussen, K. and Lindegaard, C., 1988. Effects of iron compounds on macroinvertebrate communities in a Danish lowland river system, *Water Research* 22:1101–1108.

Smith, E. J. and Sykora, J. L., 1976. Early developmental effects of lime-neutralized iron hydroxide suspensions on brook trout and coho salmon, *Transactions of the American Fishery Society* 2:308–312.

Starkel, W. M., 1985. Predicting the effect of macrobenthos on the sediment/water flux of metals and phosphorus, *Canadian Journal of Fisheries and Aquatic Science* 42:95–100.

Sykora, J. L., Smith, E. J. and Synak, M., 1972. Effect of lime neutralized iron hydroxide suspensions on juvenile brook trout (*Salvelinus fontinalis*, Mitchell), *Water Research* 6:935–950.

Sykora, J. L., Smith, E. J., Synak, M., and Shapiro, M. A., 1975. Some observations on spawning of brook trout (*Salvelinus fontinalis*, Mitchill) in lime neutralized iron hydroxide suspensions, *Water Research* 9:451–458.

U.S. Department of Commerce, National Technical Information Service, 1984. Production and Management of Leachate from Municipal Landfills: Summary and Assessment, Calscience Research, Inc., Huntington Beach, CA, PB84-187913.

U.S. Environmental Protection Agency, 1988. Constructed Wetlands and Aquatic Plant Systems for Municipal Wastewater Treatment, Center for Environmental Research Information, Cincinnati, OH, 45268.

Venugopal, B. and Luckey, T. D., 1978. *Metal Toxicity in Mammals — 2, Chemical Toxicity of Metals and Metalloids*, Plenum Press, New York.

Weller, M. W., 1987. *Freshwater Marshes: Ecology and Wildlife Management*, University of Minnesota Press, Minneapolis.

Wellnitz, T. A., Grief, K. A., and Sheldon, S. P., 1994. Response of macroinvertebrates to blooms of iron-depositing bacteria, *Hydrobiologia* 281:1–17.

CHAPTER 17

A Cost Comparison
of Leachate Treatment Alternatives

Sharon Rew and George Mulamoottil

CONTENTS

ABSTRACT: This paper is a cost comparison of three treatment alternatives for treating leachate from the Glanbrook landfill site in the Regional Municipality of Hamilton-Wentworth (RMHW), namely, a conventional pumping-and-hauling scheme, a forcemain installation, and a constructed wetland. The site opened in 1980 is currently the only active landfill site in the region. The landfill is designed for a capacity of approximately 11,800,000 m³ of domestic, commercial, and nonhazardous industrial wastes. Current estimates indicate that the active life of the landfill should extend to the year 2030. The current practice of the RMHW involves pumping leachate from an excavated pond in the active portion of the landfill and removing it from the site by tanker truck. The leachate is then transported to the sewage treatment plant and treated at a cost of approximately 5.2¢/gal. According to the region, the annual cost for such a treatment is $170,000. The forcemain alternative eliminates the costly practice of pumping and hauling of the leachate. In this situation, the leachate is conveyed by the use of a forcemain to the nearest sanitary sewer approximately 14 km away. The leachate is then transported to the sewage treatment plant. Installation of this system is estimated to cost between $2 and 3.5 million, with operation and treatment costs ranging between $119,000 and $133,000/year. The third and final alternative is a constructed wetland. Analysis of this alternative shows that capital costs could range between $37,000 and $74,000, with operation costs between $5,000 and $50,000/year. All cost figures are presented in U.S. dollars. This case study clearly illustrates that a constructed wetland alternative for the Glanbrook site is considerably less expensive than the two conventional systems presented and strongly supports the claims that constructed wetlands are less expensive than conventional treatment alternatives.

17.1 INTRODUCTION

The need for efficient collection and treatment of leachates to avoid surface water and groundwater contamination is widely recognized. However, conventional approaches to leachate treatment, such as on-site treatment or conveyance to off-site facilities, usually municipal sewage treatment plants, are considered undesirable because of the high on-site operation and maintenance costs. Further, operation and maintenance must continue after landfill closing (Staubitz et al., 1993; Surface et al., 1993). It is important to note that the addition of leachate to municipal sewage may disrupt the normal biological processes upon which treatment depends (Surface et al., 1993; Bulc et al., 1997). As well, the transport of concentrated wastes on public roads is costly and dangerous (Bulc et al., 1997). Conventional on-site treatment is usually avoided because of high construction costs and the need for continuous monitoring by a licensed operator (Surface et al., 1993).

Generally, leachates are considered anoxic with high concentrations of biochemical oxygen demand (BOD), organic carbon, nitrogen, chlorides, iron, manganese, and phenols with little or no phosphorus (Staubitz et al., 1989). The leachates may also contain high concentrations of heavy metals, pesticides, chlorinated and aromatic hydrocarbons, and other toxic chemicals (Cross and Metry, 1976; Thirumurthi, 1991; Surface et al., 1993; Aarts, 1994). As a result, groundwater and surface water contamination may result without proper collection and treatment of leachates. An ideal leachate treatment system should have the ability to treat a wide range of chemical constituents, accept varying quantities and concentrations of leachates, and be inexpensive to construct and easy to maintain with low energy and personnel requirements. In many situations the treatment of wastes is considered an unproductive process that incurs additional cost to an otherwise productive system (Martin, 1991).

In many circumstances the use of constructed wetlands for wastewater treatment has become an accepted alternative to conventional treatment systems. Constructed wetlands of differing designs have demonstrated the ability to reduce pollutant loads to acceptable levels (Staubitz et al., 1989; Urbanic-Bercic, 1994; Sanford et al., 1995; Latchum and Kangas, 1996). However, there is a lack of published information on the monetary costs of using such systems, and yet generalizations have been made that constructed wetlands are less expensive than conventional treatment technologies. For example, Hammer and Bastian (1989) state that constructed wetlands are relatively inexpensive to construct and operate. Mæhlum in Chapter 3 and Robinson et al. in Chapter 6 have also stated that the cost of constructed wetlands is much less when compared with conventional treatment systems. Yet, Peverly et al. (1993) suggest that constructed wetlands have similar energy costs to an activated sludge treatment plant, and Sartaj et al. in Chapter 10 consider the cost of a constructed wetland competitive with other treatment systems. Latchum and Kangas (1996) point out that a review of the current literature does not support the claim that constructed wetlands are less expensive than conventional treatment systems and that more-detailed studies are warranted.

Many of the constructed wetland systems have been supported on an experimental basis by public grants and therefore dollar figures may not be available (Latchum and Kangas, 1996). However, these types of data are necessary for widespread application if constructed wetland technology is to become popular (Latchum and Kangas, 1996). It is also interesting to note that a study to assess the attitudes toward artificial wetlands in Ontario has demonstrated that cost was one of the overlying concerns expressed by local municipalities and other government and non-government organizations (Carlisle et al., 1991). As funding to regional governments in Ontario continues to be cut, planning officials are now looking for more-cost-effective ways of dealing with the current practice of treating landfill leachates.

According to Kadlec and Knight (1996) a cost comparison of alternatives, including constructed wetlands, must be completed in the initial development phases to determine if the estimated costs of alternatives fall within the economic constraints. This paper is a cost comparison of treatment alternatives using a case study at the Glanbrook landfill site in the Regional Municipality of Hamilton-Wentworth (RMHW). It compares the costs of three treatment alternatives to handle leachates from the Glanbrook site, namely, a conventional pumping-and-hauling scheme, a force-main alternative, and a constructed wetland. All figures are presented in U.S. dollars. The results presented in the paper are expected to be useful for comparison with other North American situations.

17.2 SITE DESCRIPTION

The site was opened in 1980 and is currently the only active municipally owned landfill site in the region. The landfill occupies 219 ha and is situated on parts of lots 26, 27, 28, concession 9 and parts of lots 26, 27, 28, concession 10 in the township of Glanbrook. Of this land, 115 ha serves as a buffer zone comprising of floodplain and lowland and upland forest habitats.

The landfill was designed for a capacity of approximately 11,800,000 m^3 of domestic, commercial, and nonhazardous solid industrial wastes. Current regional estimates indicate that the active life of the landfill should extend to approximately the year 2030. The landfill is proceeding in two stages. Stage One comprises four waste cells (1 to 4) and covers roughly the southern portion of the landfill as shown in Figure 17.1. Stage One has been completed to final grade and capped. Stage Two also has four waste cells (5 to 8) and occupies the remaining northern portion of the landfill where there is still active landfilling.

The relatively low permeable *in situ* clay and silty soils underlying the site are the only means of leachate containment within the landfill. Leachate collection is facilitated by a toe drain collector system at the base of each cell along its outside edge, thereby creating a single perimeter collector loop around the landfill (see Figure 17.1). A single toe drain also crosses the base of the fill centrally through the landfill site from west to east. These drains discharge into a perimeter transmission pipe system, which, in turn, conveys leachates through gravity to three designated manholes.

The existing system is not effectively managing the leachates from the site and has created certain leachate-related problems. These include leachate buildup (mounding) within the facility that has resulted in leachate sideslope seepage and suspected migration of leachates out of the landfill. Leachate mounding is suspected in both stages of the site. The mound refers to the volume of liquid that would result if the refuse were allowed to drain (Salvato, 1992). Based on modeling exercises performed at the site in 1995, the total free-draining liquid in the leachate mound at the Glanbrook site is approximately 263,000 m^3 (MacLaren, 1995). Leachate mounding at the perimeter of the landfill increases the potential for contaminant migration out of the landfill toward adjacent watercourses. The upper surficial soils, which are likely fractured, provide a potential pathway for migration away from the fill area (MacLaren, 1995).

17.3 TREATMENT ALTERNATIVES

17.3.1 Present Practice: Pumping and Hauling

Until the end of 1994, the cost of leachate disposal has been a function of the time requirements for filling tanker trucks and transporting the leachates to the local sewage treatment plant for

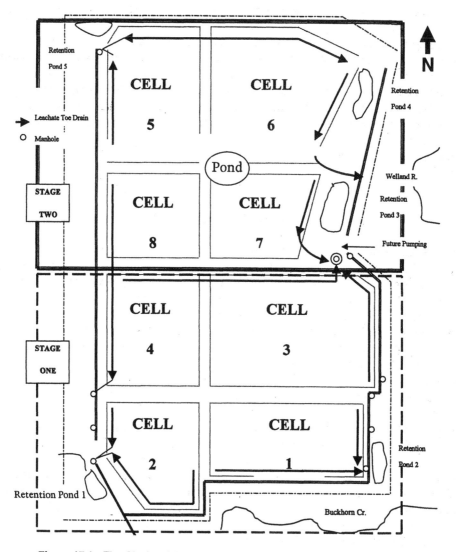

Figure 17.1 The Glanbrook landfill site layout. (Adapted from MacLaren, 1995.)

treatment. Tanker trucks pump the leachates from one of the designated manholes. It is followed by a long waiting period for the manholes to fill after quantities of leachate are pumped out. This procedure proved to be rather onerous and costly as the contractor was billing the RMHW on an hourly basis. At the present time, leachates are pumped from a collection pond excavated into the refuse in Stage Two (see Figure 17.1) and removed in a larger tanker truck. The contractor then bills the RMHW on a volumetric basis rather than the hourly rate. This new approach to pumping and hauling has led to a reduction in the disposal cost per unit volume (MacLaren, 1995).

The RMHW is still responsible for the cost of treating the leachates at the sewage treatment plant. The current cost of treating the leachates is estimated at 5.2¢/gal (Terluk, 1995; 1996). As a result, the total cost of hauling and treating approximately 2,500,000 gal of leachates per year is $170,000, out of which $40,000 is paid to haul the leachate (Terluk, 1996). The amount of leachates generated in a year is expected to double upon site closure, and the total costs will considerably increase (MacLaren, 1995).

17.3.2 Forcemain Alternative

The treatment alternative put forth by MacLaren is the installation of a forcemain system, which has been accepted by the RMHW. This system is designed to convey leachates collected in ground holding tanks to the nearest sanitary sewer approximately 14 km away. A forcemain serves to pump leachate from the holding tanks eliminating the cost associated with pumping and hauling. The sanitary sewer will, in turn, convey leachates to the sewage treatment plant. The total construction cost of this project, including engineering works, has been estimated by the RMHW to range between $2 and $3.5 million (Terluk, 1996). Annual operating costs, taking into account maintenance and hydro are estimated between $20,000 and $34,000. An estimated 600,000 gal of leachate are to be recirculated each year through the pond excavated in the active area of the landfill.

This scheme will still leave the RMHW responsible for the cost of treating approximately 1,900,000 gal of leachate per year at the sewage treatment plant. At the current cost of 5.2¢/gal, this will result in expenses of approximately $99,000. Considering the annual operating and treatment costs, the total annual costs are estimated to range between $119,000 and $133,000. When these costs are compared with the current pumping and hauling costs, the RMHW could save between $37,000 and $51,000/year.

17.3.3 Constructed Wetland Alternative

A few studies show that microbial populations in constructed wetlands are adapted to a wide range of environmental conditions and often perform biochemical transformations of wastewater constituents similar to those that take place in conventional wastewater treatment plants (Surface et al., 1993; Sanford et al., 1995). Plants and soils in wetland systems are responsible for transformations including adsorption of organics, nutrients, and metals constituents (Staubitz et al., 1989; Urbanic-Bercic, 1994; Sanford et al., 1995). It has been shown that nutrients, heavy metals, pesticides, and other natural and anthropogenic organic chemicals have been significantly reduced in wastewater applied to constructed wetlands (Staubitz et al., 1989). In addition, it has been demonstrated that constructed wetlands systems, if properly designed and maintained, can perform as well or better than the conventional treatment systems (Trautmann et al., 1989).

The performance of a constructed wetland for nutrient removal is often dependent on the proper interaction between residence time and flow, vegetation, substrate, and seasonal temperature fluctuations (Martin and Moshiri, 1994). The area of a wetland system is therefore fundamental to allow for the proper interaction between residence time and flow, and should be considered carefully in any design procedure (Surface et al., 1993; Sanford et al., 1995). However, it is difficult to determine the area needed for effective treatment of leachates in general terms since specific hydraulic and pollutant fluctuations, as well as varying climatic conditions have to be taken into consideration (Bulc et al., 1997).

Because constructed wetlands are generally land intensive, land area may be one of the most important factors when considering the cost of using such a system. If the land area required to treat the wastewater effectively is greater than the land available, land acquisition is required, adding to the total costs. Prior to final design of any wetland system, a general estimate of the land area requirements based on wastewater constituents and desired outputs can be ascertained to estimate the cost of a project. A preliminary area can be determined by using the model outlined by Kadlec and Knight (1996). This model uses desired effluent quality, first areal rate constants, and background limits to determine the preliminary sizing requirements. To achieve a conservative estimate of the land area required, modeling should be conducted on all potential pollutants. The determination of wetland area is based on the pollutant constituent that yields the highest land area requirements. The objective of the calculations is to give a general view of the size of wetland that would be required for the wastewater treatment and to determine whether or not the water quality objectives can be met (Kadlec and Knight, 1996).

The sizing model of Kadlec and Knight (1996) was used to determine the area requirements for leachate treatment at the Glanbrook site. The results of the May 1996 chemical analysis (RMHW, Environmental Laboratory, 1996) from the Glanbrook landfill were used for calculating the area. Four different pollutants were used in the preliminary sizing model, namely, BOD, total suspended solids (TSS), ammonia (NH_3), and total nitrogen (TN). The desired effluent quality was determined from the Provincial Water Quality Objectives and the Canadian Water Quality Guidelines. By using the four water quality parameters BOD, TSS, NH_3, and TN, the areas required according to the model are 0.373, 0.0064, 0.44, and 0.078 ha, respectively. In this case study, the largest area requirement for treatment was based on ammonia. Therefore, the constructed wetland leachate treatment alternative was sized for this pollutant at 0.44 ha for a volume of 73.97 m^3/day. This area is well within the recommendation of 1.7 ha for 100 m^3/day (Pries, 1994) and 3 to 4 ha for 1000 m^3 suggested by the Water Pollution Control Federation (1990). The size requirement of 0.44 ha fall very well within the large available land area designated as buffer space.

The total costs for a constructed wetland are determined by the cumulative cost of land, design, earthwork, planting, monitoring, and maintenance (Pries, 1994). By using the area of the wetland, it is possible to estimate the cost or range of costs for the capital investment and potential monitoring and maintenance cost of the system. In addition, expenditures on structures, liners, and specially engineered devices are to be added. Table 17.1 outlines different costing estimates available from the literature.

Table 17.1   Cost Estimates for a Constructed Wetland System

Source	Cost Estimates
Capital Cost	
Pries (1994)	$6800–34,000/ha
Kadlec and Knight (1996)	$10,000–50,000/ha
Water Pollution Control Federation (1990)	$500–1000/$m^3$/day (approx. $84,000–168,000/ha)
U.S. EPA Treatment Wetland Database (Knight et al., 1993)	$50,000/ha
Whalen et al. (1989)	$500/$m^3$/day ($84,000/ha)
Operation and Maintenance Costs	
Water Pollution Control Federation (1990)	$0.03–0.09/$m^3$/day (approx. $1400–5500/ha)
U.S. EPA Treatment Wetland Database (Knight et al., 1993)	$400/acre (0.4047 ha)
Kadlec and Knight (1996)	$400–2000/acre (0.4047 ha)

In general, smaller systems tend to incur higher costs than larger systems. Capital cost for a 0.44-ha system at the Glanbrook site, based on Pries (1994) and Kadlec and Knight (1996), would range between $4400 and $22,000. A better estimate of the cost is possible using the Water Pollution Control Federation values of $500 to 1000/$m^3$/day. By using these figures, the capital cost of the project is $37,000 to 74,000. These figures will increase with the inclusion of final design parameters, such as monitoring wells, engineering services, linear requirements, and permits. Whalen et al. (1989) outline the breakdown of different construction components in terms of percentage of construction costs. They suggest that, on average, piping and valves comprise 13.7% of construction costs, site work 16.4%, gravel 27.2%, structures 23.0%, mobilization 13.6%, and miscellaneous 6.1%.

Kadlec and Knight (1996) demonstrate that the base cost of a constructed wetland system is considered economical at $2000/acre (0.4047 ha); however, the addition of pumps, linears, planting, and complex structures can increase the cost to almost $80,000/acre. Kadlec and Knight (1996) suggest that the extra costs are a direct result of overengineering and that successful projects often use passive, low-tech, self-sustaining designs.

In Canada, treatment and disposal of wastes is regulated by provincial and local agencies. Each provincial environment ministry sets treatment and discharge policies that directly affect the permitting and implementation of wetland wastewater treatment costs (Pries, 1994). For example, regulations may require that building permits be issued for all on-site structures and a certificate of approval be required for discharge of treated wastewater into neighboring water bodies. Expenditures for constructed wetlands will also be influenced by provincial regulations and other regional bylaws.

Operation and maintenance cost will be dependent on personnel, energy, monitoring, and maintenance items (Kadlec and Knight, 1996). These costs will depend on the scale of monitoring, data collection, exotic plant and vector control, and water management (Pries, 1994). The Water Pollution Control Federation (1990) recommends allowing $0.03 to $0.09 operation and maintenance costs per m^3/day, and would add up to $800 to $2400/year for a wetland at the Glanbrook site. Values presented by Kadlec and Knight (1996) range between $400 and $2000/acre. However, Kadlec and Knight (1996) point out that this cost is usually not representative of the actual situation because of supplementary research costs, such as environmental monitoring and trained personnel. They go on to suggest that typical operation and maintenance costs range between $5000 and $50,000/year.

17.4 DISCUSSION

A cost comparison of the three alternatives — pumping and hauling, forcemain conveyance, and a constructed wetland — is outlined in Table 17.2. The constructed wetland figures represent costs for a 0.44-ha constructed wetland on the Glanbrook site treating an annual volume of 27,000 m^3 of landfill leachate. Figures for the constructed wetland costs are derived from Kadlec and Knight (1996), unless otherwise indicated. All figures for the conventional pumping-and-hauling scheme and the forcemain system are provided by Terluk (1995; 1996) and are presented in U.S. dollars.

Table 17.2   Cost Comparison of Three Alternative Treatment Systems

	Conventional Pumping and Hauling	Forcemain System	Constructed Wetland
Capital cost	Not available	$2–3.5 million	$37,000–74,000 (WPCF, 1990)
Operation and maintenance costs	$40,000	$20,000–34,000	$5000–50,000
Treatment costs	$130,000	$99,000	$0
Total annual	$170,000	$119,000–133,000	$5000–50,000

Table 17.2 clearly shows that the constructed wetland alternative is considerably less capital intensive than the forcemain alternative. The capital investment for a constructed wetland on the Glanbrook site is equal to 2% of the capital investment for a forcemain system. Unfortunately, no capital investment figures were available for the pumping-and-hauling scheme. Table 17.2 compares the operation and maintenance, treatment, and total annual costs of the three treatment alternatives. The total annual costs of the constructed wetland system are substantively lower when compared with the other two treatment alternatives. When comparing the operation and maintenance costs of the current pumping and hauling scheme to the constructed wetland system, the constructed wetland

costs range between 13 and 125% of the costs of the current scheme (see Table 17.2). These values suggest that a constructed wetland alternative has the potential to incur greater operation and maintenance costs than the current pumping-and-hauling scheme. However, the constructed wetland does not incur additional treatment costs, where as the current pumping-and-hauling system requires additional treatment costs of $130,000. If one considers treatment as a component of the total annual costs, a constructed wetland is much less expensive, incurring 3 to 30% of the pumping and hauling annual costs.

Similar conclusions can be drawn when the constructed wetland alternative is compared with the forcemain alternative. Operation and maintenance costs of the constructed wetland system are 25 to 147% of the costs of the forcemain system. However, as with the pumping-and-hauling scheme, the forcemain alternative is dependent on additional treatment. This treatment raises annual costs, and, as a result, the constructed wetland alternative becomes more economical representing only 4 to 38% of the overall annual forcemain costs. The specific engineering and special design features, such as impermeable liners and polyvinyl chloride (PVC) piping, may increase the capital investment required in the constructed wetland system. However, the total costs will still be much less when compared with the other two alternatives.

Ongoing monitoring and maintenance of an onsite leachate treatment system is often expensive and requires licensed operators onsite long after site closure. However, the ongoing monitoring and maintenance of a constructed wetland is minimal, and constructed wetlands perform their function even after the closing of the landfill site (Bulc et al., 1997). It is also interesting to note that the value of a wetland property is typically much higher than the site of a mechanical plant value of which will depreciate over time, and this value should be considered in any economic evaluation (Kadlec and Knight, 1996).

The treatment of wastes, although costly, is required to protect the environment. However, through the use of ecotechnologies using natural systems, these costs can be significantly reduced. The cost comparison of treatment alternatives presented illustrates the enormous cost difference between the accepted alternative and the constructed wetland alternative at the Glanbrook site. The study strongly supports the claim that constructed wetlands are less expensive than other treatment alternatives. Because these systems can help bring down the cost of treatment of landfill leachate and further contribute to the design of more-efficient and ecologically sustainable systems, similar studies in other geographic regions are warranted.

REFERENCES

Aarts, T., 1994. Alternative approaches for leachate treatment, *World Wastes* 37:28.

Bulc, T., Vrhovšek, D., and Kukanja, V., 1997. The use of constructed wetlands for landfill leachate treatment, *Water, Science and Technology* 35:301–306.

Carlisle, T., Mulamoottil, G., and Mitchell, B., 1991. Attitudes towards artificial wetlands in Ontario for stormwater control and waterfowl habitat, *Water Resources Bulletin* 27:419–427.

Cross, F. and Metry, A., 1976. Leachate control and treatment, in *Environmental Monograph Series* 7:1–30, Techonic Publishers, Westport, CT.

Hammer, D. A. and Bastian, R. K., 1989. Natural Water Purifiers? in Hammer, D. A., Ed., *Constructed Wetlands for Wastewater Treatment: Municipal, Industrial and Agricultural*, Lewis Publishers, Chelsea, MI.

Kadlec, R. and Knight, R., 1996. *Treatment Wetlands*, Lewis Publishers, Chelsea, MI, 893 pp.

Knight, R. H., Ruble, R. W., Kadlec, R. H., and Reed, S. C., 1993. Wetlands for wastewater treatment performance database, in Moshiri, G. A., Ed., *Constructed Wetlands for Water Quality Improvement*, Lewis Publishers, Boca Raton, FL, 35–58.

Latchum, J. A. and Kangas, P. C., 1996. The economic basis of a wetland wastewater treatment plant in Maryland, in *Proceedings from Constructed Wetlands in Cold Climates: Design, Operation, Performance Symposium*, June 4–5. Friends of Fort George, Niagara-on-the-Lake, Ontario.

MacLaren, F., 1995. Assessment of Existing Leachate Collection System and Other Leachate Related Issues at the Glanbrook Landfill, Draft report to the Regional Municipality of Hamilton-Wentworth.

Martin, A. M., Ed., 1991. *Biological Degradation of Wastes*, Elsvier Science, New York, Preface.

Martin, C. D. and Moshiri, G. A., 1994. Nutrient reduction in an in-series constructed wetland system treating landfill leachate, *Water, Science and Technology* 29:67–272.

Ministry of Environment and Energy, 1994. Policies Guidelines Provincial Water Quality Objectives of the Ministry of Environment and Energy, Ministry of Environment and Energy, Ontario.

Peverly, J. H., Sanford, W. E., Steenhuis, T. S., and Surface, J., 1993. Constructed Wetlands for Municipal Solid Waste Landfill Leachate Treatment: Final Report for the New York State Energy Research and Development Authority and Tompkins County.

Pries, J., 1994. Wastewater and Stormwater Applications of Wetlands in Canada, Issue Paper No. 1994-1 Sustaining Wetlands, Environment Canada, Canadian Wildlife Service, North American Wetlands Conservation Council (Canada), Ottawa, Ontario.

Regional Municipality of Hamiliton-Wentworth (RMHW), 1996. Environmental Department, Regional Environmental Laboratory.

Salvato, J. A., 1992. *Environmental Engineering and Sanitation*, 4th ed., John Wiley & Sons, New York, 1418 pp.

Sanford, W. E., Steenhuis, T. S., Surface, J. M., and Peverly, J. H., 1995. Flow characteristics of rock-reed filters for treatment of landfill leachate, *Ecological Engineering* 5:37–50.

Staubitz, W. W., Surface, J. M., Steenhuis, T. S., Peverly, J. H., Lavine, M. J., Weeks, N. C., Sanford, W. E., and Kopka, R. J., 1989. Potential use of constructed wetlands to treat landfill leachate, in Hammer, D. A., Ed., *Constructed Wetlands for Wastewater Treatment: Municipal, Industrial and Agricultural*, Lewis Publishers, Chelsea, MI, 735–741.

Surface, J. M., Peverly, J. A., Steenhuis, T. S., and Sanford, W. E., 1993. Effect of season, substrate composition and operation guidelines for small constructed wetland wastewater treatment systems, in Hammer, D. A., Ed., *Constructed Wetlands for Wastewater Treatment*, Lewis Publishers, Chelsea, MI, 461–472.

Task Force on Water Quality Guidelines of the Canadian Council of Ministers of the Environment, 1992. *Canadian Water Quality Guidelines*, Water Quality Branch Inland Waters Directorate, Ottawa, Ontario.

Terluk, V., Supervisor, 1995. Solid Waste Operations, Environmental Services Group, Regional Municipality of Hamilton-Wentworth, personal communication, November,

Terluk, V., Supervisor, 1996. Solid Waste Operations, Environmental Services Group, Regional Municipality of Hamilton-Wentworth, personal communication, December.

Thirumurthi, D., 1976. Biodegradation of sanitary landfill leachate, in Martin, A. M., Ed., *Biological Degradation of Wastes*, Elsvier Science, New York, 207–230.

Trautmann, N. M., Martin J. H., Jr., Porter, K. S., and Hawk, K. C., Jr., 1989. Use of artificial wetlands for treatment of municipal solid wastes landfill leachate, in Hammer, D. A., Ed., *Constructed Wetlands for Wastewater Treatment: Municipal, Industrial, and Agricultural*, Lewis Publishers, Chelsea, MI, 245–251.

Urbanic-Bercic, O., 1994. Investigation into the use of constructed reedbeds for municipal waste dump leachate treatment, *Water, Science and Technology* 29:289–294.

Water Pollution Control Federation, 1990. *Natural Systems for Wastewater Treatment: Manual of Practice*, Alexandria, VA, 270 pp.

Whalen, K. J., Lombardo, P. S., Wile, D. B., and Neel, T. H., 1989. Constructed wetlands: design, construction and costs, in Hammer, D. A., Ed., *Constructed Wetlands for Wastewater Treatment: Municipal, Industrial, and Agricultural*, Lewis Publishers, Chelsea, MI, p. 590–598.

Summary of Panel Discussion

The panel discussion provided an opportunity for the exchange of ideas openly and candidly for the participants with differing experiences and backgrounds. There was considerable discussion and debate on a range of issues and concerns. The conclusions and recommendations are summarized as follows.

There was general agreement on the success of using constructed wetlands (CWs) to treat different types of wastewaters. The use of CWs for treating landfill leachates was considered a logical extension of the evolving use of this ecotechnology. A few participants felt that the wetland treatment technology is not yet a proven one, and to avoid potential problems due to system failure, emphasis needs to be placed to incorporate new and improved design features in establishing CWs.

An interesting point that emerged during the discussion was how the public perceived the soundness of wastewater technologies and landfill technology in particular. The majority of the public seems to think that the current technology in wastewater treatment generally involves a closed system. In reality, it is an open system since the removal of influent contaminants is not 100%. In the case of landfill operations in solid waste management, the public seems to have significant mistrust because it thinks of these facilities as very open systems. Indeed it should be pointed out that landfills are open systems because of the release into the environment of products of continued waste deposition. However, if appropriate controls are placed on the open system and if the discharge is treated by technologies that include constructed wetlands, a landfill can be an effectively designed facility.

Landfill leachates have entrained constituents necessitating the use of a treatment system, and an integrated wetland treatment system is considered one of the favorable options. The ongoing trend in the design and implementation of integrated systems was noted. It is important that to incorporate such systems, a clear understanding is required of the characteristics of the leachate that dictate the design parameters for the creation of the wetland. Different opportunities in this approach may involve the use of engineered treatment systems, such as activated sludge units, to be directly incorporated into the CWs. Different aspects of the overall treatment could then be accomplished by the wetlands, including, for example, planning for the denitrification that will occur in the root zone of plants. Further studies are warranted to understand the optimum pretreatment required by engineered treatment facilities and the appropriate sequence of aerobic and anaerobic zones within the constructed wetlands so that the two portions of the treatment system work effectively.

Research projects should be initiated to determine the long-term pollutant removal potential of different types of substrate including, for example, peat. Investigations of the merits of different substrates to remove different constituents entrained in the leachate are urgently required.

The potential for bioaccumulation of pollutants, especially heavy metals, in the root systems of plants is a matter of concern. There is a need to initiate long-term monitoring of this potential problem in different geographic regions. The harvesting of the biomass is considered an important strategy at this time to avoid liability of treatment areas that may later be characterized as potentially contaminated.

Organic-rich substrates in CWs will retain heavy metals, in which case it could be argued that the CW treatment system may be simply transferring at least some of the pollutants from one area to another, creating a long-term liability. The uncertainties associated with the bioaccumulation of contaminants in CWs were discussed.

Pretreatment of leachates using sedimentation ponds is a useful step, but the sediment disposal could pose a potential problem that might require placement of the contaminated sediment back into the landfill.

Ammonia is a major concern in the design of constructed wetlands because of its toxicity to plant and animal life. Reed beds can handle relatively high concentrations of ammonia and recycling in reed beds of leachates with high ammonia content appears to be an appropriate strategy.

High levels of nitrogen in leachate can have a negative effect on plants. Therefore, a careful selection of plant species is vital to the success of CW projects.

CWs, like natural wetlands, provide an environment for insect breeding, especially mosquitoes. Monitoring of CWs and provision of a contingency plan for spraying in summer to control insects is warranted, especially in proximity to urban areas. Consideration should be given to possible biological control of insects.

Information is required on the concentration levels that different plant species can tolerate without interfering with their metabolism to improve our understanding of the physiology of plants in CWs.

Provision of adequate time for acclimatization of plants to various concentrations of leachate is considered essential before actual treatment is initiated.

Certain plant species are able to concentrate specific heavy metals and the use of such species in CWs is desirable. Harvesting the plants and incinerating them in an appropriate manner is one approach, but the resulting ash will likely have to be landfilled.

Genetic engineering is needed to evolve more suitable plant species that can assimilate contaminants. Concerted research efforts are required to explore this exciting possibility.

There is an urgent need to improve our understanding of rate kinetics of biological and chemical reactions. Equilibrium conditions are not necessarily achieved, and the need may exist, for example, for hydraulic containment to ensure that reactions that occur in theory are actually completed in practice.

Ongoing changes in the design aspects of landfills are being made, and the changes in design will decrease infiltration of water into landfills. These designs act to create landfills that are "dry," thereby increasing concentrations of pollutants in leachates. The implications of such high concentrations in the performance of CWs must be considered. There may be a need, for example, to recycle leachate passed through the constructed wetland to decrease the toxicity of the contaminants prior to discharge to the ambient environment.

Provision of adequate storage of the leachate is essential, since leachate is generated year-round and the effectiveness of constructed wetlands during the winter may be less.

The performance of CWs for treatment of landfill leachate in cold climates has been a matter of concern and is sometimes viewed with skepticism. Several examples where leachate has been treated in cold climatic conditions were presented. Research to prevent deep freezing by water-level manipulation and the use of various types of synthetic and natural insulating materials were considered beneficial in leachate treatment in cold regions.

The longevity of CWs to treat landfill leachate is ill-defined because of the relative newness of this ecotechnology. Modeling studies should be initiated to understand better the quality and quantity of sediments that will accumulate over time.

CWs as a treatment system, even after 20 years of use, are still considered by some regulatory agencies as an experimental technology. This negative attitude, although not warranted, is due to lack of adequate information on their performance and exaggerated views on the potential failure of CWs. It is up to scientists and engineers to provide this needed information and to the funding agencies to support CWs research.

Educating the public on the usefulness of ecotechnologies and their sustainability should be very much a top priority. The monetary advantages and the financial viability of CWs, as opposed to other leachate treatment alternatives, should be further explored.

CWs are considered a totally passive technology with little fossil fuel energy inputs and are attractive for treating large volumes of leachates. However, they will require long-term management of pollutant residuals. To improve the knowledge base of the CW system, long-term monitoring of all aspects of this ecotechnology is essential. The monitoring results will provide valuable inputs in the design of CWs and provide further improvements in natural treatment technologies. Long-term planning strategies of governments and other regulatory agencies are needed.

Index

A

Accumulation. *See* Bioaccumulation
Acetic acid, 9
Acetogenic bacteria, 8
Acid fermentation, 8
Acid mine drainage, 166
Aeration
  BOD and, 140
  cascade, 194–195
  in cold climates, 42, 44
  step aerators, 187
  temperature reduction by, 36
Aeration lagoons
  ammonia in, 162
  in cold climates, 151–153
  thermal regimes, 161
  treatment efficiencies, 157–158, 161
Aerenchyma, 226
Air pollution, 28
Air stripping of volatiles, 21
Alice and Fraser municipal landfill, Ontario, 141–150
Alkalinity, 130, 134, 212, 216–217
Aluminum, 10, 227
Ammonia
  in aeration lagoons, 162
  in leachate, 179
  in Monroe County wetland, 212, 214–216
  in Orange County Landfill wetland, 114, 129, 133
  in peat filters, 172
  in Perdido Landfill wetland, 182–184
  in reed beds, 81–82, 88, 272
Anaerobic decomposition, 8
Anaerobic ponds, 154
Areal loading, definition, 20
Arrowhead (*Sagittaria latifolia*), 66, 178
Arsenic, 200
Asellus aquaticus (sow bug), 257
Aspinwall and Company, 75
Azusa landfill (California), 15

B

Bacteria, 8, 256
Baltic isopod (*Saduria [Mesidotea] entomon*), 256
Barium, 212, 217–218
Beaver Meadow marsh system
  attenuation modeling, 147–150
  characterization, 142–147
  site description, 142
Benthic invertebrates, 253. *See also* Macroinvertebrates
Benzene, 111, 213, 217
Bicarbonate, 10
Bioaccumulation, 26–28, 255, 271–272
Bioaugmentation, 236
Biodegradation, 4, 236–241

Biohazards from leachate, 26–28
Biological oxygen demand (BOD)
  background levels, 19, 121
  in Breitenau landfill, 11–12
  in Esval and Bølstad landfills, 156, 159–160
  in Greenlane landfill, 13–15
  in Huneault Landfill, 172
  during leachate decomposition, 7–8, 10–11, 72, 115
  in Monroe County wetland, 212, 214–215
  in Orange County Landfill wetland, 124, 127, 129, 133, 136–137, 140
  overview, 181–182
  in Perdido Landfill wetland, 179–180, 182
  temperature and, 39
Biomagnification, definition, 27
Biomass in wetlands, 229–230, 271
Bioremediation, 236–241, 265
Biostimulation, 236
Bioturbation, 253
Blue flag iris (*Iris virginica*), 178
Bluegill, 120
Blue-spotted sunfish, 120
BNAEs, 110, 112, 137
BOD. *See* Biological oxygen demand
Boron, 168, 171–172, 227
Borrow pit wetlands, 188
Bølstad landfill, 152–162
Breitenau landfill leachate, 11–13
Bulrush. *See Scirpus* spp.
Butyric acid, 9
Bypass, in winter, 43

C

Caddisfly, 256
Cadmium
  criteria for, 131, 135
  in Orange County Landfill wetland, 122, 124, 126, 130, 134
  in plant tissue, 218
  treatment efficiencies, 203, 213
Calcium
  in column studies, 145–146
  in leachate, 212
  modeling results for, 149
  temporal variation, 10—11, 13
Canada, constructed wetlands in, 13–15, 35, 141–150, 165–173, 261–268. *See also* Sunoco refinery, Sarnia, Ontario
Carex spp. (sedges), 224, 228–229
Carex gracilis, 227
Carex lacustris, 226, 228, 230
Carex rostrata, 225, 229
Carex vescaria, 205, 227
Cascade aeration, 194–195. *See also* Aeration
Cattails. *See Typha* spp.
Channeling, 5–6, 171–172